Biological Magnetic Resonance

Volume 34

More information about this series at http://www.springer.com/series/5693

Lawrence J. Berliner • Narasimham L. Parinandi
Editors

Measuring Oxidants and Oxidative Stress in Biological Systems

Springer

Editors
Lawrence J. Berliner
Department of Chemistry and Biochemistry
University of Denver
Centennial, CO, USA

Narasimham L. Parinandi
Department of Internal Medicine, Division of Pulmonary, Critical Care, and Sleep Medicine
Ohio State University College of Medicine, Dorothy M. Davis Heart and Lung Research Institute
Columbus, OH, USA

ISSN 0192-6020 ISSN 2512-2215 (electronic)
Biological Magnetic Resonance
ISBN 978-3-030-47317-4 ISBN 978-3-030-47318-1 (eBook)
https://doi.org/10.1007/978-3-030-47318-1

This Springer imprint is published by the registered company Springer Nature Switzerland AG
The registered company address is: Gewerbestrasse 11, 6330 Cham, Switzerland

This volume is dedicated to the editors' precious grandchildren:

Brendan Makoto Berliner
Verity Beatrice Lester
Evangeline Grace Lester
Grier Kiran Parinandi
Graley Karthik Parinandi

Preface

It is without question that oxidants (reactive oxygen species [ROS] and reactive nitrogen species [RNS]) mediate oxidative stress and are at the epicenter of biological functions and pathophysiological events. Thus, ROS, RNS, oxygen levels, and the redox state of the biological system (e.g., cells, tissues, the whole organism) should be measured precisely by analytical methods that offer greater sensitivities and suitability for each and every reactive species, including ROS, RNS, oxygen, and redox molecules. In addition, the determination or assessment of biological, biochemical, and molecular indices of oxidative stress induced by ROS and RNS is complex and cumbersome. Overall, the methods of analysis/determination of oxidants and oxidative stress in biological systems have become challenging as they have rapidly evolved over the last five decades.

Hence, there is a pressing need to compile currently available advances in cutting-edge methods of analysis of oxidants (ROS and RNS), oxygen, redox state, and oxidative stress in biological systems, which is sought after by both novices and experts in oxidative stress biology. The book is divided into four parts: (1) Introduction (2) Methods and Reagents, (3) Clinically Related Models and Approaches, and (4) Instrumental Methods, consisting of 11 chapters by experts in those fields. We are quite fortunate to have two renowned experts in the field, Dr. Balaraman Kalyanaraman and Dr. Henry Forman, write the introductory chapters, providing in-depth analyses of measurements of mitochondrial ROS and thiol redox processes. Under Methods and Reagents, Zamora and Villamena discuss *Clinical Probes for ROS and Oxidative Stress;* Khan and coworkers review and critique *Measurement of Oxidative Stress Markers In Vitro Using Commercially Available Kits*; and Parinandi and coworkers review *Oxidative Lipidomics: Analysis of Oxidized Lipids and Lipid Peroxidation in Biological Systems with Relevance to Health and Disease.* The Clinically Related Models and Approaches part includes chapters by Veeraraghavan, Parinandi, and Hund on *Oxidant-Induced Models of Vascular Leak*; this is followed with a chapter by Parinandi and coauthors on *Ozone-Specific Oxysterols and Neuronal Cell Signaling*; finally Uppu, Woods, and Parinandi discuss *Measurement Oxidative Stress Status in Human Populations: A Critical Need for a Metabolomic Approach.* The Instrumental Methods part contains work by Kuppusamy on *Sense and Sensibility of Oxygen in Pathophysiology Using EPR Oximetry*; then Hirata and coworkers discuss *Resonators for Clinical EPR*; lastly Ichikawa presents a comprehensive chapter on *Biomedical Overhauser Magnetic Resonance Imaging (OMRI): Noninvasive Imaging of Redox Processes.* The authors are all leaders in their respective fields. Each chapter is both pedagogical and critical with an intended audience not only of workers in the field but also new students and researchers entering the area. We envision that this book will be a highly useful resource for both novices and experts in the field of oxidant and oxidative stress biology.

Centennial, CO, USA Lawrence J. Berliner
Columbus, OH, USA Narasimham L. Parinandi

Contents

About the Editors

Lawrence J. Berliner received his BS in chemistry at UCLA, then gained his PhD in (bio)physical chemistry as one of the first Stanford University graduate students under H. M. McConnell. This era signaled the beginning of the spin labeling technique, wherein he was involved from the beginning to this day. He was then offered the first British-American Heart Postdoctoral Fellowship sponsored jointly by the American Heart Association and British Heart Fund in 1967 to work in protein crystallography at the Laboratory of Molecular Biophysics at Oxford University under the tutelage of Sir David Phillips. He joined the Ohio State University in 1969 as Assistant Professor of Chemistry. He was promoted to Full Professor in 1981 and then became Emeritus Professor in 2001, following a move to become Chair of the Department of Chemistry and Biochemistry at the University of Denver, where he was an adjunct member of the University of Colorado Toxicology Graduate Program in the School of Pharmacy. He is a biophysical chemist with interests in magnetic resonance studies of proteins and enzymes, in vivo MR studies, and free radical chemistry. He is well known in the EPR field for his contributions to spin labeling for over five decades, in addition to his edited book, *Spin Labeling: Theory and Applications* (Academic Press, 1976). The "series" has resulted in four volumes, with a fifth planned. The first 1976 tattered volume remains today on many lab bookshelves around the world, with the Russian translation "selling" more volumes than the English version. He conceived the series "Biological Magnetic Resonance" in 1979 with NMR pioneer Jacques Reuben, which is now in its 30th volume, and he now coordinates new volumes with guest coeditors. His work has covered proteolytic enzymes, including key coagulation proteins involved in thrombosis, protein-protein interactions in lactation and other key biological processes, in vivo EPR on small animals, and studies of free radical processes in mammalian system. Some of the accomplishments that he is best known for are his collaborative work with Kalman Hideg in developing the MTSL spin label and first demonstrating in vivo

EPR imaging on a plant species and small animals. He was recognized with the 2000 IES Silver Medal in Biology and Medicine. He takes great pride in being inducted as a charter member of the IES as well as the Society of Magnetic Resonance in Medicine. He has received numerous honors and awards, including fellow of the American Association for the Advancement of Science. He was further recognized, after his move to Denver, with the Lifetime Achievement Award in Biological EPR Spectroscopy at EPR 2005 in Columbus, Ohio, and as the only Fellow of the American Chemical Society at the University of Denver. In 2016 he was the Rudi Lemberg Traveling Fellow of the Australian Academy of Science. He was nominated as Fellow of the International EPR Society in 2019. His other activities, particularly in the American Chemical Society, involve educating and interacting with US legislators about the importance of funding basic scientific research and training. His goal, after stepping down as Chair in late 2008, was learning and pursuing new biological EPR and biophysical problems. Hence, his most recent sabbatical was spent as the Joseph Meyerhoff Visiting Professorship at the Department of Chemical Physics, Weizmann Institute of Science, and International Visiting Research Scientist at the University of Western Sydney. His other professional responsibility is as Editor in Chief of two Springer-Nature journals, *Cell Biochemistry and Biophysics* and *The Protein Journal.*

He spends a great deal of his time mentoring young scientists on career decisions and aspirations. For those who know him personally, he is someone who sees the world in a bigger perspective. While he had more than once concluded that spin labeling had run its course and had reached a dead end, yet he admits that spin labeling continues to flourish and has probably met a renaissance in the 2000s with site-directed spin labeling approaches and the many new techniques that colleagues in high-field EPR have developed. He summarized his perspective of the History of Spin Labeling and In Vivo EPR in a festschrift issue of the *European Biophysics Journal* (2010) and *Biomedical Spectroscopy and Imaging* (2016).

Narasimham L. Parinandi (pAri) is an Associate Professor in the Department of Internal Medicine at the Ohio State University College of Medicine. Parinandi received his BSc (Hons) in botany with chemistry, zoology, and English and MSc in botany with environmental biology in Berhampur University, India, in 1975–1977. From 1977 to 1980, he was a research fellow in Environmental Sciences at Andhra University, India. He earned his PhD (1986) at the University of Toledo, Toledo, OH, in biology and toxicology under the tutelage of Prof. Woon H. Jyung, an established zinc metabolism expert and aging biologist. During his graduate training at Toledo, he was exposed to the field of lipids by Prof. Max Funk, an expert lipoxygenase enzymologist from the lineage of Prof. Ned Porter. He did his postdoctoral fellowship (1986–1990) at the Hormel Institute, University of Minnesota, the premier lipid institute in the USA where he was trained with Prof. Harald

Schmid, a celebrity in the area of ether lipids and a pioneer in anandamide chemistry. At the Hormel Institute of the University of Minnesota, Parinandi was associated with Prof. Ralph T. Holman (member of the National Academy of Sciences and pioneer in fatty acid and lipoxygenase biochemistry who also coined the name "Omega-3 Fatty Acid") and conducted studies on omega-3 fatty acid dynamics in humans. He was also a research scientist/junior faculty at the Johns Hopkins University School of Medicine (1998–2002) under the mentorship of Prof. V. Natarajan, renowned lipid signaling expert, and Prof. Joe G.N. (Skip) Garcia, a celebrated lung vascular biologist. Parinandi has published nearly 125 peer-reviewed original scientific papers, reviews, and book chapters and edited books on *Free Radicals and Antioxidant Protocols* with Prof. William Pryor, the legendary free radical and lipid peroxidation scientist, and *Mitochondrial Function in Lung Health and Disease* with Prof. Natarajan, a lipid signaling celebrity. Parinandi collaborated and published original research papers with the Nobel Laureate Prof. Louis Ignarro on the pharmacology of NO donor drug. Currently, Parinandi is editing a Springer book on the *Methods of Determination of Oxidative Stress* with Prof. Lawrence Berliner of Denver University, a legendary scientist of biological EPR spectroscopy and imaging. Parinandi has given more than 50 invited scientific lectures at the national level in the USA and international institutions and conferences. He has also conducted and chaired several scientific conferences and symposia in the areas of oxidative stress and lipidology. He has teaching and mentoring experience of more than 35 years and mentored over 75 students, technicians, fellows, and junior faculty in his laboratory. He served as an editor of the Chemical Abstracts of the American Chemical Society. He has been a reviewer of nearly 70 peer-reviewed journals in the area of biochemistry, molecular biology, cell biology, and lipidomics. Parinandi has been on the editorial board of the *Molecular Biology Reports* (Springer), *Frontiers of Pharmacology*, *World Journal of GI Pharmacology*, *Current Chemical Research*, *Cell Biophysics and Biochemistry* (Associate Editor), and *The Protein Journal*. He has also received extramural funding from the National Institutes of Health (NIH), Department of Defense (DOD), American Thoracic Society (ATS), and International Academy of Oral Medicine and Toxicology (IAOMT) as a principal investigator (PI) and co-investigator (Co-I). Parinandi also serves as a reviewer of grant proposals for the NIH, AHA, DOD, US Universities, Government of Israel, Government of Austria, and Government of South Africa. Parinandi has received awards including the Gold Medal for securing the highest GPA in the MS class of 1975–1977 from Berhampur University, India, the Outstanding Teaching Assistant Award of the Biology Department of the University of Toledo in 1986, Distinguished Mentor Award of the Davis Heart and Lung Research Institute of the Ohio State University Wexner Medical Center in 2008, and the Distinguished Undergraduate Mentor Award of the Ohio State Undergraduate Research Program in 2009.

Teaching, Mentoring, and Research Interests

Teaching and Mentoring: He teaches a wide variety of topics in life sciences and biochemistry and environmental sciences. He is passionate and committed to mentoring undergraduate and graduate students on independent projects in fundamental research in biological sciences and biochemistry.

Research: Biochemistry and Toxicology; Lipidomics of Phospholipases and Bioactive Lipids; Lipid Signaling in Regulation of Secretory Phenomena; Heavy Metal and Lipid Bilayer Interactions; Oxidized Lipidome; Translational Lipidomics of Chronic Diseases; Analytical Biochemistry; Methods of Teaching and Mentoring Research.

Part I
Introduction

Chapter 1
Assays for Thiols and Modifications

Henry Jay Forman

Abstract This chapter comments on how to assay thiols and their modifications. Several biological compounds are derived from cysteine (the main cellular thiol), cysteamine, glutathione (GSH, gamma-glutamyl-cysteinyl-glycine), coenzyme A, and their disulfide and mixed disulfide forms, such as cystine. In addition, there are other biological sulfur compounds such as hydrogen sulfide, methionine, and polysulfur molecules. This chapter focuses on cysteine, glutathione, and their oxidized and S-conjugated forms including the S-nitrosated products of cysteine and glutathione.

Keywords Thiols · Analysis of thiols · Assay of GSH and modified thiols · Cysteine and glutathione assay · GSSG assay · Oxidized and nitrosated thiol assay

This is a brief commentary on the constantly evolving subject of how to assay thiols and their modifications. There are many biological compounds derived from cysteine, the core thiol in cells, cysteamine, glutathione (GSH, gamma-glutamyl-cysteinyl-glycine), coenzyme A, and their disulfide and mixed disulfide forms, such as cystine. There are also other biological sulfur compounds that include hydrogen sulfide, methionine, and polysulfur compounds. Here, however, we focus on cysteine, glutathione, and their oxidized and S-conjugated forms including S-nitrosated products of cysteine and glutathione. Frequently, glutathione disulfide (GSSG) is referred to as oxidized glutathione, but it is only one of several oxidized forms. The oxidized forms of a thiol in which addition of oxygen occurs are the sulfenic, sulfinic, and sulfonic forms, which have one, two, and three oxygens, respectively. The way to remember this sequence is to sing the "Old MacDonald had a farm" song, "e, i, o."

H. J. Forman (✉)
Leonard Davis School of Gerontology, University of Southern California, Los Angeles, CA, USA

L. J. Berliner, N. L. Parinandi (eds.), *Measuring Oxidants and Oxidative Stress in Biological Systems*, Biological Magnetic Resonance 34,
https://doi.org/10.1007/978-3-030-47318-1_1

GSH, GSSG, and Mixed Disulfides

Interestingly, some of the old assays are still the best in terms of accuracy of specificity. Among the oldies but goodies are Sies and Akerboom measurements for GSH and GSSG used spectrophotometric and fluorometric assays [1]. The principles involved kinetic recycling, which allowed low concentrations of GSH and GSSG to be measured by allowing GSH to reduce 5,5′-dithiobis(2-nitrobenzoic acid) (DTNB):

(a) 2 GSH + DTNB → 2 TNB + GSSG

which generates GSSG and then reducing GSSG with glutathione reductase:

(b) GSSG + NADPH → 2 GS− + NADP+ + H+

To determine GSSG in a sample, GSH is initially modified with N-ethylmaleimide (NEM) so that only the GSSG remains to be recycled. But that requires removing the excess NEM. Later, a more direct method using high-performance liquid chromatography to detect fluorescently labeled GSH and GSSG was devised by Fariss and Reed [2]. Although other methods have been devised, this remains the gold standard. To measure disulfides formed between GSH and protein cysteines or with small thiols, Sies and Akerboom suggested reduction with sodium borohydride, followed by measurement of GSH [1].

Conjugates Formed with GSH

There are many glutathione S-transferases (GSTs) in cells that catalyze the conjugation to the sulfur of GSH in an SN_2 reaction (a):

(c) $GSH + RX \rightarrow GSX + H^+ + X^-$

where R is an alkane and X is a halide (Cl^-, Br^-, I^-), nitrile (CN^-), or nitro (NO_2^-) moiety or by Michael addition to an alpha, beta-unsaturated carbonyl such as 4-hydroxy-2-nonenal (HNE)

(d) $GSH + RC{=}CHC{=}OR' \rightarrow RC(SG)CH_2C{=}OR'$

where R′ can be an alkyl group of a hydrogen.

For the (c) reactions with GSH, Habig and Jakoby described spectrophotometric and titrimetric assays [3] that accurately determine the reaction products of GSTs. Protein thiols can also react nonenzymatically in reaction (c) and (d). Nonenzymatic reactions of thiols may have similar kinetics to the GST-catalyzed reactions. But, although adducts can form with protein cysteine, competition with millimolar GSH in cells generally allows only the proteins with particularly nucleophilic cysteine to form a conjugate under normal conditions. This is why it is essential to consider the concentration of an electrophile like HNE in exposure experiments [4].

The specific determination of the adducts to cysteine by alpha, beta-unsaturated carbonyls in reaction (d) is best done using mass spectrometry [5]. Although kinetics favor addition to cysteine rather than other amino acids, antibodies to HNE adducts to protein generally detect adducts to lysine or histidine rather than to cysteine [6].

Oxidation products of protein cysteine The oxidation of cysteine in proteins by H_2O_2 and other hydroperoxides underlies the mechanism of much of redox signaling [7]. The question arose as to how the oxidation of protein cysteines could occur under conditions of physiologically generated hydroperoxides (ROOH), particularly as physiology demands that this be specific and regulated. Cysteine can be nonenzymatically oxidized by H_2O_2 to the sulfenic acid but the reaction is too slow, even for the better nucleophilic cysteine thiolate (S^-), to account for the oxidation that occurs in signaling proteins. Furthermore, the oxidation of the vast majority of protein cysteines by H_2O_2 is no faster than the oxidation of GS^- by H_2O_2 [8]. And, as the concentration of GS^- is thousands to millions of times the concentration of protein thiolates, the nonenzymatic reaction to form a sulfenic acid is extremely unlikely. The exceptions are in the active site of the sulfur-containing glutathione peroxidases and peroxiredoxin 6 in which a proton donating amino acid accelerates the splitting of the O–O bond in ROOH by a million fold [9] or in having zinc bound to cysteine, as in Keap1 [10]. Nonetheless, claims were made that sulfenic acid could be measured in proteins by forming adducts to dimedone [11]. The problem with that is that dimedone reacts far too slowly with sulfenic acid to account for the observations.

Realizing that the data was real but the interpretation incorrect, we demonstrated that a reaction with a sulfenylamide rather than sulfenic acid can account for those measurements [12]. The formation of the sulfenylamide can be accounted for by the equilibrium that exists between a glutathionylated protein (protein-glutathione disulfide) and the sulfenylamide while the glutathionylated proteins can be enzymatically produced [12]. There are four potential enzyme-catalyzed mechanisms for the formation of a glutathionylated protein (see [12] for detailed explanation). The measurement of glutathionylated proteins is described earlier. Ascertaining which cysteine in a protein is modified, however, requires mass spectrometry. Similarly, measurement of protein sulfenylamide is best done by mass spectrometry.

S-Nitroso Proteins

Finally, the measurement of proteins that have been S-nitrosated will be considered. A big issue is how the formation of an S-nitroso group occurs on proteins or GSH under physiological conditions, as any reaction mechanism requires the oxidation of NO [13]. While the formation of S-nitrosated proteins is most often called S-nitrosylation to make it seem like other posttranslational modifications, nitrosation is the correct chemical term. In the original biotin-switch assay used to detect S-nitroso functions on proteins [14], unmodified cysteines are blocked by modification with methyl methane thiosulfonate, which is then removed by chromatography or with acetone. Ascorbic acid is then added to reduce the S-nitroso function to the thiol form which reacts with HPDP-biotin (*N*-[6-(Biotinamido)hexyl]-3′-(2′-pyridyldithio)-propionamide). The biotin-conjugated protein can then be detected with a streptavidin-horseradish peroxidase after separation by electrophoresis. A modified method that appears to increase accuracy involves the use of added copper [14].

With the discovery that the H_2S is used in cell signaling and that polysulfur compounds are formed in biological systems, new assays are being developed to address this area. If a similar history occurs in this area as in thiol oxidation, it could be decades before it is sorted out. The take away message from this brief description is that it is important to use quantitatively accurate and specific assays to understand what thiol modifications are relevant to physiology.

References

1. Akerboom TPM, Sies H. Assay of glutathione, glutathione disulfide, and glutathione mixed disulfides in biological samples. Methods Enzymol. 1981;77:373–82.
2. Fariss M, Reed DJ. High-performance liquid chromatography of thiols and disulfides: dinitrophenol derivatives. Methods Enzymol. 1987;143:101–9.
3. Habig WH, Jakoby WB. Assays for differentiation of glutathione s-transferases. Methods Enzym. 1981;77:398–405.
4. Zhang H, Forman HJ. Signaling by 4-hydroxy-2-nonenal: exposure protocols, target selectivity and degradation. Arch Biochem Biophys. 2017;617:145–54.
5. Colzani M, Aldini G, Carini M. Mass spectrometric approaches for the identification and quantification of reactive carbonyl species protein adducts. J Proteome. 2013;92:28–50.
6. Just J, Jung T, Friis NA, Lykkemark S, Drasbek K, Siboska G, Grune T, Kristensen P. Identification of an unstable 4-hydroxynoneal modification on the 20s proteasome subunit α7 by recombinant antibody technology. Free Radic Biol Med. 2015;89:786–92.
7. Forman HJ, Maiorino M, Ursini F. Signaling functions of reactive oxygen species. Biochemistry. 2010;49:835–42.
8. Winterbourn CC, Metodiewa D. Reactivity of biologically important thiol compounds with superoxide and hydrogen peroxide. Free Radic Biol Med. 1999;27:322–8.
9. Forman HJ, Ursini F, Maiorino M. An overview of mechanisms of redox signaling. J Mol Cell Cardiol. 2014;73:2–9.
10. Dinkova-Kostova AT, Holtzclaw WD, Wakabayashi N. Keap1, the sensor for electrophiles and oxidants that regulates the phase 2 response, is a zinc metalloprotein. Biochemistry. 2005;44:6889–99.
11. Yang J, Gupta V, Tallman K, Porter N, Carroll KS, Liebler DC. Global, in situ, site-specific analysis of protein s-sulfenylation. Nat Protoc. 2015;10:1022–37.
12. Forman HJ, Davies MJ, Krämer AC, Miotto G, Zaccarin M, Zhang H, Ursini F. Protein cysteine oxidation in redox signaling: caveats on sulfenic acid detection and quantification. Arch Biochem Biophys. 2017;617:26–37.
13. Li Q, Lancaster JR Jr. A conspectus of cellular mechanisms of nitrosothiol formation from nitric oxide. Immunopathol Dis Ther. 2012;3:183–91.
14. Jaffrey SR, Snyder SH. The biotin switch method for the detection of s-nitrosylated proteins. Sci STKE. 2001;2001(86):PL1.

Chapter 2
Pitfalls of Reactive Oxygen Species (ROS) Measurements by Fluorescent Probes and Mitochondrial Superoxide Determination Using MitoSOX

Balaraman Kalyanaraman

Abstract Intracellular and mitochondrial superoxide formation is detected using phenanthrene-based dyes such as hydroethidine, mitochondria-targeted hydroethidine, or MitoSOX. HE and MitoSOX are redox probes, which undergo two-electron oxidation forming ethidium (E^+) and Mito-ethidium (Mito-E^+). The two-electron oxidation products derived from these probes exhibit the characteristic fluorescence that aids in fluorescence microscopy or flow cytometry or related techniques that are used to detect and determine superoxide (sometimes referred to as ROS, mitochondrial ROS, or mROS). This chapter briefly addresses the pitfalls of fluorescence-based techniques for detecting the intracellular superoxide.

Keywords Reactive oxygen species measurement · Pitfalls of ROS detection and determination · Fluorescent probes for ROS · Mitochondrial superoxide determination · MitoSOX ROS determination pitfall · Hydroethidine mitochondrial ROS

Phenanthrene-based dyes (e.g., hydroethidine, mitochondria-targeted hydroethidine, or MitoSOX) have been used to detect intracellular and mitochondrial superoxide formation [1, 2]. Both HE and MitoSOX are redox probes and undergo two-electron oxidation forming ethidium (E^+) and Mito-ethidium (Mito-E^+). The two-electron oxidation products derived from these probes exhibit characteristic fluorescence, and therefore, fluorescence microscopy or flow cytometry or related techniques have been used to detect and quantitate superoxide (sometimes referred to as ROS, mitochondrial ROS, or mROS). Most recently, MitoSOX-derived fluorescence was used to detect mitochondrial ROS formed in activated T cells [3]. This short commentary addresses the pitfalls of fluorescence-based techniques for

B. Kalyanaraman (✉)
Department of Biophysics, Medical College of Wisconsin, Milwaukee, Milwaukee, WI, USA
e-mail: balarama@mcw.edu

L. J. Berliner, N. L. Parinandi (eds.), *Measuring Oxidants and Oxidative Stress in Biological Systems*, Biological Magnetic Resonance 34,
https://doi.org/10.1007/978-3-030-47318-1_2

detecting intracellular superoxide. Readers are referred to the previous reviews on this topic [4–8].

Many years ago, we showed that superoxide reacts with hydroethidine and other analogs including hydropropidine (HPr) and MitoSOX, forming a characteristic hydroxylated product (e.g., 2-hydroxyhydroethidium or 2-OH-E^+ and Mito-2-OH-E^+), but not the corresponding ethidium (E^+ or Mito-E^+) [9, 10]. The fluorescence parameters of 2-OH-E^+ and Mito-2-OH-E^+ are significantly different, and therefore, monitoring the red fluorescence of HE or MitoSOX in cells will not measure intracellular superoxide formation, and the increase in fluorescence intensity is merely indicative of increased oxidation of HE and MitoSOX. Although the exact mechanism of oxidation is not determined, it is conceivable that redox metal ions (iron, for example) and/or peroxidatic mechanism is responsible for the oxidation of HE to E^+ and MitoSOX to Mito-E^+. Therefore, the use of MitoSOX to measure mitochondrial superoxide formation, using the fluorescence technique, is incorrect and flawed.

Evidence for one-electron oxidant formation in extracellular and intracellular settings was obtained by determining dimeric product formation (e.g., E^+-E^+, Mito-E^+-Mito-E^+) using HPLC and LC-MS techniques [11]. The dimeric products are not fluorescent.

Another caveat is the uptake of redox dyes into cells. The intracellular uptake varies depending upon the experimental conditions (changes in oxidative profile, membrane potential, apoptosis). Measuring the intracellular concentration of the fluorescent probes is critical for interpreting the results. At the same rate of intracellular oxidant generation, an increase or decrease in probe uptake will alter the amount of product formation.

Improper use of these probes during extraction or incubation procedure can induce hydrolysis, like in other assays, and give rise to confounding results [12]. We have used isotopically labeled oxidants (e.g., O-18-labeled superoxide) and unequivocally shown that the oxygen atom in 2-OH-E^+ is incorporated from molecular oxygen and not from water (unpublished results).

In conclusion, irrespective of the probes (HE, Mito-SOX or HPr) used, it is important to obtain a global profile of probe uptake and oxidation products in order to fully assess the extracellular, intracellular, and mitochondrial oxidant formation.

References

1. Zielonka J, Kalyanaraman B. Hydroethidine- and Mito-SOX-derived red fluorescence is not a reliable indicator of intracellular superoxide formation: another inconvenient truth. Free Radic Biol Med. 2010;48:983–1001. Review.
2. Kalyanaraman B. What effects do antioxidants have on tumors? Sci Trends. 2018. https://sciencetrends.com/what-effects-do-antioxidants-have-on-tumors/.
3. De Biasi S, Gibellini L, Bianchini E, Nasi M, Pinti M, Salvioli S, Cossarizza A. Quantification of mitochondrial reactive oxygen species in living cells by using multi-laser polychromatic flow cytometry. Cytometry A. 2016;89(12):1106–10.

4. Zielonka J, Joseph J, Sikora A, Kalyanaraman B. Real-time monitoring of reactive oxygen and nitrogen species in a multiwell plate using the diagnostic marker products of specific probes. Methods Enzymol. 2013;526:145–57.
5. Zielonka J, Hardy M, Kalyanaraman B. HPLC study of oxidation products of hydroethidine in chemical and biological systems: ramifications in superoxide measurements. Free Radic Biol Med. 2009;46:329–38.
6. Zielonka J, Vásquez-Vivar J, Kalyanaraman B. Detection of 2-hydroxyethidium in cellular systems: a unique marker product of superoxide and hydroethidine. Nat Protoc. 2008;3:8–21.
7. Kalyanaraman B, Hardy M, Zielonka J. A critical review of methodologies to detect reactive oxygen and nitrogen species by NADPH oxidase enzymes: implications in pesticide toxicity. Curr Pharmacol Rep. 2016;2:193–201.
8. Kalyanaraman B, Hardy M, Podsiadly R, Cheng G, Zielonka J. Recent developments in detection of superoxide radical anion and hydrogen peroxide: opportunities, challenges, and implications in redox signaling. Arch Biochem Biophys. 2017;617:38–47.
9. Zhao H, Kalivendi S, Zhang H, Joseph J, Nithipatikom K, Vásquez-Vivar J, Kalyanaraman B. Superoxide reacts with hydroethidine but forms a fluorescent product that is distinctly different from ethidium: potential implications in intracellular fluorescence detection of superoxide. Free Radic Biol Med. 2003;34:1359–68.
10. Zhao H, Joseph J, Fales HM, Sokoloski EA, Levine RL, Vásquez-Vivar J, Kalyanaraman B. Detection and characterization of the product of hydroethidine and intracellular superoxide by HPLC and limitations of fluorescence. Proc Natl Acad Sci U S A. 2005;102:5727–32. [Erratum: Proc Natl Acad Sci USA. 102:9086, 2005].
11. Zielonka J, Srinivasan S, Hardy M, Ouari O, Lopez M, Vásquez-Vivar J, Avadhani NG, Kalyanaraman B. Cytochrome c-mediated oxidation of hydroethidine and mito-hydroethidine in mitochondria: identification of homo- and heterodimers. Free Radic Biol Med. 2008;44:835–46.
12. Xiao Y, Meierhofer D. Are hydroethidine-based probes reliable for reactive oxygen species detection? Antioxid Redox Signal. 2018;31(4):359–67. https://doi.org/10.1089/ars.2018.7535.

Part II
Methods and Reagents

Chapter 3
Clinical Probes for ROS and Oxidative Stress

Pedro L. Zamora and Frederick A. Villamena

Abstract Electron paramagnetic resonance (EPR) spectroscopy is considered as a valuable tool in the determination and characterization of free radicals in vitro and in animal models; however, its use in humans presents technical challenges. While spin traps and spin probes have their own advantages and disadvantages, there are several factors that need to be considered for the appropriateness of any possible applications, which is a topic that will be discussed in this chapter. Besides the use of exogenous probes for the detection of free radicals, several endogenous molecules are used to determine the redox status in patients using the EPR techniques. In this chapter, both endogenous and exogenous agents for clinical studies of oxidative stress will be discussed. EPR signal formation or disappearance is a measure of the extent of ROS production, but changes in the spectral profile are also exploited, such as in the case of some trityl radical probes. The mechanisms of signal formation, disappearance, or changes thereof will be mentioned in detail in this chapter along with the limitations in their applications and cautionary notes in their interpretation.

Keywords Clinical EPR ROS probes · Electron paramagnetic resonance spectroscopy clinical probes · Oxidative stress clinical EPR · Reactive oxygen species clinical EPR spectroscopy

---

P. L. Zamora · F. A. Villamena (✉)
Department of Biological Chemistry and Pharmacology, The Ohio State University, Columbus, OH, USA
e-mail: frederick.villamena@osumc.edu

L. J. Berliner, N. L. Parinandi (eds.), *Measuring Oxidants and Oxidative Stress in Biological Systems*, Biological Magnetic Resonance 34,
https://doi.org/10.1007/978-3-030-47318-1_3

Introduction

Free radicals are critical mediators of normal cellular function, and they also play a major role in the pathogenesis of many human diseases. The term *oxidative stress* refers to a pathogenic burden of free radicals as they perturb the delicate balance between antioxidant defense mechanisms and pro-oxidative events. Unfortunately, there remains a disconnect between our understanding of this process as a mediator of human disease and our ability to use this knowledge in the clinical setting. It is conceivable that the development of techniques to measure these molecular processes in patients would provide unique and clinically actionable information. Electron paramagnetic resonance (EPR) spectroscopy is the gold standard for studying free radicals in biological systems, and EPR-based technologies may provide such a platform for analyzing oxidative stress in the clinical setting. EPR spectroscopy has proven to be a valuable tool in the measurement and speciation of free radicals in vitro and in animal models; however, its use in humans presents technical challenges. Factors limiting the direct-detection of radicals in most biological systems include the relatively low rate of endogenous radical production and the short half-lives of radicals. The result is a low steady-state concentration of radical species that falls short of the EPR detection threshold. To overcome the constraint of low radical concentrations, spin probes or spin traps are introduced to the system under investigation. While spin probes such as trityl or aminoxyl (nitroxides) radicals have been employed to detect ROS, they most often failed to identify specific radical species although current advances in the synthesis of spin trap and trityl radicals could overcome such limitation. Spin traps, however, have been used to study small molecule radicals such as superoxide (e.g., $O_2^{\cdot-}$ or $HO^{\cdot}$) as well as radical formation on larger biomolecules such as lipids, proteins, and nucleotides, which unlike most of the spin probes, spin traps can identify the nature of the ROS other than just detect them. While spin traps and spin probes have their own advantages and disadvantages, there are several factors that need to be considered for the appropriateness of any possible applications, which is a topic that is discussed in this chapter. Aside from the exogenous probes used for the detection of free radicals, several endogenous agents have also been used to assess redox status in patients using EPR techniques. In this chapter, both endogenous and exogenous agents for clinical studies of oxidative stress will be presented. Shown in Fig. 3.1 are characteristic EPR-detectable species for various probes in response to ROS production. EPR signal formation or disappearance is a measure of the extent of ROS production, but changes in the spectral profile are also exploited such as in the case of some trityl radical probes. The mechanisms of signal formation, disappearance, or changes thereof will be mentioned in detail in this chapter as well as limitations in their applications and cautionary notes in their interpretation.

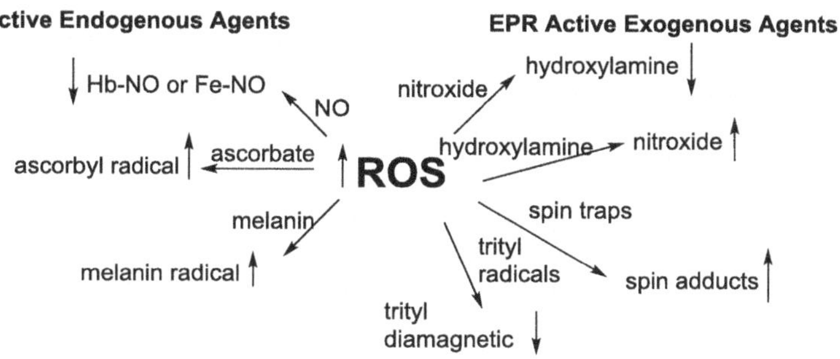

Fig. 3.1 Endogenous and exogenous probes used in ROS detection. Arrows indicate formation (↑) or quenching (↓) of EPR signal upon ROS production

Endogenous Probes

Hemoglobin- and Dinitrosyl Iron Complexes

Although nitric oxide (NO) is not under the category of ROS, levels in biological system could be correlated with the overall oxidative stress status, since NO bioavailability depends on the level of ROS production. This is due in part to the fact that production by the enzyme nitric oxide synthase (NOS) is redox-sensitive and that NO reacts with $O_2^{\cdot -}$, thereby lowering NO bioavailability. Nitric oxide is a colorless, odorless gas, and a free radical that is involved in a myriad of mammalian physiologic processes including maintenance of vascular and airway tones, platelet aggregation, angiogenesis, and immune-function [1] and is also implicated in the pathogenesis of important human diseases [2]. Due to its role as a potent vasodilatory signaling molecule, excessive generation of NO, whether due to overmedication with NO-donating nitrites or as a consequence of inducible nitric oxide synthase (iNOS) activation in sepsis [3]—leads to profound hypotension and organ hypoperfusion. There is also evidence that NO is involved in mediating transplant rejection. Dysfunction of endothelial nitric oxide synthase (eNOS) has been demonstrated to contribute to the development of cardiovascular disease, diabetes mellitus, and stroke.

The half-life of NO in serum is estimated to be in the range of 0.05–1.8 ms [4], and thus direct measurement of serum NO concentration is not practical. Therefore, there is a need for their detection using endogenous probes. Such probes are hemoglobin (Hb) and iron-nitrosyls. Hemoglobin is a tetrameric protein with four iron-containing heme groups. Each heme group is capable of binding a single O_2 molecule, and the entire protein undergoes conformational changes according to the number of O_2binding sites that are occupied. The oxygenation saturation status of the protein also determines the EPR spectra for NO-bound Hb or HbNO. As the oxygen tension of the blood determines the number of heme sites occupied and thus the conformation of Hb, the oxygen tension will also determine the predominant EPR species. During periods of oxidative insult, NO reacts readily with $O_2^{\cdot-}$ to form peroxynitrite ($ONOO^-$) which rapidly decomposes into nitrogen dioxide ($^{\cdot}NO_2$) and $HO^{\cdot}$—two powerful oxidants. Nitric oxide can also interact with Hb or with much smaller non-heme complexes to form heme-nitrosyls (HbNO) [5] and dinitrosyl-iron complexes (DNIC) [6], respectively, which produce characteristic EPR spectra [7, 8]. Because a stable paramagnetic adduct is formed with NO, these iron–heme complexes are functioning as endogenous spin traps. It is conceivable that using EPR spectroscopy to measure concentrations of heme- or iron-nitrosyl species in whole blood or solid tissues can provide useful information on various physiologic processes and disease states.

Iron-dinitrosyl complexes (DINCs) have a general stoichiometry of $[Fe(NO)_2L_2]^-$ where L typically corresponds to small molecule thiols such as cysteine ($CysS^-$) or glutathione (GS^-) and has been a reliable EPR probe for NO with $S = ½$ signal at $g = 2.03$, although this was observed in other non-thiol ligands [8]. Iron–sulfur cluster proteins have been shown to be a sensor for NO, and although the major [4Fe-4S] cluster nitrosyl products do not exhibit EPR spectra [9, 10], there are a variety of NO-Fe complexes that do produce EPR signals. For example, extra- and intracellular NO in rat liver biopsies after LPS injection was quantified using HbNO and DINCs EPR signals. In nonperfused liver tissue, HbNO was observed in both blood and liver, and only Fe-NO in liver homogenate. In perfused liver tissue, only the HbNO was observed in blood and Fe-NO in tissue with concentrations ranging from 9 to 30 nmol/cm^3 and detection limits of 0.61 and 0.52 nmol/cm^3 [11].

Hogg and colleagues [12] showed that the EPR signal in arterial and venous blood collected from human subjects after NO inhalation reflects the intravascular oxygen tension, with the mixed venous blood displaying predominately the triplet hyperfine structure of the 5-coordinate T-state signal, one of at least three different HbNO conformers. In organ transplantation, early studies by Lancester et al. demonstrated EPR signals attributable to HbNO in cytotoxic activated macrophages [13] and in the heart tissue of rats during acute allograft rejection [14]. In the later study, the characteristic EPR signal was absent in syngeneic grafts as well as in allografts that had been treated with immune suppressants. HbNO EPR signal in venous blood peaked on postoperative day seven in pancreas allografts in rats, while the signal was absent in syngeneic grafts [15]. The gradual development of the HbNO EPR signal with a peak at 1 week post-op follows the natural history of disease for acute transplant rejection, suggesting that HbNO can be used as a proxy for NO as a serum marker of graft rejection.

EPR signal changes during neoplastic transformation were first observed in the 1960s [16]. The characteristic triplet hyperfine splitting was attributed to the unpaired electron on a nitrogen atom in a species initially termed an "iron-complex with a nitrogen-containing ligand," later identified as 5-coordinate HbNO [16, 17]. It is now known that the increased NO production in tumor cells is a result of upregulated inducible NOS (iNOS) expression [18]. It appears that increased iNOS expression is a universal occurrence during neoplastic transformation, regardless of the tissue of origin [19–21]. The role of increased NO production during tumorogenesis is still not entirely clear—but possible contributions to evasion of immune surveillance and neoangiogenesis cannot be discounted [22]. Increased NO expressed has been positively correlated with P53 tumor suppressor mutations in cancers of the lung, colon, and oropharynx [23]. It is unclear if NO (or a reactive NO derivative such as $ONOO^-$) is contributing to the rate of P53 mutations by damaging DNA, or if the increased NO is a downstream effect of P53 loss of function. If increased NO expression does in fact reflect mutation burden in tumor cells, it is possible that detection of HbNO by EPR could play a role in tumor prognosis and treatment response. The HbNO EPR signal has also been shown to increase in photodynamically stressed human lung adenocarcinoma [24].

As a critical attenuator of vascular tone, NO can easily traverse to the lipid bilayer of endothelial cells to elicit relaxation of the surrounding vascular smooth muscle. Using EPR spectroscopy, the HbNO signal can serve as a reliable surrogate for NO activity in whole blood [25], and therefore, could be a viable tool in the diagnosis of cardiovascular diseases. Nitrites in circulation are reduced to NO in the presence of deoxyhemoglobin to yield methemoglobin [26]. In hypoxic/anaerobic states, such as during cardiac arrest, intracellular nitrites are reduced to NO in reactions catalyzed by myoglobin [27, 28], eNOS [29], xanthine oxidoreductase [30, 31], and mitochondrial cytochrome *c* oxidase [32]. Intracellular nitrites can also be reduced to NO at low pH via a nonenzymatic mechanism [33]. In ischemic heart, nitrite-mediated nitrosyl-heme complexes are formed [34] and whole-body EPR image of the nitrosyl-heme complexes in mice subjected to cardiopulmonary arrest was obtained using L-band EPR, demonstrating this process in vivo [35]. Of note, the nitrite-derived NO produced in critically ischemic tissue creates a radical burden during reperfusion, which reacts with superoxide to produce peroxynitrite—a key mediator of the reperfusion injury [36]. Detection of the HbNO signal by EPR has been applied to measure NO levels in hypertension models [37], as well as to demonstrate reactive hyperemia following brachial artery occlusion [38].

Ascorbyl Radical

Ascorbic acid (vitamin C) is an essential human nutrient that is a cofactor in multiple enzymatic reactions including those necessary for collagen synthesis. Due to its low redox potential, ascorbic acid also serves as a potent scavenger of superoxide and other free radicals, forming the stable ascorbyl radical in a one-electron oxidation step [39]. As such, ascorbyl radical formation is a sensitive indicator of oxidative insult. EPR studies in animal models show steady formation of ascorbyl radical in coronary effluent during periods of myocardial ischemia, with spikes in production following coronary reperfusion [40, 41]. Similar patterns of ascorbyl radical production have been demonstrated by EPR in models of pulmonary ischemia-reperfusion injury [42]. In diseases where the relationship between free radical injury and pathophysiology is not as immediately evident as it is in ischemia-reperfusion events, ascorbyl radical detection is perhaps even more important. In the cerebrospinal fluid of patients suffering from amyotrophic lateral sclerosis (ALS) an ascorbyl radical EPR signal is present, while it is absent in fluid collected from controls [43]. As we begin to understand the scope of free radical behavior and oxidative stress in various human disease states, the utility of ascorbyl radical detection by EPR as a catch-all marker for oxidative stress should be further investigated.

Melanins

Melanins are a group of heterogeneous polymeric pigments synthesized from oxidized tyrosine residues. These pigments, including eumelanin and pheomelanin, give color to hair and skin, and protect against UV light. The eumelanin structure includes multiple ortho-quinones which can be reduced under a variety of conditions to yield metastable semiquinone radicals—the main paramagnetic centers detected in melanin EPR studies [44]. The magnitude of melanin radical generation has been demonstrated to be directly proportional to the square-root of UVA irradiation intensity in EPR studies [45]. Animal models have shown that the same wavelength within the UV spectrum at 303 nm is the "action spectrum" for both peak melanin radical production and malignant transformation in melanocytes [45]. These results suggest that the melanin radical is a causal agent or key mediator in the pathogenesis of malignant melanoma and that EPR may be deployed as a clinical tool to better characterize this disease. The most essential prognostic factors in melanoma that is not metastatic are the primary lesion's spatial characteristics: the thickness of the lesion and depth of invasion into the underlying tissue. EPR imaging techniques based on detection of melanin radical spectra are being developed and hold promise to become a rapid, noninvasive method of revealing these key lesion parameters [46].

Exogenous or Synthetic Probes

To date, only a few exogenous EPR probes have been employed for the assessment of oxidative stress in clinical settings. The main roadblocks for their clinical applications are stability; due to the ubiquity of oxido-reductants in biological systems, and specificity; their susceptibility to redox reactions limits their ability to discern various ROS with similar redox potentials. Discussion of clinical oximetry probes will be presented in other chapters, while this chapter focuses on the application of synthetic probes for the assessment of total redox statuses. A more detailed discussion on the clinical applications of trityl probes is done in the succeeding chapters.

Nitroxides (Aminoxyls)

Redox sensitive probes such as aminoxyls, better-known as nitroxides, provide an overall assessment of the redox status of biological samples whether in vivo, in vitro, or ex vivo. The redox-sensitive imaging probe 3-methoxycarbonyl-2,2,5,5-tetramethylpiperidine-1-oxyl (MCP) has been widely used to

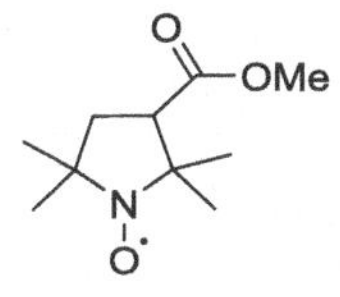

MCP

study brain pathology due to its membrane permeability. The rate of decay of EPR signal intensity has been correlated with increased oxidative stress status whereby fast disappearance of EPR signal is indicative of increased ROS levels or prooxidant state. In transgenic mouse models of Alzheimer's disease overexpressing amyloid-β peptides, MCP was employed as an EPR imaging agent to assess brain tissue oxidative stress status [47, 48]. Using L-band EPR imaging, orally administered lipophilic nitroxides (e.g., MCP and hydroxy-TEMPO) showed higher decay rates than those of nitroxides with low partition coefficients such as carboxy- and trimethylammonium analogs in the gastric mucosa from indomethacin-induced gastric ulcers in rats. In this study, cell membrane-permeable nitroxides were also found to inhibit ulcer formation demonstrating their potential utility as theranostic agents [49]. The EPR active MCP is reduced by ROS to form the EPR-silent hydroxylamine, but this is an over simplification and is not mechanistically accurate. Six-membered ring nitroxides reaction with $O_2^{\cdot-}$ exhibit a second order rate constant of 10^4–10^5 M^{-1} s^{-1} [50]. For example, nitroxides have a SOD mimetic property where upon reaction with $O_2^{\cdot-}$ [51], nitroxide is oxidized to oxoammonium cation with formation of H_2O_2, and the formed oxoammonium further reacts with $O_2^{\cdot-}$ to regenerate the nitroxide according to Eqs. 3.1 and 3.2:

$$RR'NO^{\cdot} + O_2^{\cdot-} + 2H^+ \rightarrow \left[RR'NO^+\right] + H_2O_2 \quad (3.1)$$

$$\left[RR'NO^+\right] + O_2^{\cdot-} \rightarrow RR'NO^{\cdot} + O_2 \quad (3.2)$$

Therefore, reduction of nitroxide to hydroxylamine by $O_2^{\cdot-}$ is only possible through the presence of a 2-electron reducing agent such as NADH or NADPH according to Eq. 3.3:

$$NADH + \left[RR'NO^+\right] \rightarrow RR'NOH + NAD^+ \quad (3.3)$$

With HO$^\cdot$, addition to the aminoxyl was a typical reaction with concomitant formation of oxoammonium cation. The formation of hydroxylamine typically occurs via HO$^\cdot$ addition to the nitroxyl group (40% in the case of 4-oxo-2,2,6,6-tetramethylpiperidine-*N*-oxyl) to form the oxoammonium cation via an *N*-peroxyl intermediate (see Eq. 3.4) with the rest undergoing H-abstraction. H-atom abstraction by the nitroxide from a secondary HO-adduct also occurs according to Eqs. 3.4 and 3.5 [52]:

$$\mathrm{RR'NO^{\cdot} + HO^{\cdot} \rightarrow [RR'NOOH] \rightarrow [RR'NO^{+}] + HO^{-}} \tag{3.4}$$

$$\mathrm{RR'}NO^{\cdot} + \mathrm{H} - Ar - OH^{\cdot} \rightarrow \mathrm{RR'}NOH + Ar - OH \tag{3.5}$$

Besides the SOD mimetic property of nitroxides, they have been shown to inhibit lipid peroxidation through reaction with alkyl or alkoxyl radical initiating species according to Eq. 3.6 [53]:

$$\mathrm{RR'NO^{\cdot} + L^{\cdot} \rightarrow RR'NOL} \tag{3.6}$$

Nitroxide EPR signal decay can be mediated indirectly by other nonradical species, therefore, data interpretation needs special precaution. For example, in the presence of hemoglobin, nitroxide is oxidized to its oxoammonium form in the presence of H_2O_2 (Eq. 3.7) [54];

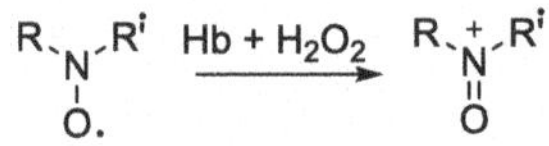

(3.7)

Peroxidase-like activity of metmyoglobin ($MbFe^{3+}$) with H_2O_2 to form O_2 and ferrylmyoglobin ($MbFe^{4+}$) was found to be enhanced fourfold to sixfold by nitroxides via reduction of $MbFe^{4+}$ by nitroxide to $MbFe^{3+}$ with formation of oxoammonium cation. Oxoammonium cation can further react with H_2O_2 or $O_2^{\cdot-}$ to form $O_2^{\cdot-}$ or O_2, respectively, with the regeneration of the nitroxide [55]. Additionally, nitroxide can react with labile Fe^{2+} to form hydroxylamine and Fe^{3+} in a reversible manner [56] where the equilibrium is shifted toward the reduction of hydroxylamine to nitroxide, and is catalyzed at basic pH. Also of note, in the presence of phosphate, six-membered ring nitroxides showed higher reactivity to Fe^{2+} than the five-membered ring nitroxides [57].

Aside from the redox properties of nitroxides the line shape of the EPR spectrum as a function of their degree of tumbling rate or motion in solution had been exploited for the diagnosis of oxidative stress and as an indicator of disease status. Albumin has the ability to bind to disease-associated molecules such as fatty acid transporters and changes in albumin conformation could provide utility in the diagnosis or prognosis of disease, certain cancers for example. Using 16-doxyl stearic acid (2-(14-carboxytetradecyl)-2-ethyl-4,4-dimethyl-3-oxazolidinyloxy) as a spin probe and peripheral blood samples, albumin conformation in cancer patients and healthy controls showed significant differences as demonstrated by EPR spectra, where the unbound probe was ~2 times greater concentrations and with narrower linewidths in cancer patients than in healthy controls showing high diagnostic sensitivity and specificity between groups [58]. The same technique was investigated for the potential diagnosis of oral squamous cell carcinoma from peripheral blood samples [59]. In studies of colorectal tissues in healthy controls and patients with colorectal neoplasms the results are not as conclusive as in blood albumin using lipophilic spin probes [60]. Membrane fluidity on noncultured lung cancer tissues was also assessed through the EPR spectra of lipophilic spin probes and showed that membranes of tumor tissues were more fluid than those of normal lungs [61]. Utilizing a dianionic spin-label probe, [1-*N*-(2,2,6,6-tetramethyl-1-oxyl-4-piperidinyl)-5-*N*-(1-aspartate)-2,4-dinitrobenzene], changes in serum albumin conformation upon binding with bilirubin was demonstrated, a finding which could lead to clinical applications in the diagnosis of hyperbilirubinemia or jaundice in neonates [62].

In summary, nitroxides are employed to measure total ROS production and membrane fluidity, both as indicators of oxidative stress and disease state, respectively. While significant differences between controls and oxidatively challenged samples have been observed, the use of nitroxides suffers from lack-of specificity due to a variety of redox events that could lead to the formation of the EPR-silent hydroxylamine. Moreover, these redox reactions can be mediated directly or indirectly by hemeproteins, labile transition metal ions or reducing substrates such as NADH or NADPH. In the diagnosis of diseases through EPR line shape analysis, the source of the biological sample has been shown to be critical for this application; for example, the analysis of albumin conformation is most robust in samples of peripheral blood (compared to other tissues). Perhaps design of new nitroxide spin probes and labels that have high specificity to certain types of ROS, inertness to non-ROS oxidoreductants as well as high affinity and specificity to specific proteins could improve the diagnostic potential of nitroxides.

Hydroxylamines [63]

Like nitroxides, hydroxylamines are redox-sensitive probes but while the EPR application of nitroxides works through spin quenching of the paramagnetic probe by ROS, the mechanism under study with hydroxylamine is opposite, that is, the EPR-silent hydroxylamine is transformed to the EPR-active nitroxide upon reaction with certain ROS. The hydroxylamine, 1-hydroxy-3-methoxycarbonyl-2.2.5.5-tetramethylpyrrolidine (CMH) which is a one-electron reduced

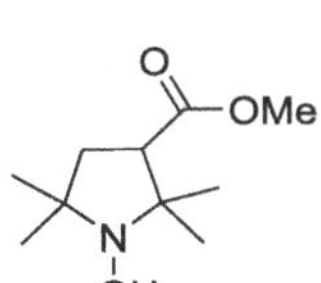

CMH

form of MCP has been widely employed as an EPR probe in the assessments of oxidative stress in clinical studies. For example, CMH was used to study the effect of a dietary supplement on the overall redox status of whole blood from human subjects using EPR with CMH as a probe. This study showed attenuation of ROS generation after 1 h of supplement intake [64]. The CMH probe was also used in the assessment of ROS production from blood samples in patients with aneurysmal subarachnoid hemorrhage [65]. Human liver biopsies from patients with liver disease (both viral and nonviral) shows higher nitroxide formation compared to liver samples from healthy controls using a highly lipophilic spin scavenger bis(1-hydroxy-2,2,6,6-tetramethyl-4-piperidinyl)decandioate [66]. The same bis-hydroxylamine

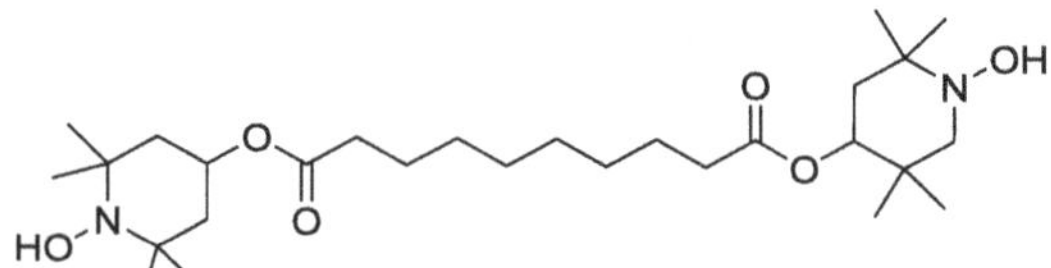

bis-hydroxylamine

probe was employed in the investigation of redox state in peripheral blood samples from conditioned athletes and normal-activity adults during a 60 min exercise challenge [67]. Another exercise-mediated ROS production study was also carried out using CMH as the spin probe [68]. Total oxidative stress status was assessed in blood from patients affected by thalassemia [69] and from rat renal tissues upon administration of vitamin E [70] using the bis-hydroxylamine probe, both showing heightened oxidative stress status.

Hydroxylamine reacts with both $O_2^{\cdot -}$ and $ONOO^-$ with considerably high rate constants of 10^3–10^4 M^{-1} s^{-1} and 6×10^9 M^{-1} s^{-1}, respectively [71, 72], to form nitroxide according to Eqs. 3.8 and 3.9:

$$RR'NOH + O_2^{\cdot -} \rightarrow RR'NO^{\cdot} + H_2O_2 \quad (3.8)$$

$$RR'NOH + ONOO^- \rightarrow RR'NO^{\cdot} + {}^{\cdot}NO_2 + H_2O_2 \quad (3.9)$$

The rate constant of $O_2^{\cdot -}$ oxidation of hydroxylamine was found to be pH-dependent with the rate of nitroxide regeneration shown to decrease at lower pHs [72]. Also, five-membered ring hydroxylamines show higher reactivity to $O_2^{\cdot -}$ compared to the six-membered ring species. In comparison of the lipophilic positively charged spin probe mito-TEMPO and its hydroxylamine analog (mito-TEMPO-H), lower $O_2^{\cdot -}$ scavenging ability was observed for the latter species, consistent with the lower rate constant of $O_2^{\cdot -}$ reaction with hydroxylamine compared to nitroxide [73]. Whether this is due to direct inactivation of hydroxylamine of the enzyme is not clear since for example hydroxylamine could be susceptible to hydrogen atom transfer as catalyzed by metal porphyrins or other heme cofactors in biological systems resulting in hydroxylamine disproportionation to TEMPO and amine [74]. In Alzheimer's disease, generation of free radicals from certain Aβ peptides can enhance the metal-catalyzed oxidation of hydroxylamine derivatives [75]. While H_2O_2 has slow reactivity with hydroxylamine, the oxidation reaction of hydroxylamine by H_2O_2 is enhanced by cupric-Aβ peptide [76]. In living cells under aerobic conditions, lipid-soluble hydroxylamines are oxidized to their respective nitroxide more than the water-soluble analogs in a reaction found to be enzyme-dependent, as oxidation is inhibited by heat, TCA or by cyanide with cytochrome oxidase being the major oxidizing enzyme [77]. Furthermore, hydroxylamines are susceptible to oxidation to nitroxide by ascorbyl radicals that could be generated from the ascorbate/ascorbate oxidase system and dehydroascorbic acid. The rate constant for the oxidation of HA by ascorbyl radical is comparable to that of $O_2^{\cdot-}$ and since ascorbyl radical concentration in the blood could exceed that of $O_2^{\cdot-}$ during vitamin C supplementation or in tissues with high ascorbate contents, this may contribute to the overall kinetics of nitroxide reduction and hydroxylamine oxidation, but it is worth noting that GSH also plays a major role in the reduction of ascorbate radicals in biological systems [78]. Aside from the non-ROS-mediated oxidation of hydroxylamine, nitroxides could comproportionate [72] with oxoammonium cation to form a nitroxide with equilibrium

$$2RR'N\dot{O} + 2H^+ \underset{}{\overset{K_4}{\rightleftharpoons}} RR'NO^+ + RR'N^+(H)OH \tag{3.10}$$

constant (K_4) being highly dependent on the structure of the hydroxylamine (Eq. 3.10) [79]. Figure 3.2 shows the oxoammonium-nitroxide-hydroxylamine redox triad. Careful interpretation of data is necessary when considering the actual source or mechanism of hydroxylamine oxidation or nitroxide reduction.

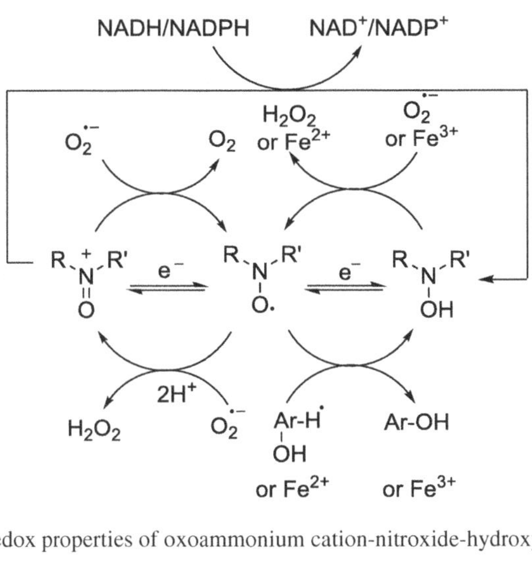

Fig. 3.2 General redox properties of oxoammonium cation-nitroxide-hydroxylamine triad

Nitrone Spin Traps

Similar to the mode of radical detection using hydroxylamine, spin traps produce an EPR spectrum upon reaction with free radicals. But unlike hydroxylamine that typically gives a *single* triplet spectral profile, spin trapping imparts a *multiple* spectral profile that can provide a wealth of information on the nature of radical species formed due to the distinctiveness of their respective spectra. The ability to discern between radicals and characterize various radicals based on their distinctive spectra is one of the hallmarks of EPR spin trapping detection.

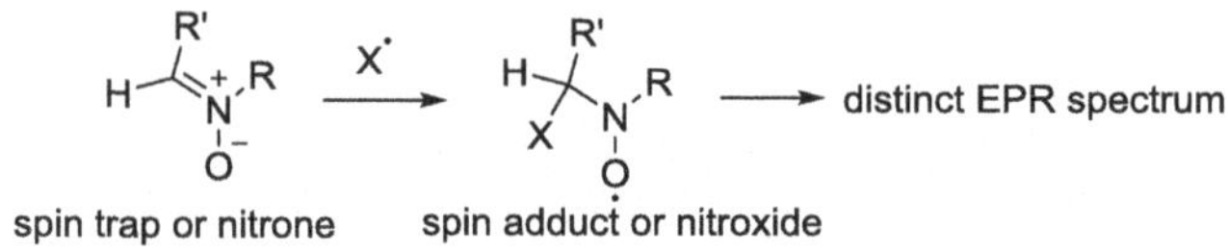

Depending on the type of radical or spin trap, the rate of spin adduct formation can vary significantly. For example, in cell-free systems the rate of reaction of $HO^{\cdot}$ with nitrones can reach the diffusion-controlled rate of $10^9\ M^{-1}\ s^{-1}$, while with $O_2^{\cdot-}$ it is a dismally slow rate of <1–100 $M^{-1}\ s^{-1}$ [80]. Half-life of $O_2^{\cdot-}$ adducts varies for different spin traps as well, which ranges from 1 to 14 min with 5,5-dimethyl-1-pyrroline *N*-oxide (DMPO) exhibiting the shortest half-life for its $O_2^{\cdot-}$ adduct of ~1 min followed by 5-ethylcarbonyl-5-methyl-1-pyrroline *N*-oxide (EMPO) of 7 min. and 5-diethoxyphosphoryl-5-methyl-1-pyrroline *N*-oxide (DEPMPO) of 14 min [81]. Protein radicals were also demonstrated to be trapped by DMPO in metmyoglobin-H_2O_2 reaction indicating the formation of Tyr-103 and Cys-110 DMPO adducts [82]. In vitro, 5-(2,2-dimethyl-1,3-propoxy cyclophosphoryl)-5-methyl-1-pyrroline *N*-oxide (CYPMPO) was found to be superior to DMPO and DEPMPO in scavenging $O_2^{\cdot-}$ from suspensions of human oral polymorphonuclear leukocytes (OPMNs) stimulated by phorbol 12-myristate 13-acetate and with the lowest toxicity [83]. DMPO and POBN was shown to be the safest spin trap for human skin studies including the NO spin traps, iron complexes of dithiocarbamates, which gave the least irritation up to 500 mM compared to PBN and DEPMPO [84].

Application of spin trapping yielded some important findings and insights on the antioxidant activity as well as pathogenesis of some diseases. In the investigation of the antioxidative action of certain compounds such as Edaravone in the treatment of acute ischemic brain disorder, EPR spin trapping was employed to assess the burst of radical production in rat neonatal brain before, during and after hypoxia using in vivo brain microdialysis technique and α-(4-pyridyl-*N*-oxide)-*N*-tert-butylnitrone (POBN) as a spin trap to demonstrate formation of lipid radicals during and after the ischemic event [85]. Also, the protective property of tea extract on normal lymphocytes but not clonal blasts in leukemia under UV-induced ROS production was investigated using 5-*tert*-butoxycarbonyl-5-methyl-1-pyrroline *N*-oxide (BMPO) spin trap [86].

In body fluids, ROS detection in blood drawn from patients during cardiopulmonary bypass procedures revealed that PBN radical adduct concentration increased in blood during preaortic cross-unclamping, and could reach a maxima during aortic cross-unclamping making a case for administering an antioxidant therapeutic agent prior to coronary artery bypass grafting for possible improved patient outcomes after surgery [87]. In the context of lung disease, xanthine oxidase activity was investigated from bronchoalveolar lavage fluid of patients with chronic obstructive pulmonary disease through ROS detection using DMPO [88]. There are instances that spin trapping may not be a suitable technique for a particular biological sample, such as in the case of human seminal plasma [89] or venous blood from type 2 diabetic patients [90] where endogenous ascorbyl radical was shown to be a more appropriate probe than DMPO or DEPMPO for ROS detection.

In cells, radical formation during the synthesis of prostaglandin H synthase from human platelets using DMPO and PBN was investigated and showed EPR signal formation indicative of a glutathyl radical adduct, however, with poor signal-to-noise ratio [91]. Spin trapping could also be used to assess the antioxidant capacity of cells other than just in the assessment of ROS production, for example, antioxidant potential of red blood cells in uremic patients was assessed using DMPO and in the presence of butylhydroperoxide as the oxidant. The study suggests that GSH plays a significant role in peroxide detoxification in red blood cells [92]. Formation of $O_2^{\cdot -}$ from human neutrophil granulocytes was shown to be induced by amyloid beta protein, βA(25–35) using the spin trap DEPMPO [93]. Using human endothelial cells, which have lesser propensity to generate radicals, were exposed to hypoxia and reoxygenation, and ROS generation was successfully detected using DMPO [94, 95].

In tissues, using human coronary artery tissue, flow-induced dilation-mediated ROS production was investigated using BMPO and was shown to be formed from the mitochondria but not NOS or cyclooxygenase [96]. ROS production was assessed from the ventricular myocardium from failing hearts versus nonfailing hearts using DEPMPO as spin trap showing increased $O_2^{\cdot -}$ concentration by two-fold in failing myocardium after treatment of the homogenized tissue with the spin trap and NADPH [97]. Intratracheal installation of asbestos in rats showed carbon-centered radical formation from chloroform extracts of lungs after 24 h exposure to the particles using POBN [98].

Since spin traps exhibit different rates of reaction with various radicals and that their respective spin adducts have varying half-life, improper choice of spin trap may give false negative results. For example, the linear nitrones, PBN and POBN are more appropriate spin traps for the detection of C-centered radicals in lipophilic environments while the cyclic nitrones, DMPO, DEPMPO, or EMPO are more suitable for the detection of O-centered radicals due to their ability to discern between $HO^{\cdot}$ and $O_2^{\cdot -}$ radicals. So how does one choose the right spin trap? A reasonable first consideration would be to choose a spin trap is already proven to work for a particular system. For example, ex-vivo studies using bile, plasma, whole blood, or urine used PBN or POBN, although they may not be able to detect O-centered radicals due to the O-centered adducts short half-lives, the C-centered adducts which are relatively stable could provide an indirect evidence for ROS production but could only give information on the overall redox status of the system. Considering cyclic nitrones, their respective C-centered adducts are quite stable as well and could provide as much information as the linear nitrones but with the added advantage of providing additional information on the nature of primary O-centered radicals, since cyclic nitrones have longer O-centered adduct half-lives. In considering a trap for an exploratory study, it is recommended to at least try both PBN and DMPO as starting spin traps since these two spin traps have different polarity where the former is more lipophilic than the latter. In most instances, only the HO-adduct as well as C-centered adduct are observable so a more robust spin trap such as DEPMPO or EMPO can then be used to investigate further if $O_2^{\cdot -}$ was indeed formed as a primary species. A cocktail of spin traps composed of DEPMPO, EMPO, PBN, and POBN has been proposed as a way to screen the best spin trap to use for a particular study [99]. One other important factor to consider is that nitrones, like nitroxides and hydroxylamine could participate in the redox regulation of a system under investigation. For example, nitrones have been shown to release NO upon reaction with $O_2^{\cdot -}$ that could further sequester ROS [100] or form protein-radical adducts that may have implications for enzyme inactivation or enervation [101–103]. In fact, spin traps have exhibited pharmacological activity in micromolar concentrations in some diseases such as certain cancers, ischemia and reperfusion injury, and stroke [104, 105]. Therefore, all of these factors must be considered in the interpretation of the levels of ROS produced in a particular system through the use of proper controls and inhibition studies as well as independent experiments for ROS detection using other nonspin trapping-based techniques.

Newly developed spin traps have shown increased $O_2^{\cdot -}$ adduct stability through their conjugation to permethylated (CD-DEPMPO) [106] or nonmethylated cyclodextrin (CDNMPO) [107, 108] with a $t_{1/2}$ ~ 55 min and ~30 min, respectively. Anion recognition host molecules such as calixpyrrole have been used to conjugate to the nitrone (CalixMPO) via an ester linker. Due to CalixMPO poor solubility in water, half-life for $O_2^{\cdot -}$ adduct of ~25 min was determined in 85% DMSO in PBS [109]. DMSO could annihilate the intramolecular H-bond interaction which could lead to a short adduct half-life, and this was evident for CDNMPO-O_2H with $t_{1/2}$ of 6 min in the same DMSO/PBS solvent system compared to ~30 min in PBS. In spite of these conditions, CalixMPO-O_2H in DMSO/PBS still exhibited a longer half-life due to the strong H-bond stabilizing interaction by the calixpyrrole. Another class of spin traps are those conjugated to the lipophilic tryphenylphosphnium cation (Mito-DEPMPO) which shows mitochondrial specificity with a long $O_2^{\cdot -}$ adduct half-life of $t_{1/2}$ ~ 73 min [110, 111]. Although these new nitrone spin traps hold promise, more clinical applications in the investigation of oxidative stress have yet to be realized and this is perhaps due to their availability to biomedical investigators which has been the limiting factor. Large scale synthesis and making these compounds widely available should be the next step toward their clinical applications for disease diagnosis and prognosis.

Trityl Radicals

Trityl radicals have been gaining popularity as the newest spin probe but most of their applications have been focused on oximetry and imaging due to the narrow linewidth they exhibit [112]. Some have employed trityl for pH [113] and inorganic phosphates [114] measurements in the cell, but these applications for trityl are explored in succeeding chapters while this chapter only briefly focuses on the detection of ROS. Since trityl clinical applications are still not widely employed, this section focuses only on the various trityl probes that have been developed for ROS detection.

The first demonstration of trityl radical application to measure $O_2^{\cdot-}$ production was done using the water-soluble analog of tetrathiatriarylmethyl radical, TAM OX06 [115] exhibiting stability in the presence of oxidoreductants. The rate constant for $O_2^{\cdot-}$ was several orders of magnitude faster than a typical spin trap of $3 \times 10^3\ M^{-1}\ s^{-1}$, which is comparable with the hydroxylamine reactivity to $O_2^{\cdot-}$. However, unlike with hydroxylamines, $O_2^{\cdot-}$ production is measured via loss of the singlet EPR signal. Using perchlorotritylmethyl (PTM) radical, this rate constant is even much higher than the TAM analogs with $k = 8 \times 10^8\ M^{-1}\ s^{-1}$ to form oxygen and an EPR-silent perchlorotritylphenylmethane as products [116–118].

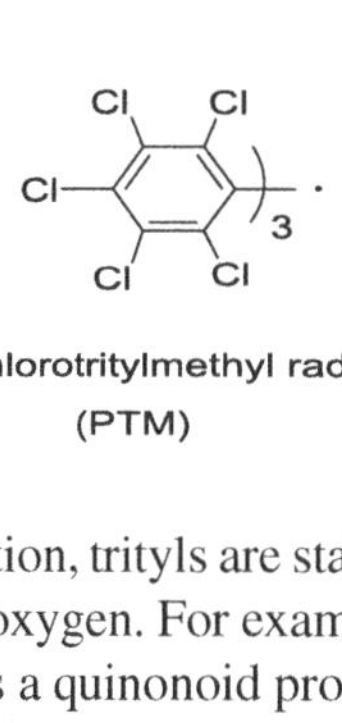

perchlorotritylmethyl radical
(PTM)

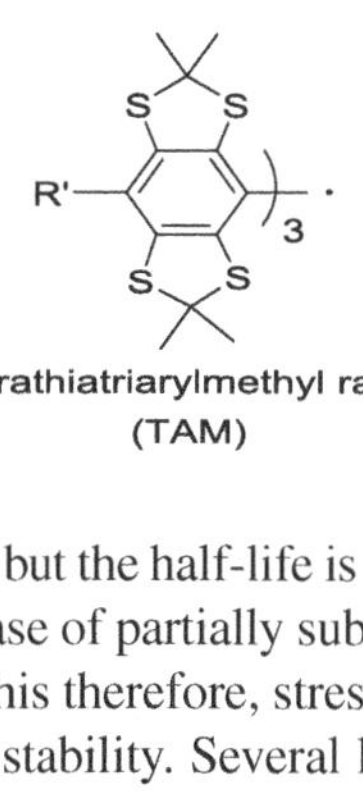

tetrathiatriarylmethyl radical
(TAM)

In anoxic condition, trityls are stable for days, but the half-life is greatly diminished in the presence of oxygen. For example, in the case of partially substituted TAM radical, 3-CO_2Et forms a quinonoid product [119]; this therefore, stresses the importance of fully substituting trityl radicals for increased stability. Several limitations of trityl have been addressed as far as their solubility in water, spectral profile, sensitivity, specificity, and stability. The water solubility of TAM probes was achieved through either partial or full para-substitution of the aryl group with carboxylate moieties, such as in the case of CT-03 [119]. However, carboxylate groups play a major role

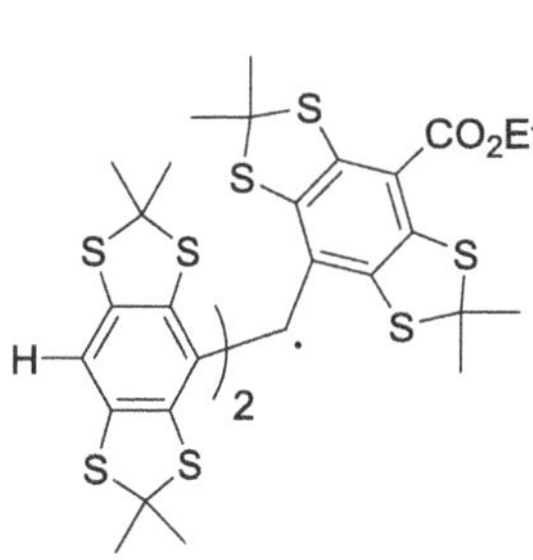

3-CO_2Et

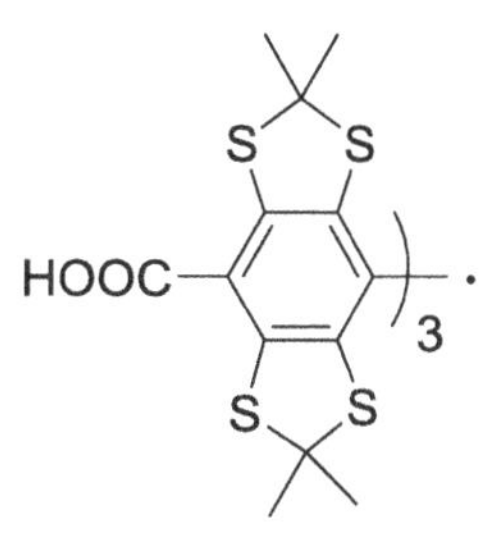

CT-03

in the TAM probe's selectivity to $O_2^{\cdot -}$/ROO$^{\cdot}$, which is due to their ability to undergo oxidation to form the quinone-methide products via decarboxylation reaction according to Eq. 3.11 [120].

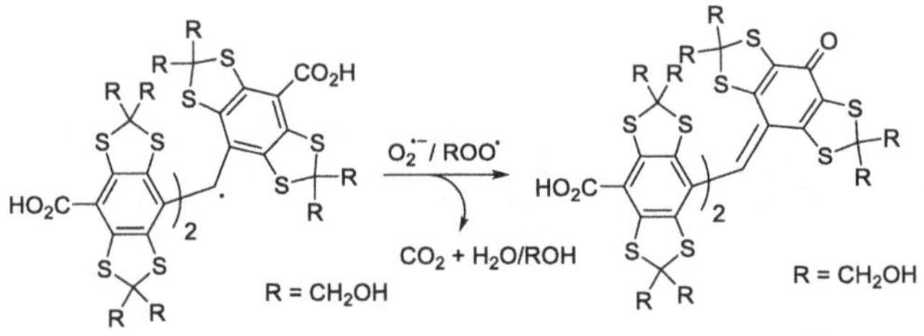

(3.11)

To improve the spectral profile of trityl other than from spin quenching of a singlet spectrum, various probes were designed that impart an initial and final distinctive spectral profile. TAM radical conjugation with the diamagnetic hydroxylamine (TN1) gave a singlet signal but upon oxidation of the hydroxylamine to nitroxide, gave a distinctive triplet spectrum. The triplet signal formation could be observed over a period

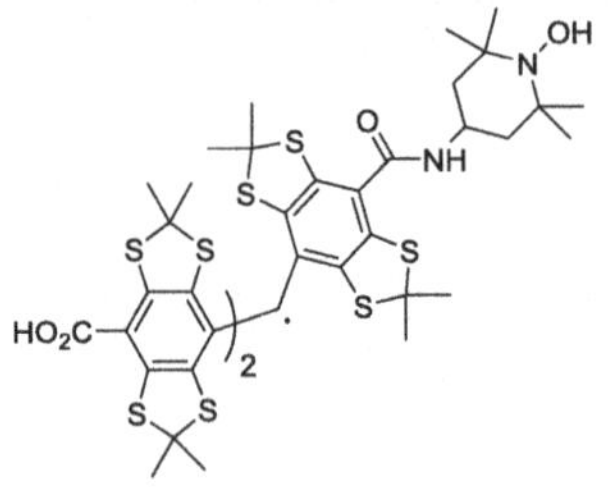

TN1

of 30 min from fresh rat liver homogenates upon succinate addition and this was coupled with simultaneous O_2 consumption measurement from the changes in the linewidth [121]. The spectral profile could also be manipulated into a doublet by isotopically labeling the amide-N with ^{15}N as TNN15 with better sensitivity than the ^{14}N (TNN14)

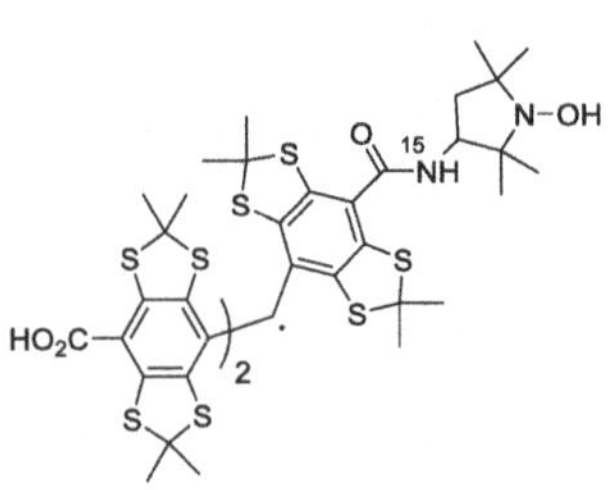

TNN15

analog due to lesser number of peaks [122]. A biradical design, that is, trityl-trityl (TSST) and trityl-nitroxide (TSSN), linked with a disulfide group was accomplished with a potential utility as thiol probe. The broad signal from the biradicals is converted to a signal corresponding to the monoradicals upon cleavage by GSH with $k \sim 0.3\ M^{-1}\ s^{-1}$ for TSSN and $0.2\ M^{-1}\ s^{-1}$ for TSST with the reduction of the nitroxide moiety in the case of TSSN. These probes were also employed in rat liver homogenates showing similar results [123].

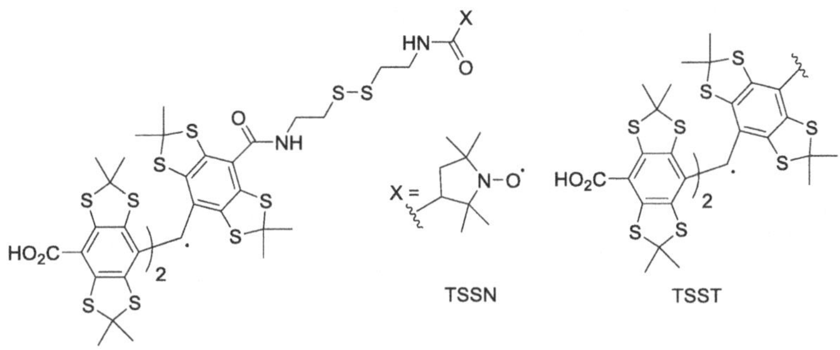

A noncleavable vicinal dihydroxy ester linker group was used to design a TAM biradical (TAM-TAM), which initially imparts a broad signal and forms a sharp singlet signal upon reaction with $O_2^{\cdot -}$ due to the conversion of one of the trityls to the diamagnetic quinone-methide (see Eq. 3.11) exhibiting a rate constants that is expected for TAM radicals of $6.7 \times 10^3\ M^{-1}\ s^{-1}$ [124].

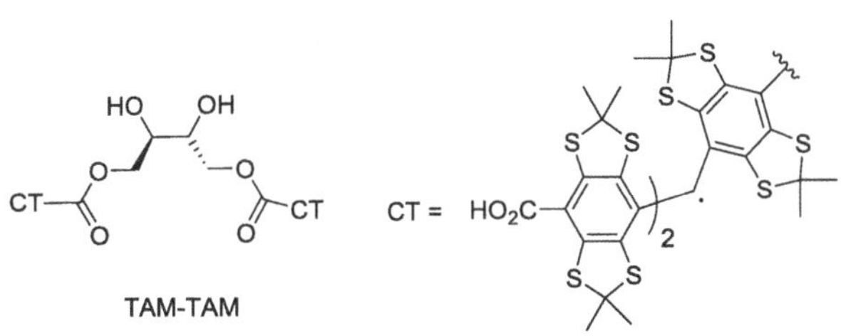

High sensitivity and specificity to $O_2^{\cdot -}$ was achieved through a TAM design bearing one unsubstituted aryl group as CT02-H, and gave a second rate constant of $1.7 \times 10^4\ M^{-1}\ s^{-1}$ which is higher than that observed for the fully carboxylate-substituted TAM with $k = 3.1 \times 10^3\ M^{-1}\ s^{-1}$ [125]. The doublet signal decays as a function of $O_2^{\cdot -}$ and to some extent $ROO^{\cdot}$, but not with GSH, Asc, $Fe^{2+/3+}$, H_2O_2, $HO^{\cdot}$ and $ONOO^{-}$.

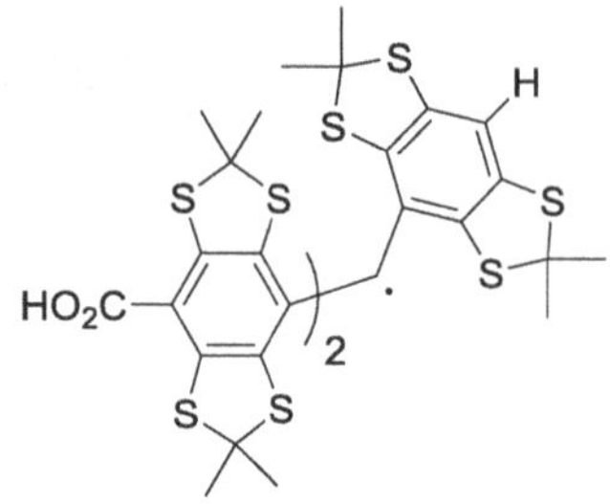

CT02-H

An ^{15}N-nitro-conjugated TAM was synthesized and showed higher rate of reaction with $O_2^{\cdot-}$ compared to other previously synthesized trityls with rate constant of 7×10^5 M^{-1} s^{-1} to form a the EPR-silent nitronate quinone methide [126].

In spite of the promising properties that trityl radicals have to offer in the realm of ROS detection, there are only few clinical applications of trityl outside of the fields of oximetry and EPR imaging where they are widely employed. Caution is urged, however, in the interpretation of data since some trityls could interfere in ROS production. For example, para-trisubstituted ester- and amide-carboxylate derivatives of TAM could be reduced by thiols such as GSH and cysteine to generate the corresponding trityl carbanions that subsequently reduces molecular O_2 to $O_2^{\cdot-}$ and with the regeneration of the trityl radical. However, the rate constant of thiol reaction with trityl is low with k ranging from 0.03 to 0.33 M^{-1} s^{-1} [127]. Also worth noting is that the fully carboxylated TAM (CT-03) can be metabolized by rat, human, and pig liver microsomes by cytochromes P450 and cytochrome P450 reductase in the presence of NADPH, both aerobically and anaerobically to form quinone-methide and triarylmethane, respectively [128]. Oxidation of TAM by microsome/NADPH, or by horseradish peroxidase/H_2O_2 systems could lead to the formation of carbocation intermediates which then can undergo adduction reaction with α-amino acids containing a strong nucleophilic residue such as thiols and other less nucleophilic groups [129]. Moreover, CT-03 was shown to bind to albumin [130] with high binding constant of 2×10^5 M^{-1} which could have implications for its clinical applications due to the ubiquity of albumin in blood. TAM binding to albumin could result in line-broadening of the trityl signal that could compromise its detection sensitivity; however, PEGylated trityl-derivatives were synthetized to overcome this problem of albumin-dependent linewidth broadening [131]. This is further supported in vivo where isolation of OX-063 from animal effluents only yields 25–47% recovery from crude lyophilized trityl with 98% purity suggesting that protein adduction of trityl in actual animal models could occur [132]. This also leads to the development of dendritic TAM radicals in which the TAM is covalently linked to dendrons containing three (PST-TA) or nine (PST-NA) carboxylic acids exhibiting improved biostabilitty against various oxidoreductancts without compromising their high rate of reactivity to $O_2^{\cdot-}$ with a detection limit of 2.1 nM/min [133]. Both dendrimeric trityls were oxidized to the EPR-silent quinone methide upon reaction with $O_2^{\cdot-}$ and was shown to measure $O_2^{\cdot-}$ generation from RAW 264.7 macrophages.

Conclusion

One possible clinical application of EPR would fulfill a similar role to current PET scans in marrying information on cellular physiology with structural information from superimposed conventional imaging studies. The future lies not only on imaging tissue oxygenation levels but also spectro-spatial imaging of ROS production in real time in tissues or whole animals. The ability to image $O_2^{\cdot-}$ specifically, for example, instead of just the total redox status as demonstrated through spin loss from nitroxide and trityl probes, would be more informative mechanistically, and could only be achieved through the use of spin traps and superoxide-specific trityl probes. However, this would entail the use of probes with $O_2^{\cdot-}$ adduct half-lives that last for hours, with high biostability and narrow linewidth. Perhaps a probe in the form of a nitrone-trityl conjugate could fulfill all of these requirements for a suitable ROS imaging probe. Other applications could extend to ex vivo EPR performed on in vivo spin-trapping samples obtained from peripheral blood or other tissues to provide bedside analysis of tissue perfusion or end organ damage.

The holy grail of EPR ROS detection is a biorthogonal approach using spin traps and redox probes, wherein the reaction of ROS with these probes does not interfere with the native biochemical processes that take place in the living system. So far, currently used nitroxides, trityls, and spin traps have the capacity to perturb the redox balance upon reaction with ROS through either redox cycling, NO release, or thiol sequestration through Michael adduction. Therefore, host–guest chemistry could be a viable approach in the development of such biorthogonal probes.

References

1. Ignarro LJ. Nitric oxide: a unique endogenous signaling molecule in vascular biology (Nobel lecture). Angew Chem Int Ed. 1999;38:1882–92.
2. Moncada S, Higgs EA. The discovery of nitric oxide and its role in vascular biology. Br J Pharmacol. 2006;147:S193–201.
3. Titheradge MA. Nitric oxide in septic shock. Biochim Biophys Acta. 1999;1411:437–55.
4. Liu X, Miller MJ, Joshi MS, Sadowska-Krowicka H, Clark DA, Lancaster JR Jr. Diffusion-limited reaction of free nitric oxide with erythrocytes. J Biol Chem. 1998;273:18709–13.
5. Fago A, Crumbliss AL, Peterson J, Pearce LL, Bonaventura C. The case of the missing NO-hemoglobin: spectral changes suggestive of heme redox reactions reflect changes in NO-heme geometry. Proc Natl Acad Sci U S A. 2003;100:12087–92.
6. Butler AR, Megson IL. Non-heme iron nitrosyls in biology. Chem Rev. 2002;102:1155–66.
7. Hogg N. Detection of nitric oxide by electron paramagnetic resonance spectroscopy. Free Radic Biol Med. 2010;49:122–9.
8. Vanin AF, Serezhenkov VA, Mikoyan VD, Genkin MV. The 2.03 signal as an indicator of dinitrosyl–iron complexes with thiol-containing ligands. Nitric Oxide. 1998;2:224–34.

9. Crack JC, Green J, Thomson AJ, Le Brun NE. Iron-sulfur clusters as biological sensors: the chemistry of reactions with molecular oxygen and nitric oxide. Acc Chem Res. 2014;47:3196–205.
10. Vanin AF. Dinitrosyl iron complexes with thiol-containing ligands as a "working form" of endogenous nitric oxide. Nitric Oxide. 2016;54:15–29.
11. Dumitrescu Sergiu D, Meszaros Andras T, Puchner S, Weidinger A, Boros M, Redl H, Kozlov Andrey V. EPR analysis of extra- and intracellular nitric oxide in liver biopsies. Magn Reson Med. 2017;77:2372–80.
12. Piknova B, Gladwin MT, Schechter AN, Hogg N. Electron paramagnetic resonance analysis of nitrosylhemoglobin in humans during NO inhalation. J Biol Chem. 2005;280:40583–8.
13. Lancaster JR Jr, Hibbs JB Jr. EPR demonstration of iron-nitrosyl complex formation by cytotoxic activated macrophages. Proc Natl Acad Sci U S A. 1990;87:1223–7.
14. Lancaster JR Jr, Langrehr JM, Bergonia HA, Murase N, Simmons RL, Hoffman RA. EPR detection of heme and nonheme iron-containing protein nitrosylation by nitric oxide during rejection of rat heart allograft. J Biol Chem. 1992;267:10994–8.
15. Tanaka S, Kamiike W, Ito T, Uchikoshi F, Matsuda H, Nozawa M, Kumura E, Shiga T, Kosaka H. Generation of nitric oxide as a rejection marker in rat pancreas transplantation. Transplantation. 1995;60:713–7.
16. Emanuel NM, Saprin AN, Shabalkin VA, Kozlova LE, Krugljakova KE. Detection and investigation of a new type of ESR signal characteristic of some tumour tissues. Nature. 1969;222:165–7.
17. Brennan MJ, Cole T, Singley JA. A unique hyperfine ESR spectrum in mouse neoplasms analyzed by computer simulation. Proc Soc Exp Biol Med. 1966;123:715–8.
18. Jenkins DC, Charles IG, Thomsen LL, Moss DW, Holmes LS, Baylis SA, Rhodes P, Westmore K, Emson PC, Moncada S. Roles of nitric oxide in tumor growth. Proc Natl Acad Sci U S A. 1995;92:4392–6.
19. Thomsen LL, Lawton FG, Knowles RG, Beesley JE, Riveros-Moreno V, Moncada S. Nitric oxide synthase activity in human gynecological cancer. Cancer Res. 1994;54:1352–4.
20. Thomsen LL, Miles DW, Happerfield L, Bobrow LG, Knowles RG, Moncada S. Nitric oxide synthase activity in human breast cancer. Br J Cancer. 1995;72:41–4.
21. Wilson KT, Fu S, Ramanujam KS, Meltzer SJ. Increased expression of inducible nitric oxide synthase and cyclooxygenase-2 in Barrett's esophagus and associated adenocarcinomas. Cancer Res. 1998;58:2929–34.
22. Huang B, Zhao J, Li H, He KL, Chen Y, Chen SH, Mayer L, Unkeless JC, Xiong H. Toll-like receptors on tumor cells facilitate evasion of immune surveillance. Cancer Res. 2005;65:5009–14.
23. Lala PK, Chakraborty C. Role of nitric oxide in carcinogenesis and tumour progression. Lancet Oncol. 2001;2:149–56.
24. Jakubowska M, Michalczyk-Wetula D, Pyka J, Susz A, Urbanska K, Plonka BK, Kuleta P, Lacki P, Krzykawska-Serda M, Fiedor L, Plonka PM. Nitrosylhemoglobin in photodynamically stressed human tumors growing in nude mice. Nitric Oxide. 2013;35:79–88.
25. Kirima K, Tsuchiya K, Sei H, Hasegawa T, Shikishima M, Motobayashi Y, Morita K, Yoshizumi M, Tamaki T. Evaluation of systemic blood NO dynamics by EPR spectroscopy: HbNO as an endogenous index of NO. Am J Physiol Heart Circ Physiol. 2003;285:H589–96.
26. Cosby K, Partovi KS, Crawford JH, Patel RP, Reiter CD, Martyr S, Yang BK, Waclawiw MA, Zalos G, Xu X, Huang KT, Shields H, Kim-Shapiro DB, Schechter AN, Cannon RO III, Gladwin MT. Nitrite reduction to nitric oxide by deoxyhemoglobin vasodilates the human circulation. Nat Med. 2003;9:1498–505.
27. Hendgen-Cotta UB, Merx MW, Shiva S, Schmitz J, Becher S, Klare JP, Steinhoff HJ, Goedecke A, Schrader J, Gladwin MT, Kelm M, Rassaf T. Nitrite reductase activity of myoglobin regulates respiration and cellular viability in myocardial ischemia-reperfusion injury. Proc Natl Acad Sci U S A. 2008;105:10256–61.

28. Rassaf T, Flogel U, Drexhage C, Hendgen-Cotta U, Kelm M, Schrader J. Nitrite reductase function of deoxymyoglobin: oxygen sensor and regulator of cardiac energetics and function. Circ Res. 2007;100:1749–54.
29. Webb AJ, Milsom AB, Rathod KS, Chu WL, Qureshi S, Lovell MJ, Lecomte FM, Perrett D, Raimondo C, Khoshbin E, Ahmed Z, Uppal R, Benjamin N, Hobbs AJ, Ahluwalia A. Mechanisms underlying erythrocyte and endothelial nitrite reduction to nitric oxide in hypoxia: role for xanthine oxidoreductase and endothelial nitric oxide synthase. Circ Res. 2008;103:957–64.
30. Li H, Samouilov A, Liu X, Zweier JL. Characterization of the magnitude and kinetics of xanthine oxidase-catalyzed nitrate reduction: evaluation of its role in nitrite and nitric oxide generation in anoxic tissues. Biochemistry. 2003;42:1150–9.
31. Cantu-Medellin N, Kelley EE. Xanthine oxidoreductase-catalyzed reduction of nitrite to nitric oxide: insights regarding where, when and how. Nitric Oxide. 2013;34:19–26.
32. Castello PR, David PS, McClure T, Crook Z, Poyton RO. Mitochondrial cytochrome oxidase produces nitric oxide under hypoxic conditions: implications for oxygen sensing and hypoxic signaling in eukaryotes. Cell Metab. 2006;3:277–87.
33. Zweier JL, Wang P, Samouilov A, Kuppusamy P. Enzyme-independent formation of nitric oxide in biological tissues. Nat Med. 1995;1:804–9.
34. Tiravanti E, Samouilov A, Zweier JL. Nitrosyl-heme complexes are formed in the ischemic heart: evidence of nitrite-derived nitric oxide formation, storage, and signaling in post-ischemic tissues. J Biol Chem. 2004;279:11065–73.
35. Kuppusamy P, Shankar RA, Roubaud VM, Zweier JL. Whole body detection and imaging of nitric oxide generation in mice following cardiopulmonary arrest: detection of intrinsic nitrosoheme complexes. Magn Reson Med. 2001;45:700–7.
36. Wang P, Zweier JL. Measurement of nitric oxide and peroxynitrite generation in the post-ischemic heart. Evidence for peroxynitrite-mediated reperfusion injury. J Biol Chem. 1996;271:29223–30.
37. Kanematsu Y, Tsuchiya K, Ohnishi H, Motobayashi Y, Izawa Y, Ishihara M, Ishizawa K, Abe S, Kawazoe K, Tamaki T. Effects of angiotensin II type 1 receptor blockade on the systemic blood nitric oxide dynamics in Nomega-nitro-L-arginine methyl ester-treated rats. Hypertens Res. 2006;29:369–74.
38. Lobysheva II, Biller P, Gallez B, Beauloye C, Balligand JL. Nitrosylated hemoglobin levels in human venous erythrocytes correlate with vascular endothelial function measured by digital reactive hyperemia. PLoS One. 2013;8:e76457.
39. Buettner GR, Jurkiewicz BA. Ascorbate free radical as a marker of oxidative stress: an EPR study. Free Radic Biol Med. 1993;14:49–55.
40. Vergely C, Maupoil V, Benderitter M, Rochette L. Influence of the severity of myocardial ischemia on the intensity of ascorbyl free radical release and on postischemic recovery during reperfusion. Free Radic Biol Med. 1998;24:470–9.
41. Pietri S, Culcasi M, Stella L, Cozzone PJ. Ascorbyl free radical as a reliable indicator of free-radical-mediated myocardial ischemic and post-ischemic injury. A real-time continuous-flow ESR study. Eur J Biochem. 1990;193:845–54.
42. Gielis JF, Boulet GA, Briede JJ, Horemans T, Debergh T, Kusse M, Cos P, Van Schil PE. Longitudinal quantification of radical bursts during pulmonary ischaemia and reperfusion. Eur J Cardiothorac Surg. 2015;48:622–9.
43. Spasojevic I, Stevic Z, Nikolic-Kokic A, Jones DR, Blagojevic D, Spasic MB. Different roles of radical scavengers—ascorbate and urate in the cerebrospinal fluid of amyotrophic lateral sclerosis patients. Redox Rep. 2010;15:81–6.
44. Enochs WS, Nilges MJ, Swartz HM. A standardized test for the identification and characterization of melanins using electron paramagnetic resonance (EPR) spectroscopy. Pigment Cell Res. 1993;6:91–9.

45. Wood SR, Berwick M, Ley RD, Walter RB, Setlow RB, Timmins GS. UV causation of melanoma in Xiphophorus is dominated by melanin photosensitized oxidant production. Proc Natl Acad Sci U S A. 2006;103:4111–5.
46. Vanea E, Charlier N, Dewever J, Dinguizli M, Feron O, Baurain JF, Gallez B. Molecular electron paramagnetic resonance imaging of melanin in melanomas: a proof-of-concept. NMR Biomed. 2008;21:296–300.
47. Fang D, Zhang Z, Li H, Yu Q, Douglas JT, Bratasz A, Kuppusamy P, Yan SSD, Reddy H. Increased electron paramagnetic resonance signal correlates with mitochondrial dysfunction and oxidative stress in an Alzheimer's disease mouse brain. J Alzheimers Dis. 2016;51:571–80.
48. Matsumura A, Emoto MC, Suzuki S, Iwahara N, Hisahara S, Kawamata J, Suzuki H, Yamauchi A, Sato-Akaba H, Fujii HG, Shimohama S. Evaluation of oxidative stress in the brain of a transgenic mouse model of Alzheimer disease by in vivo electron paramagnetic resonance imaging. Free Radic Biol Med. 2015;85:165–73.
49. Utsumi H, Yasukawa K, Soeda T, Yamada K, Shigemi R, Yao T, Tsuneyoshi M. Noninvasive mapping of reactive oxygen species by in vivo electron spin resonance spectroscopy in indomethacin-induced gastric ulcers in rats. J Pharmacol Exp Ther. 2006;317:228–35.
50. Samuni A, Krishna CM, Mitchell JB, Collins CR, Russo A. Superoxide reaction with nitroxides. Free Radic Res Commun. 1990;9:241–9.
51. Krishna MC, Russo A, Mitchell JB, Goldstein S, Dafni H, Samuni A. Do nitroxide antioxidants act as scavengers of $O_2^{\cdot-}$ or as SOD mimics? J Biol Chem. 1996;271:26026–31.
52. Samuni A, Goldstein S, Russo A, Mitchell JB, Krishna MC, Neta P. Kinetics and mechanism of hydroxyl radical and OH-adduct radical reactions with nitroxides and with their hydroxylamines. J Am Chem Soc. 2002;124:8719–24.
53. Miura Y, Utsumi H, Hamada A. Antioxidant activity of nitroxide radicals in lipid peroxidation of rat liver microsomes. Arch Biochem Biophys. 1993;300:148–56.
54. Yamaguchi T, Nakano T, Kimoto E. Oxidation of nitroxide radicals by the reaction of hemoglobin with hydrogen peroxide. Biochem Biophys Res Commun. 1984;120:534–9.
55. Krishna MC, Samuni A, Taira J, Goldstein S, Mitchell JB, Russo A. Stimulation by nitroxides of catalase-like activity of hemeproteins. Kinetics and mechanism. J Biol Chem. 1996;271:26018–25.
56. Mitchell JB, Samuni A, Krishna MC, DeGraff WG, Ahn MS, Samuni U, Russo A. Biologically active metal-independent superoxide dismutase mimics. Biochemistry. 1990;29:2802–7.
57. Bar-On P, Mohsen M, Zhang R, Feigin E, Chevion M, Samuni A. Kinetics of nitroxide reaction with iron(II). J Am Chem Soc. 1999;121:8070–3.
58. Kazmierczak SC, Gurachevsky A, Matthes G, Muravsky V. Electron spin resonance spectroscopy of serum albumin: a novel new test for cancer diagnosis and monitoring. Clin Chem. 2006;52:2129–34.
59. Moergel M, Kammerer PW, Schnurr K, Klein MO, Al-Nawas B. Spin electron paramagnetic resonance of albumin for diagnosis of oral squamous cell carcinoma (OSCC). Clin Oral Investig. 2012;16:1529–33.
60. Rossi S, Giuntini A, Balzi M, Becciolini A, Martini G. Nitroxides and malignant human tissues: electron spin resonance in colorectal neoplastic and healthy tissues. Biochim Biophys Acta. 1999;1472:1–12.
61. Sok M, Sentjurc M, Schara M. Membrane fluidity characteristics of human lung cancer. Cancer Lett. 1999;139:215–20.
62. Hsia JC, Kwan NH, Er SS, Wood DJ, Chance GW. Development of a spin assay for reserve bilirubin loading capacity of human serum. Proc Natl Acad Sci U S A. 1978;75:1542–5.
63. Dikalov SI, Polienko YF, Kirilyuk I, Polienko YF, Kirilyuk I. Electron paramagnetic resonance measurements of reactive oxygen species by cyclic hydroxylamine spin probes. Antioxid Redox Signal. 2018;28(15):1433–43.
64. Nemzer BV, Fink N, Fink B. New insights on effects of a dietary supplement on oxidative and nitrosative stress in humans. Food Sci Nutr. 2014;2:828–39.

65. Ewelina G, Krzysztof S, Marek M, Krzysztof K. Blood free radicals concentration determined by electron paramagnetic resonance spectroscopy and delayed cerebral ischemia occurrence in patients with aneurysmal subarachnoid hemorrhage. Cell Biochem Biophys. 2017;75:351–8.
66. Valgimigli L, Pedulli GF, Paolini M. Measurement of oxidative stress by EPR radical-probe technique. Free Radic Biol Med. 2001;31:708–16.
67. Paolini M, Valgimigli L, Marchesi E, Trespidi S, Pedulli GF. Taking EPR "snapshots" of the oxidative stress status in human blood. Free Radic Res. 2003;37:503–8.
68. Mrakic-Sposta S, Gussoni M, Montorsi M, Porcelli S, Vezzoli A. Assessment of a standardized ROS production profile in humans by electron paramagnetic resonance. Oxidative Med Cell Longev. 2012;2012:973927.
69. Filosa A, Valgimigli L, Pedulli GF, Sapone A, Maggio A, Renda D, Scazzone C, Malizia R, Pitrolo L, Lo Pinto C, Borsellino Z, Cuccia L, Capra M, Canistro D, Broccoli M, Soleti A, Paolini M. Quantitative evaluation of oxidative stress status on peripheral blood in beta-thalassaemic patients by means of electron paramagnetic resonance spectroscopy. Br J Haematol. 2005;131:135–40.
70. Vivarelli F, Canistro D, Franchi P, Sapone A, Vornoli A, Della Croce C, Longo V, Lucarini M, Paolini M. Disruption of redox homeostasis and carcinogen metabolizing enzymes changes by administration of vitamin E to rats. Life Sci. 2016;145:166–73.
71. Dikalov S, Skatchkov M, Bassenge E. Quantification of peroxynitrite, superoxide, and peroxyl radicals by a new spin trap hydroxylamine 1-hydroxy-2,2,6,6-tetramethyl-4-oxo-piperidine. Biochem Biophys Res Commun. 1997;230:54–7.
72. Zhang R, Goldstein S, Samuni A. Kinetics of superoxide-induced exchange among nitroxide antioxidants and their oxidized and reduced forms. Free Radic Biol Med. 1999;26:1245–52.
73. Dikalova AE, Bikineyeva AT, Budzyn K, Nazarewicz RR, McCann L, Lewis W, Harrison DG, Dikalov SI. Therapeutic targeting of mitochondrial superoxide in hypertension. Circ Res. 2010;107:106–16.
74. Porter TR, Mayer JM. Radical reactivity of the Fe(III)/(II) tetramesitylporphyrin couple: hydrogen atom transfer, oxyl radical dissociation, and catalytic disproportionation of a hydroxylamine. Chem Sci. 2014;5:372–80.
75. Dikalov SI, Vitek MP, Maples KR, Mason RP. Amyloid beta peptides do not form peptide-derived free radicals spontaneously, but can enhance metal-catalyzed oxidation of hydroxylamines to nitroxides. J Biol Chem. 1999;274:9392–9.
76. Dikalov SI, Vitek MP, Mason RP. Cupric-amyloid beta peptide complex stimulates oxidation of ascorbate and generation of hydroxyl radical. Free Radic Biol Med. 2004;36:340–7.
77. Chen K, Swartz HM. Oxidation of hydroxylamines to nitroxide spin labels in living cells. Biochim Biophys Acta. 1988;970:270–7.
78. Bobko AA, Kirilyuk IA, Grigor'ev IA, Zweier JL, Khramtsov VV. Reversible reduction of nitroxides to hydroxylamines: roles for ascorbate and glutathione. Free Radic Biol Med. 2007;42:404–12.
79. Sen VD, Tikhonov IV, Borodin LI, Pliss EM, Golubev VA, Syroeshkin MA, Rusakov AI. Kinetics and thermodynamics of reversible disproportionation-comproportionation in redox triad oxoammonium cations—nitroxyl radicals—hydroxylamines. J Phys Org Chem. 2015;28:17–24.
80. Villamena FA, Xia S, Merle JK, Lauricella R, Tuccio B, Hadad CM, Zweier JL. Reactivity of superoxide radical anion with cyclic nitrones: role of intramolecular H-bond and electrostatic effects. J Am Chem Soc. 2007;129:8177–91.
81. Lauricella R, Allouch A, Roubaud V, Bouteiller JC, Tuccio B. A new kinetic approach to the evaluation of rate constants for the spin trapping of superoxide/hydroperoxyl radical by nitrones in aqueous media. Org Biomol Chem. 2004;2:1304–9.
82. Gunther MR. Probing the free radicals formed in the metmyoglobin-hydrogen peroxide reaction. Free Radic Biol Med. 2004;36:1345–54.

83. Saito K, Takahashi M, Kamibayashi M, Ozawa T, Kohno M. Comparison of superoxide detection abilities of newly developed spin traps in the living cells. Free Radic Res. 2009;43:668–76.
84. Fuchs J, Groth N, Herrling T. In vitro and in vivo assessment of the irritation potential of different spin traps in human skin. Toxicology. 2000;151:55–63.
85. Noor JI, Ueda Y, Ikeda T, Ikenoue T. Edaravone inhibits lipid peroxidation in neonatal hypoxic-ischemic rats: an in vivo microdialysis study. Neurosci Lett. 2007;414:5–9.
86. Tepe Cam S, Polat M, Esmekaya MA, Canseven AG, Seyhan N. Tea extracts protect normal lymphocytes but not leukemia cells from UV radiation-induced ROS production: an EPR spin trap study. Int J Radiat Biol. 2015;91:673–80.
87. Goudeau J-J, Clermont G, Guillery O, Lemaire-Ewing S, Musat A, Vernet M, Vergely C, Guiguet M, Rochette L, Girard C. In high-risk patients, combination of antiinflammatory procedures during cardiopulmonary bypass can reduce incidences of inflammation and oxidative stress. J Cardiovasc Pharmacol. 2007;49:39–45.
88. Pinamonti S, Leis M, Barbieri A, Leoni D, Muzzoli M, Sostero S, Chicca MC, Carrieri A, Ravenna F, Fabbri LM, Ciaccia A. Detection of xanthine oxidase activity products by EPR and HPLC in bronchoalveolar lavage fluid from patients with chronic obstructive pulmonary disease. Free Radic Biol Med. 1998;25:771–9.
89. Ochsendorf FR, Goy C, Fuchs J, Morke W, Beschmann HA, Bromme HJ. Lucigenin chemiluminescence in human seminal plasma. Free Radic Res. 2001;34:153–65.
90. Delmas-Beauvieux M-C, Peuchant E, Thomas M-J, Dubourg L, Pinto AP, Clerc M, Gin H. The place of electron spin resonance methods in the detection of oxidative stress in type 2 diabetes with poor glycemic control. Clin Biochem. 1998;31:221–8.
91. Chou D-S, Hsiao G, Shen M-Y, Tsai Y-J, Chen T-F, Sheu J-R. ESR spin trapping of a carbon-centered free radical from agonist-stimulated human platelets. Free Radic Biol Med. 2005;39:237–48.
92. Klemm A, Voigt C, Friedrich M, Funfstuck R, Sperschneider H, Jager E-G, Stein G. Determination of erythrocyte antioxidant capacity in haemodialysis patients using electron paramagnetic resonance. Nephrol Dial Transplant. 2001;16:2166–71.
93. Andersen JM, Myhre O, Aarnes H, Vestad TA, Fonnum F. Identification of the hydroxyl radical and other reactive oxygen species in human neutrophil granulocytes exposed to a fragment of the amyloid beta peptide. Free Radic Res. 2003;37:269–79.
94. Zweier JL, Broderick R, Kuppusamy P, Thompson-Gorman S, Lutty GA. Determination of the mechanism of free radical generation in human aortic endothelial cells exposed to anoxia and reoxygenation. J Biol Chem. 1994;269:24156–62.
95. Zweier JL, Kuppusamy P, Thompson-Gorman S, Klunk D, Lutty GA. Measurement and characterization of free radical generation in reoxygenated human endothelial cells. Am J Phys. 1994;266:C700–8.
96. Liu Y, Zhao H, Li H, Kalyanaraman B, Nicolosi AC, Gutterman DD. Mitochondrial sources of H_2O_2 generation play a key role in flow-mediated dilation in human coronary resistance arteries. Circ Res. 2003;93:573–80.
97. Sam F, Kerstetter DL, Pimental DR, Mulukutla S, Tabaee A, Bristow MR, Colucci WS, Sawyer DB. Increased reactive oxygen species production and functional alterations in antioxidant enzymes in human failing myocardium. J Card Failure. 2005;11:473–80.
98. Kadiiska MB, Ghio AJ, Mason RP. ESR investigation of the oxidative damage in lungs caused by asbestos and air pollution particles. Spectrochim Acta A. 2004;60A:1371–7.
99. Marchand V, Charlier N, Verrax J, Buc-Calderon P, Leveque P, Gallez B. Use of a cocktail of spin traps for fingerprinting large range of free radicals in biological systems. PLoS One. 2017;12:e0172998/0172991–15.
100. Locigno EJ, Zweier JL, Villamena FA. Nitric oxide release from the unimolecular decomposition of the superoxide radical anion adduct of cyclic nitrones in aqueous medium. Org Biomol Chem. 2005;3:3220–7.

101. Kumar A, Ganini D, Deterding LJ, Ehrenshaft M, Chatterjee S, Mason RP. Immuno-spin trapping of heme-induced protein radicals: implications for heme oxygenase-1 induction and heme degradation. Free Radic Biol Med. 2013;61:265–72.
102. Ranguelova K, Rice AB, Lardinois OM, Triquigneaux M, Steinckwich N, Deterding LJ, Garantziotis S, Mason RP. Sulfite-mediated oxidation of myeloperoxidase to a free radical: immuno-spin trapping detection in human neutrophils. Free Radic Biol Med. 2013;60:98–106.
103. Chatterjee S, Ehrenshaft M, Bhattacharjee S, Deterding LJ, Bonini MG, Corbett J, Kadiiska MB, Tomer KB, Mason RP. Immuno-spin trapping of a post-translational carboxypeptidase B1 radical formed by a dual role of xanthine oxidase and endothelial nitric oxide synthase in acute septic mice. Free Radic Biol Med. 2009;46:454–61.
104. Floyd RA. Nitrones as therapeutics in age-related diseases. Aging Cell. 2006;5:51–7.
105. Villamena FA, Das A, Nash KM. Potential implication of the chemical properties and bioactivity of nitrone spin traps for therapeutics. Future Med Chem. 2012;4:1171–207.
106. Poulhes F, Rizzato E, Bernasconi P, Rosas R, Viel S, Jicsinszky L, Rockenbauer A, Bardelang D, Siri D, Gaudel-Siri A, Karoui H, Hardy M, Ouari O. Synthesis and properties of a series of β-cyclodextrin/nitrone spin traps for improved superoxide detection. Org Biomol Chem. 2017;15:6358–66.
107. Han Y, Liu Y, Rockenbauer A, Zweier JL, Durand G, Villamena FA. Lipophilic β-cyclodextrin cyclic-nitrone conjugate: synthesis and spin trapping studies. J Org Chem. 2009;74:5369–80.
108. Han Y, Tuccio B, Lauricella R, Villamena FA. Improved spin trapping properties by β-cyclodextrin-cyclic nitrone conjugate. J Org Chem. 2008;73:7108–17.
109. Kim S-U, Liu Y, Nash KM, Zweier JL, Rockenbauer A, Villamena FA. Fast reactivity of a cyclic nitrone-calix[4]pyrrole conjugate with superoxide radical anion: theoretical and experimental studies. J Am Chem Soc. 2010;132:17157–73.
110. Hardy M, Poulhes F, Rizzato E, Rockenbauer A, Banaszak K, Karoui H, Lopez M, Zielonka J, Vasquez-Vivar J, Sethumadhavan S, Kalyanaraman B, Tordo P, Ouari O. Mitochondria-targeted spin traps: synthesis, superoxide spin trapping, and mitochondrial uptake. Chem Res Toxicol. 2014;27:1155–65.
111. Hardy M, Rockenbauer A, Vasquez-Vivar J, Felix C, Lopez M, Srinivasan S, Avadhani N, Tordo P, Kalyanaraman B. Detection, characterization, and decay kinetics of ROS and thiyl adducts of mito-DEPMPO spin trap. Chem Res Toxicol. 2007;20:1053–60.
112. Ardenkjaer-Larsen JH, Laursen I, Leunbach I, Ehnholm G, Wistrand LG, Petersson JS, Golman K. EPR and DNP properties of certain novel single electron contrast agents intended for oximetric imaging. J Magn Reson. 1998;133:1–12.
113. Marchand V, Leveque P, Driesschaert B, Gallez B, Driesschaert B, Marchand-Brynaert J. In vivo EPR extracellular pH-metry in tumors using a triphosphonated trityl radical. Magn Reson Med. 2017;77:2438–43.
114. Bobko AA, Eubank TD, Driesschaert B, Dhimitruka I, Evans J, Mohammad R, Tchekneva EE, Dikov MM, Khramtsov VV. Interstitial inorganic phosphate as a tumor microenvironment marker for tumor progression. Sci Rep. 2017;7:41233.
115. Rizzi C, Samouilov A, Kutala VK, Parinandi NL, Zweier JL, Kúppusamy P. Application of a trityl-based radical probe for measuring superoxide. Free Radic Biol Med. 2003;35:1608–18.
116. Kutala VK, Villamena FA, Ilangovan G, Maspoch D, Roques N, Veciana J, Rovira C, Kuppusamy P. Reactivity of superoxide anion radical with a perchlorotriphenylmethyl (trityl) radical. J Phys Chem B. 2008;112:158–67.
117. Kutala VK, Parinandi NL, Zweier JL, Kuppusamy P. Reaction of superoxide with trityl radical: implications for the determination of superoxide by spectrophotometry. Arch Biochem Biophys. 2004;424:81–8.
118. Dang V, Wang J, Feng S, Buron C, Villamena FA, Wang PG, Kuppusamy P. Synthesis and characterization of a perchlorotriphenylmethyl (trityl) triester radical: a potential sensor for superoxide and oxygen in biological systems. Bioorg Med Chem Lett. 2007;17:4062–5.
119. Xia S, Villamena FA, Hadad CM, Kuppusamy P, Li Y, Zhu H, Zweier JL. Reactivity of molecular oxygen with ethoxycarbonyl derivatives of tetrathiatriarylmethyl radicals. J Org Chem. 2006;71:7268–79.

120. Decroos C, Li Y, Bertho G, Frapart Y, Mansuy D, Boucher JL. Oxidation of tris-(p-carboxyltetrathiaaryl)methyl radical EPR probes: evidence for their oxidative decarboxylation and molecular origin of their specific ability to react with O_2*. Chem Commun. 2009;(11):1416–8.
121. Liu Y, Villamena FA, Rockenbauer A, Zweier JL. Trityl-nitroxide biradicals as unique molecular probes for the simultaneous measurement of redox status and oxygenation. Chem Commun. 2010;46:628–30.
122. Liu Y, Villamena FA, Song Y, Sun J, Rockenbauer A, Zweier JL. Synthesis of 14N- and 15N-labeled trityl-nitroxide biradicals with strong spin-spin interaction and improved sensitivity to redox status and oxygen. J Org Chem. 2010;75:7796–802.
123. Liu Y, Song Y, Rockenbauer A, Sun J, Hemann C, Villamena FA, Zweier JL. Synthesis of trityl radical-conjugated disulfide biradicals for measurement of thiol concentration. J Org Chem. 2011;76:3853–60.
124. Poncelet M, Driesschaert B, Bobko AA, Khramtsov VV. Triarylmethyl-based biradical as a superoxide probe. Free Radic Res. 2018;52(3):373–9.
125. Liu Y, Song Y, De Pascali F, Liu X, Villamena FA, Zweier JL. Tetrathiatriarylmethyl radical with a single aromatic hydrogen as a highly sensitive and specific superoxide probe. Free Radic Biol Med. 2012;53:2081–91.
126. Driesschaert B, Bobko AA, Khramtsov VV, Driesschaert B, Bobko AA, Khramtsov VV, Zweier JL. Nitro-triarylmethyl radical as dual oxygen and superoxide probe. Cell Biochem Biophys. 2017;75:241–6.
127. Tan X, Chen L, Song Y, Rockenbauer A, Villamena FA, Zweier JL, Liu Y. Thiol-dependent reduction of the triester and triamide derivatives of Finland trityl radical triggers O_2-dependent superoxide production. Chem Res Toxicol. 2017;30:1664–72.
128. Decroos C, Li Y, Bertho G, Frapart Y, Mansuy D, Boucher JL. Oxidative and reductive metabolism of tris(p-carboxyltetrathiaaryl)methyl radicals by liver microsomes. Chem Res Toxicol. 2009;22:1342–50.
129. Decroos C, Boucher JL, Mansuy D, Xu-Li Y. Reactions of amino acids, peptides, and proteins with oxidized metabolites of tris(p-carboxyltetrathiaaryl)methyl radical EPR probes. Chem Res Toxicol. 2014;27:627–36.
130. Song Y, Liu Y, Liu W, Villamena FA, Zweier JL. Characterization of the binding of the Finland trityl radical with bovine serum albumin. RSC Adv. 2014;4:47649–56.
131. Song Y, Liu Y, Hemann C, Villamena FA, Zweier JL. Esterified dendritic TAM radicals with very high stability and enhanced oxygen sensitivity. J Org Chem. 2013;78:1371–6.
132. Serda M, Wu YK, Barth ED, Halpern HJ, Rawal VH. EPR imaging spin probe trityl radical OX063: a method for its isolation from animal effluent, redox chemistry of its quinone methide oxidation product, and in vivo application in a mouse. Chem Res Toxicol. 2016;29:2153–6.
133. Tan X, Tao S, Liu W, Rockenbauer A, Villamena FA, Zweier JL, Song Y, Liu Y. Synthesis and characterization of the perthiatriarylmethyl radical and its dendritic derivatives with high sensitivity and selectivity to superoxide radical. Chemistry. 2018;24(27).

Chapter 4
Measurement of Oxidative Stress Markers In Vitro Using Commercially Available Kits

Bryan Gardiner, Julie A. Dougherty, Devasena Ponnalagu, Harpreet Singh, Mark Angelos, Chun-An Chen, and Mahmood Khan

Abstract Reactive oxygen species (ROS) and reactive nitrogen species (RNS) are highly reactive molecules, with significant effects in human diseases including cancer and cardiovascular disease. The ability to accurately and precisely detect the formation of free radicals within cells and tissues is crucial to developing proper treatments for the problems caused by ROS/RNS. Fluorescent probes have become widely available reagents of detecting ROS/RNS within cells. Several commercially available kits have shown their specificity toward detecting the formation of ROS and RNS. In this chapter, we discuss the principle behind each kit and the benefits and shortcomings of these kits, namely dihydroethidium (DHE), dichlorohydrofluoresin diacetate (DCF-DA), 4-amino-5-methylamino-2′,7′-difluorofluorescein diacetate (DAF-FM diacetate), and 10-acetyl-3,7-dihyrdroxyphenoxazine (Amplex red). DHE is used to specifically detect superoxide, while DCF-DA readily detects hydroxyl radicals. Amplex red is used to detect hydrogen peroxide, and DAF-FM is used for measuring nitric oxide. However, due to the nature of their reactivity, the probes are not absolutely specific for the noted ROS/RNS species, and will react with others. ROS measurement may need to be made in real time, and they are short-lived within the cell, especially superoxide and nitric oxide. This chapter

B. Gardiner · J. A. Dougherty · M. Angelos · C.-A. Chen
Department of Emergency Medicine, Dorothy M. Davis Heart Lung and Research Institute, The Ohio State University Wexner Medical Center, Columbus, OH, USA

D. Ponnalagu · H. Singh
Department of Physiology and Cell Biology, The Ohio State University Wexner Medical Center, Columbus, OH, USA

M. Khan (✉)
Department of Emergency Medicine, Dorothy M. Davis Heart Lung and Research Institute, The Ohio State University Wexner Medical Center, Columbus, OH, USA

Department of Physiology and Cell Biology, The Ohio State University Wexner Medical Center, Columbus, OH, USA
e-mail: mahmood.khan@osumc.edu

L. J. Berliner, N. L. Parinandi (eds.), *Measuring Oxidants and Oxidative Stress in Biological Systems*, Biological Magnetic Resonance 34,
https://doi.org/10.1007/978-3-030-47318-1_4

explains the mechanism behind each chemical kit, the protocols used with the kit, and show typical results after imaging. Additionally, an assessment is made on the use of the kit, identifying the advantages and disadvantages of each probe.

Keywords ROS · RNS · Mitochondria · Cardiomyocytes · Cardiovascular disease

Introduction

Reactive oxygen species (ROS) and reactive nitrogen species (RNS) have dichotomous effects within cells and tissues, being beneficial or detrimental to physiological function. Responses are based on the system's ability to recognize the type, amount, and location of ROS/RNS generated, as well as the duration of its presence in the cell. At low or moderate concentrations in cells, ROS/RNS are beneficial for regulating growth, apoptosis, and cell signaling; which at the tissue level contribute to regulation of blood pressure, immune response, cognitive function, and metabolism [1]. Antioxidant mechanisms within the cell respond to and control ROS/RNS signals and maintain homeostasis. When endogenous antioxidant activity is insufficient to deal with high concentrations or longevity of ROS/RNS within cells then pathological conditions arise [1]. Oxidative stress or nitrosative stress refers to these molecules causing biological damage [2–5]. Excessive ROS/RNS damage lipids, proteins, or DNA and inhibit their normal function thus leading to their implication in the general aging process [2]. These types of stresses are also implicated in disease conditions including cardiovascular disease, cancer, neurological disorders, diabetes, and inflammatory diseases [6–9].

Superoxide ($\cdot O_2^-$) is the primary ROS, and it reacts with other molecules to produce other secondary ROS, including hydrogen peroxide, hydroxyl radicals, and alkoxyl radicals [2]. ROS also includes myriad peroxides, including those of lipids, proteins, and nucleic acids. Over 90% of ROS is generated by electrons leaking from the electron transport chain of the mitochondria [10]. Various enzymes within the cell also generate ROS, including xanthine oxidase, NAPDH oxidases (NOXs), and the uncoupled eNOS [11]. Similarly, there are enzymatic and nonenzymatic processes in the cell to eliminate ROS. Antioxidant molecules within the cell interact directly with ROS or act as cofactors for ROS-detoxifying enzymes [1, 12]. Superoxide dismutase catalyzes the reaction of superoxide with addition of protons to produce molecular oxygen and hydrogen peroxide [13]. Hydrogen peroxide is then converted to water and molecular oxygen by catalase or reduced by peroxidases [14]. A mechanism for removal of the hydroxyl radical has not yet been found, likely due to its high reactivity and short half-life [15]. Therefore, prevention of its formation is the best method for protection from its hazardous effects [1].

The primary RNS in the cell is nitric oxide (NO) and also includes peroxynitrite and related compounds [16–18]. NO is generated by nitric oxide synthases during the conversion of arginine to citrulline [19]. NO is a very important signaling molecule involved in numerous biological processes including vasodilation and immune regulation [20]. During inflammatory responses immune cells produce superoxide and NO, which react together to form peroxynitrite anion, a strong oxidizer that causes DNA fragmentation and lipid oxidation [21]. NO reacts with susceptible protein thiol groups to induce S-nitrosylation and S-gluathionylation, which regulates the function of numerous mammalian proteins [11, 22].

Reactive oxygen species (ROS) can cause detrimental damage to cells at high concentrations including DNA base damage, cell fluidity modification, and protein oxidation [23]. This cell damage can eventually lead to large-scale physiological effects, and many kinds of diseases, including cancer and cardiovascular diseases, which are the two leading causes of death in the United States [24, 25]. Some treatments for these diseases involve the use of antioxidants, compounds which quench the activity of ROS [24, 26]. When investigating the pathophysiology of disease, as well as the treatment progress, it is imperative to study the concentration of ROS.

There are many available methods for detecting ROS and RNS in vitro using fluorescent probes, most of which are readily available through chemical providers, and each having unique properties. When selecting a kit for ROS determination, it is important to consider the type of ROS being measured, the expected concentration, the importance on quantification, and the location within the cell of the ROS, among other variables. This chapter will look at four commercially available kits which can be used for detecting ROS and other free radicals in living cells. Each kit has different specificity for certain radical species, as well as differences in protocols and mechanisms. The reagents investigated are dihydroethidium (DHE), dichlorohydrofluoresin diacetate (DCF-DA), and amplex red for ROS detection and 4-amino-5-methylamino-2′,7′-difluorofluorescein diacetate (DAF-FM diacetate) for detection of NO. Becoming familiar with these kits is a vital component to any study investigating oxidative and nitrosative stress, as fluorescent imaging and measurement are convenient and reliable methods of detecting these important and often dangerous byproducts of all living cells.

Dihydroethidium (DHE) for Measuring Superoxide Generation

How It Works

Dihydroethidium (DHE) (1), a fluorogenic probe, strongly reacts with $^{\cdot}O_2^-$, and to a lesser extent with other ROS and NO (Fig. 4.1). After reacting with an $^{\cdot}O_2^-$, which hydroxylates the DHE forming 2-hydroxyethidium (2) [27], this intercalates into DNA and produces a red fluorescence (ex: 490 nm and em: 590 nm) in biological systems. DHE also reacts with other ROS such H_2O_2, it generates ethidium (3), which also produces a red fluorescence (ex: 480 nm and em: 580 nm). MitoSOX™ Red reagent (4) (Fig. 4.2) (Life Technologies, CA) is a cationic derivative (triphe-

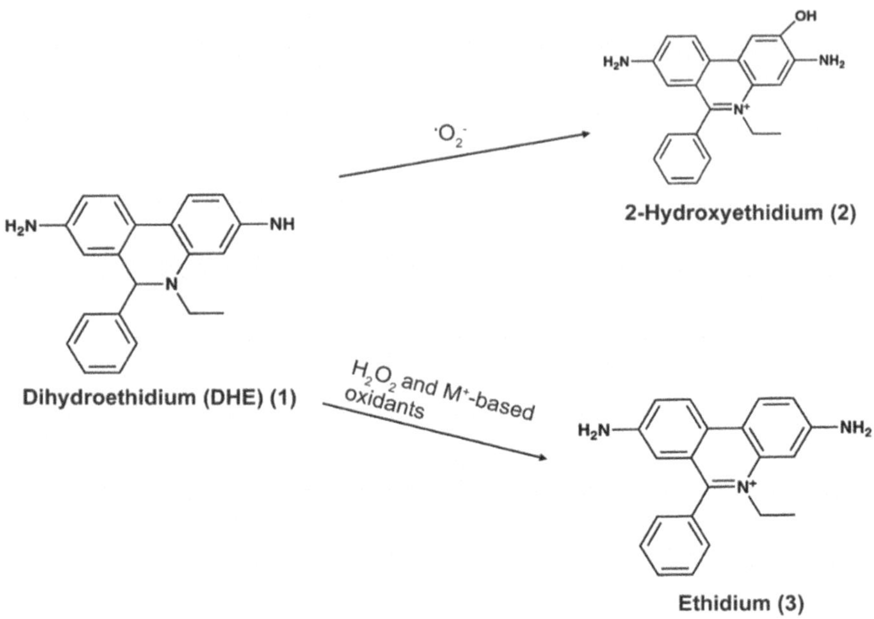

Fig. 4.1 Schematic of DHE reaction with $^{\cdot}O_2^-$ and other species of ROS

Fig. 4.2 Schematic of MitoSOX reagent chemical structure

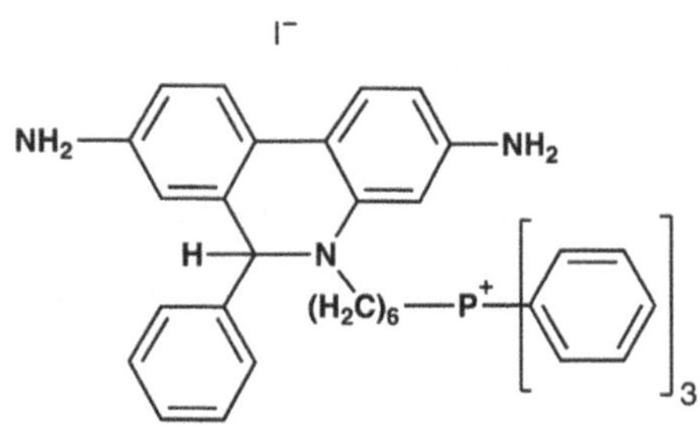

MitoSOX™ (4)

nylphosphonium) of DHE. This derivative is directly targeted to the mitochondria to specifically detect $^{\cdot}O_2^-$ generated in the mitochondria in live cells [27]. Menadione can be used to induce superoxide generation in live cells [28] and act as a positive control for the assay. MnTBAP is a SOD-mimetic [29] and scavenges $^{\cdot}O_2^-$, comparison of cells treated with or without MnTBAP can indicate the level of $^{\cdot}O_2^-$ generated by experimental treatments. The intensity of this red fluorescence is measured by a fluorescent microscope.

Methodology

1. Bovine aortic endothelial cells (BAECs) were cultured to approximately 70% confluence, human-induced pluripotent stem cell-derived cardiomyocytes (hiPSC-CMs) were grown as a confluent monolayer. The cells were washed using PBS, then treated with 10 μM DHE-containing media for 30 min at 37 °C, NucBlue™ (ThermoFisher, MA) was added to label nuclei.
2. DHE-containing media was removed and replaced with untreated growth media (specific for cell type). The cells were imaged with a fluorescent microscope with filters for red and blue fluorescence (Fig. 4.3).

Notes

- If MitoSOX™ is used instead of DHE, incubate for 10 min in 5 μM MitoSOX™-treated media rather than 30 min [30].

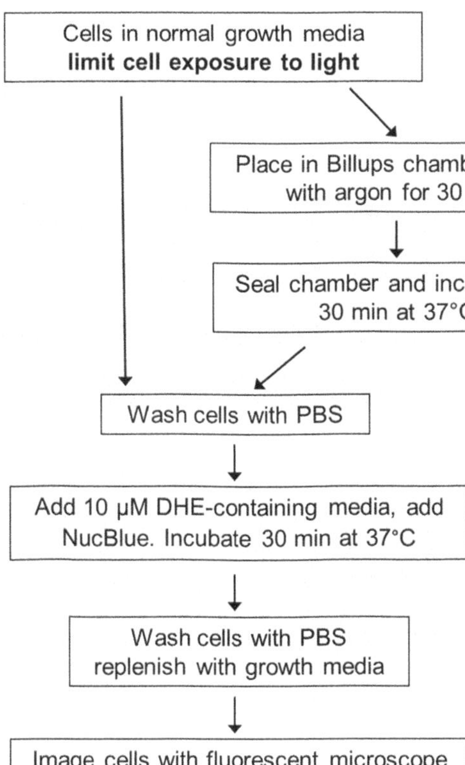

Fig. 4.3 Work flow for DHE/MitoSOX™ treatment

- Dye treatment may be staggered, but this is not necessary as the fluorescence persists for a long time, and staggering may induce unwanted stress. Added stress will give false positive signals of the red fluorescence.
- To create a positive control, treat cells with 25 μM menadione in media and incubate at 37 °C for 15 min prior to treatment with DHE/MitoSOX™.

Results

DHE and MitoSOX™ were tested on BAEC and hiPSC-CM cell cultures. All cells were prepared exactly as described in the Methodology section and a positive control and untreated cells are included. A positive control was created by treating the cells with menadione prior to probe treatment. Menadione promotes ROS generation within the cells [31]. The untreated control was treated only with the probe to assess basal ROS levels.

Image Analysis Images used were selected to have similar cell confluence from each sample, to show that fluorescent difference is due only to changes in the dye activation, and not cell confluence. Positive and untreated controls show a significant difference in fluorescent signal (Figs. 4.4 and 4.5).

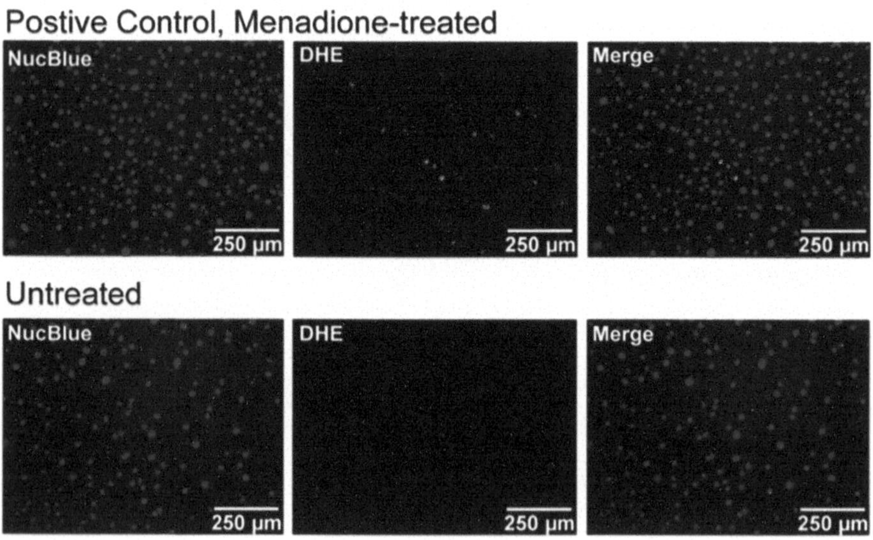

Fig. 4.4 DHE detection of ROS in BAECs in positive control, menadione-treated cells (top panel) and untreated cells (bottom panel)

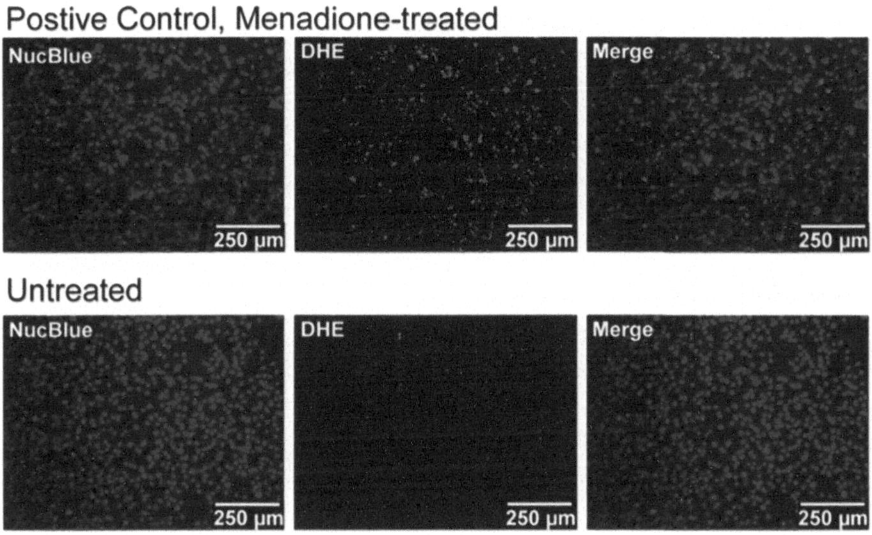

Fig. 4.5 DHE detection of ROS in hiPSC-CMs in positive control, menadione-treated cells (top panel) and untreated cells (bottom panel)

The untreated control images both show some background fluorescence, showing small amounts of ROS generation in those untreated cells. Under identical conditions, MitoSOX™ shows similar results, displaying a large difference in fluorescence between the positive and untreated controls (Fig. 4.6).

Assessment

Pros The signal produced by both DHE and MitoSOX™ is very strong and distinct, with very little background fluorescence outside of the cells. The procedure for treating cells with DHE and MitoSOX™ is very simple and does not require a long incubation period. Fluorescence produced by DHE byproducts degrades slowly, allowing a large time window for imaging and analysis. MitoSOX™ is specific to mitochondria and will not detect ROS outside of mitochondrial membranes.

Cons DHE is not completely exclusive to $^{\cdot}O_2^-$ and can react with other ROS and rarely nitric oxide species, this can make it unsuitable for studies specifically on $^{\cdot}O_2^-$, especially for quantification.

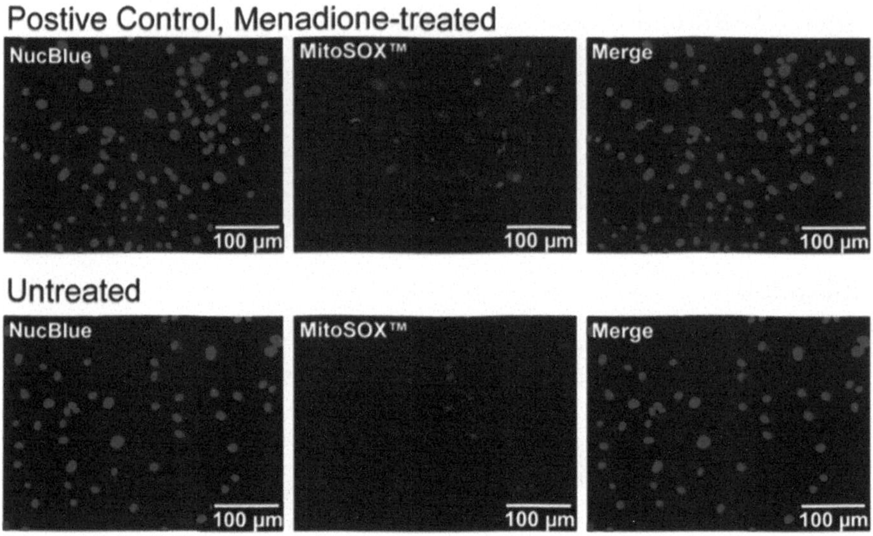

Fig. 4.6 MitoSOX™ detection of ROS in BAECs in positive control, menadione-treated cells (upper panel) and untreated cells (lower panel)

Summary

Overall, Dihydroethidium (DHE) (1) is an excellent and inexpensive fluorescent probe for detecting ROS generation in living cells, specifically for $^{\cdot}O_2^-$. With clear signals, long-lasting fluorescence, and simple procedures, it serves as a great benchmark for cellular redox status measurements. Even though it reacts most strongly with $^{\cdot}O_2^-$, other oxygen free radicals also react with DHE to form the red fluorescence, which cannot be differentiated by using a fluorescent microscope. However, with the proper control experiment, DHE is still a powerful fluorescent probe specifically measuring the level of $^{\cdot}O_2^-$ generation in living cells. In order to overcome this potential obstacle, our previous works [32, 33] and others [34] using high-performance liquid chromatography (HPLC) equipped with C18 column have shown that 2-hydroxyethidium (2), the adduct of $^{\cdot}O_2^-$ and DHE, can be easily separated from ethidium (3), a general oxidation product of DHE, by HPLC method. The level of $^{\cdot}O_2^-$ generated from living cells can be further determined by the quantification of the fluorescent intensity of 2-hydroxyethidium (Fig. 4.1). MitoSOX™ is equally useful, but for assessing the presence of superoxides exclusively in mitochondria. DHE shows responsiveness in both BAECs and hiPSC-CMs, making it a reliable ROS detection method in studies involving these cells.

Dichlorohydrofluoresin Diacetate (DCF-DA) for Measuring Hydroxyl Radicals

How It Works

DCF-DA (5) is a diacetate and nonfluorescent form of DCF, and is cell permeable. After entering the cytoplasm, intracellular esterases will cleave the two ester bonds of DCF-DA to form DCF (6), which in turn reacts with many types of ROS, most specifically hydroxyl radicals ($^{\cdot}OH$) [35]. DCFDA is nonfluorescent and will not be if it remains outside of the cell. The molecule requires de-esterification which can only be catalyzed by intracellular esterases [35]. The resulting molecule is the fluorescent derivative dichlorofluoresin (DCF) (7) (Fig. 4.7). DCF emanates a strong green fluorescent signal (ex: 485 nm and em: 530 nm), which can be easily measured using a fluorescent microscope. Without further treatment, DCFDA will not produce a fluorescent byproduct by reacting with $^{\cdot}O_2^-$, NO, or H_2O_2.

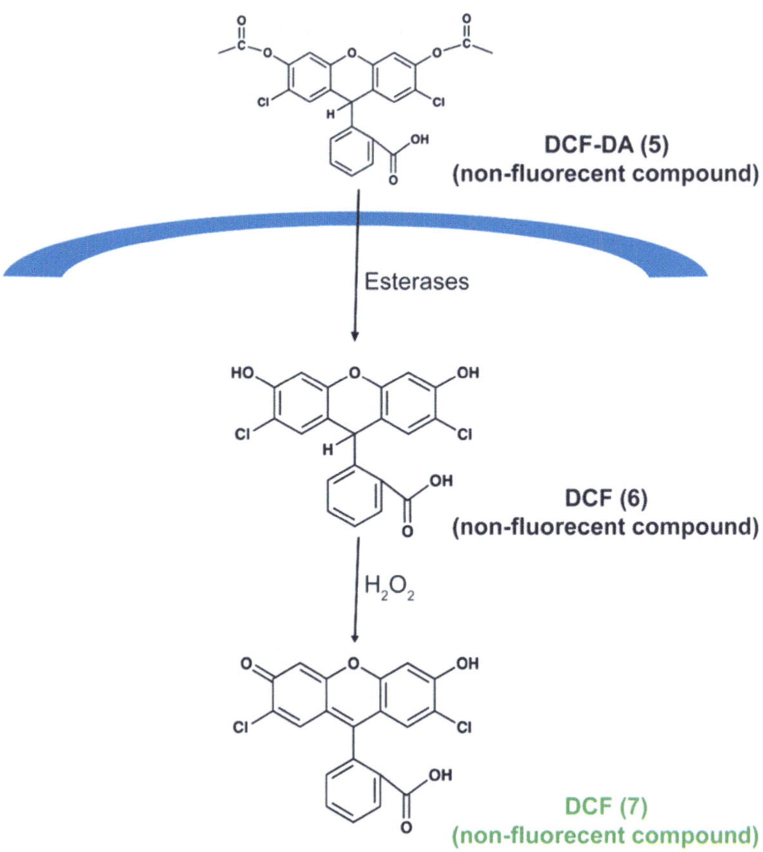

Fig. 4.7 Schematic of detection of hydroxyl radicals by DCF-DA

Methodology

1. BAECs were cultured to 70% confluence. The cells were washed with PBS, then treated with 10 μM DCF-DA-containing media, and incubated for 30 min at 37 °C.
2. DCF-DA media was removed and replaced with basal media, and NucBlue™ was added for nuclear labeling. The cells were incubated for 5–15 min at 37 °C to allow for complete de-esterification of the DCF-DA. NucBlue™ was added for nuclear labeling during this step.
3. Cells were imaged for green and blue fluorescence.

Notes

- Caution must be taken to avoid excess stress to the cells, as DCF-DA reacts very strongly to the ROS produced by cellular stress and will skew data.
- Concentration of DCF-DA will need to be empirically derived based on cell confluence, cell type, and experimental conditions (Fig. 4.8).

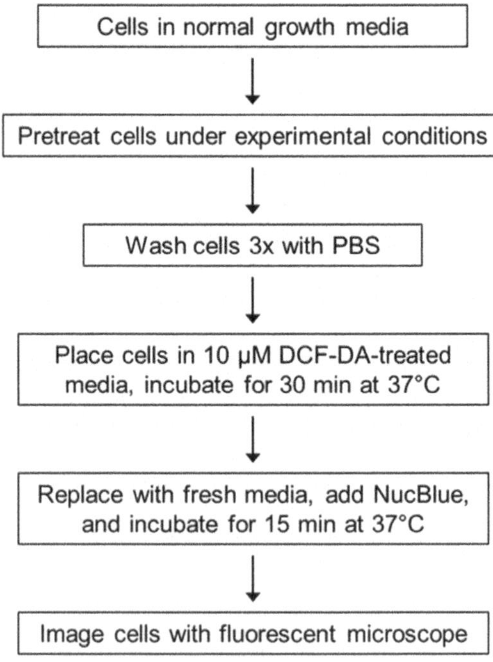

Fig. 4.8 Work flow for DCF-DA

Results

DCF-DA was tested on BAEC cultures using a positive and untreated control. Cells were prepared in a similar fashion as mentioned in the methodology section. Prior to treatment, the positive control samples were treated with menadione-containing media for 15 min to induce ROS generation [31]. There were no ROS quenchers used after the treatment to show any natural ROS presence under standard in vitro conditions (Fig. 4.9).

In addition, another experiment was performed to compare positive control (Menadione-treated) and catalase-treated cells (Fig. 4.9). Catalase decomposes H_2O_2 into H_2O and O_2 [36]. These treatment groups were compared to test whether DCF-DA reacts with H_2O_2 as is, or only with further treatment. For this comparison, the cells were placed under hypoxic conditions to induce ROS generation [37].

Image Analysis Images used were selected to have similar cell confluence, to show that fluorescent difference is due only to changes in dye activation, and not

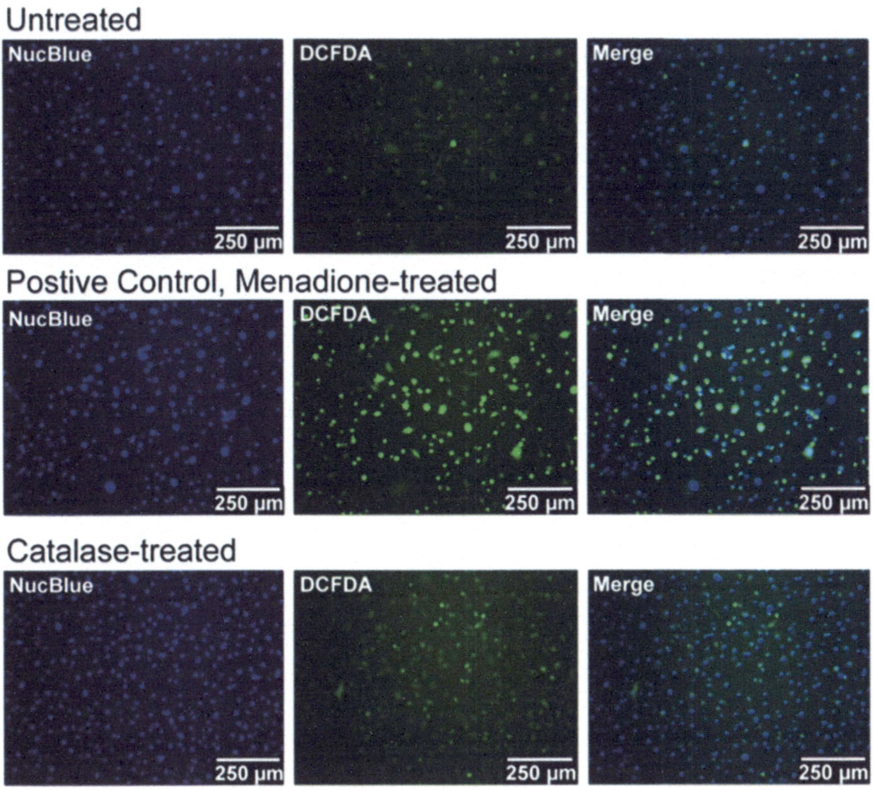

Fig. 4.9 DCF-DA detection of ROS in BAECs in positive control, menadione-treated cells (middle panel) and untreated cells (upper panel) and catalase treated cells (lower panel)

cell number. There is a noticeable difference between the fluorescent signals from the positive and untreated controls (Fig. 4.9). The untreated control shows a decent amount of basal ROS detection. This should be considered when calculating concentration of hydroxyl radicals when using DCF-DA. When comparing a positive control and a catalase-treated sample, no noticeable difference in fluorescent signal is seen (Fig. 4.9). This indicates that DCF-DA is not able to detect H_2O_2, otherwise there would be a decrease in signal. However, DCF-DA can detect H_2O_2 following additional treatment with a transition metal [35].

Assessment

Pros The fluorescent signal produced by DCF-DA is very strong and requires low concentrations for a reliable signal. The signal is also very long-lasting, up to 6 h while maintaining a strong signal [38]. This allows for larger experiments because the risk of data loss is minimal. The protocol is very simple and relatively quick, lasting 45–75 min.

Cons The working concentration varies depending on ROS concentration, as low DCF-DA concentrations are poor at measuring large amounts of ROS, and high DCF-DA concentrations are poor at measuring small amounts on ROS. This means that the concentration of DCF-DA might not be universal across experiments [38].

Summary

DCF-DA is a versatile fluorescent probe for the detection of a variety of ROS generated in cells. It can be used for specifically measuring the concentration of radicalized ROS, such as peroxyl, alkoxyl, $OH^{\cdot}$, $CO_3^{\cdot -}$, and $NO_2^{\cdot}$ as well as peroxynitrite [35]. DCF-DA can be a strong detector of H_2O_2, but only when additional treatment with a transition metal is performed [35]. DCF-DA is currently widely used as a general ROS indicator but could see better use as a method of quantifying concentrations of hydroxyl radicals and H_2O_2.

4-Amino-5-Methylamino-2′,7′-Difluorofluorescein Diacetate (DAF-FM Diacetate) for Nitric Oxide Detection

How It Works

DAF-FM diacetate (8) is a very potent NO probe, able to accurately detect NO at very low concentrations. DAF-FM diacetate is a stable nonfluorescent molecule, capable of diffusing across the cell membrane. Upon entering the cytoplasm, intracellular esterases will cleave the diacetate bond of the molecule to form DAF-FM (9) [39], which in turn will react with NO to form the fluorescent compound Bezotriazole (10) (Fig. 4.10). The molecule exhibits green fluorescence (ex: 495 nm and em: 515 nm). This reaction is irreversible, so the fluorescence will decay gradually during measurement due to photo-bleaching, dye leakage, etc. [40]. The fluorescent signal can be measured using a fluorescent microscope.

Methodology

1. HUVEC cells were cultured to 70% confluence. The cells were washed with PBS and placed in DAF-FM diacetate-treated media (final concentration of 5 μM), then incubated at 37 °C for 1 h.
2. Cells were removed from DAF-FM diacetate and placed in media to allow DAF-FM to be de-esterified by intracellular enzymes. They were incubated for 30 min at 37 °C. NucBlue™ was added for nuclear labeling upon completion of this step.
3. Cells were imaged under blue and green fluorescence (Fig. 4.11).

Notes

- Step 2 is critical, as the DAF-FM diacetate will be unable to react with NO unless this process occurs.
- DAF-FM, once it reacts with NO, decays rather quickly. Images should be acquired within about 20 min to ensure accurate signal.

Results

DAF-FM was tested on HUVEC cultures. Cells were prepared exactly as described in the Methodology section. Following treatment, the positive control samples were treated with S-Nitroso-*N*-acetyl-DL-penicillamine (SNAP), an NO donor. This was

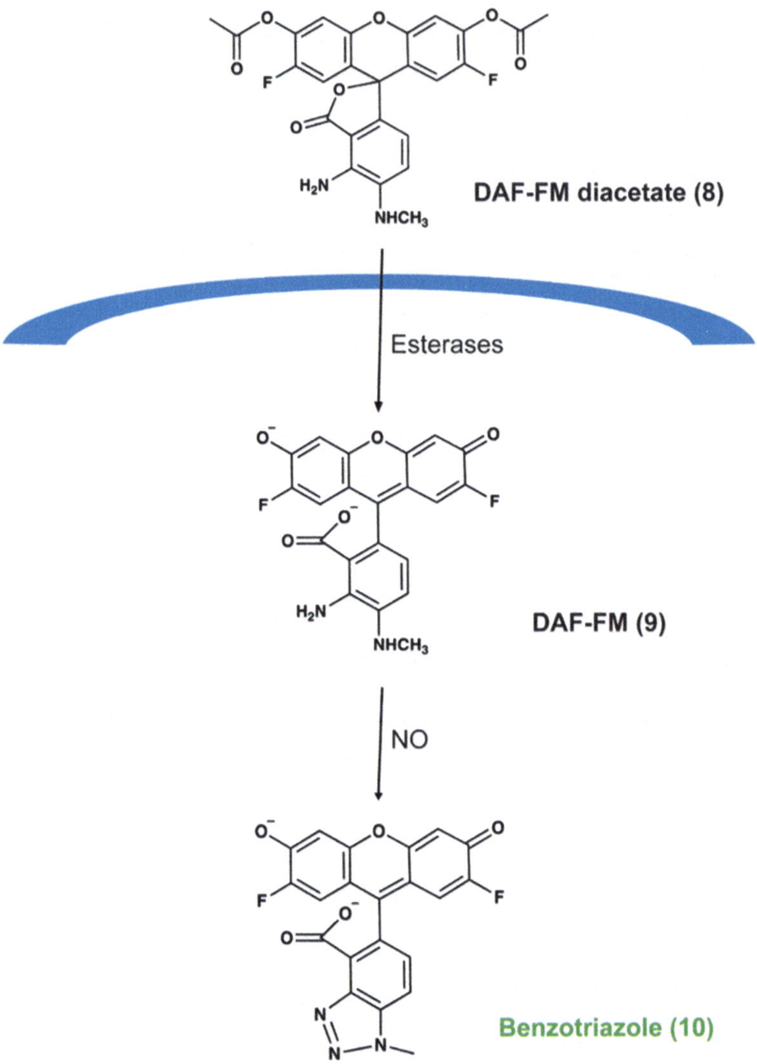

Fig. 4.10 Schematic of mechanism of NO detection by DAF-FM diacetate

to simulate a NO-rich environment for the cells, and to trigger the DAF-FM to form its fluorescent product [41]. The untreated control was not subject to any treatment aside from DAF-FM.

Image Analysis The positive and untreated controls of the HUVEC sample show a clear difference in the fluorescence of DAF-FM in NO-rich environments versus NO-deficient environments (Fig. 4.12). In addition, the complete lack of signal in the untreated control shows very little background radical detection. This shows that

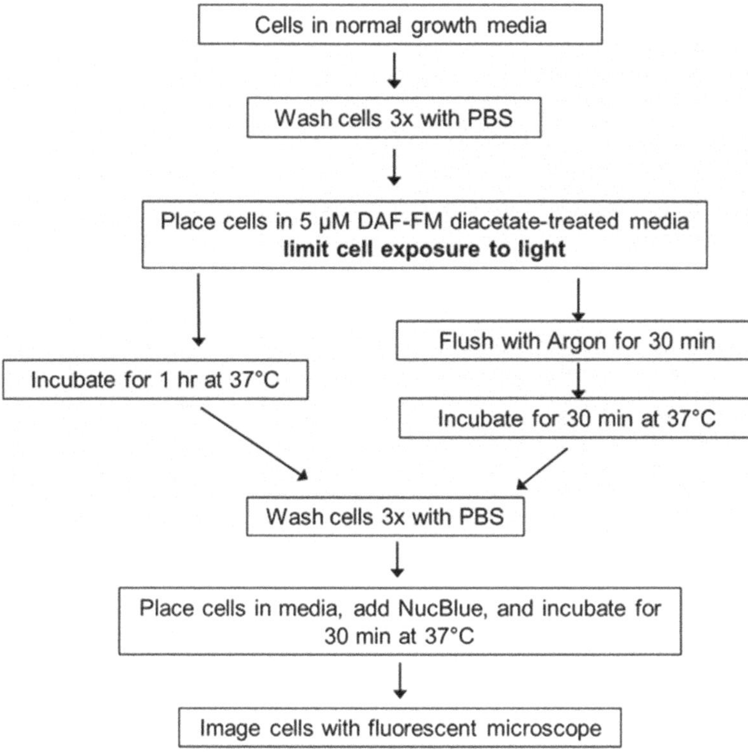

Fig. 4.11 Work flow used for DAF-FM diacetate treatment

DAF-FM is specific to NO, as no other radicals were induced into the positive control.

Assessment

Pros DAF-FM is very specific to NO, making it a great candidate for quantifying concentration of NO. An advantage specific to the diacetate variant is its quality of only reacting with NO inside the cells, due to the required de-esterification process. This leads to more accurate quantification than forms capable of reacting outside of the cell.

Cons A long protocol with limited time for imaging could make this indicator inefficient for larger projects. For a low number of samples, it is much more manageable. Another disadvantage is that this process includes more media changes than similar protocols, this can increase stress in certain cells types, which can influence the collection of data, as stressed cells will often give false positives.

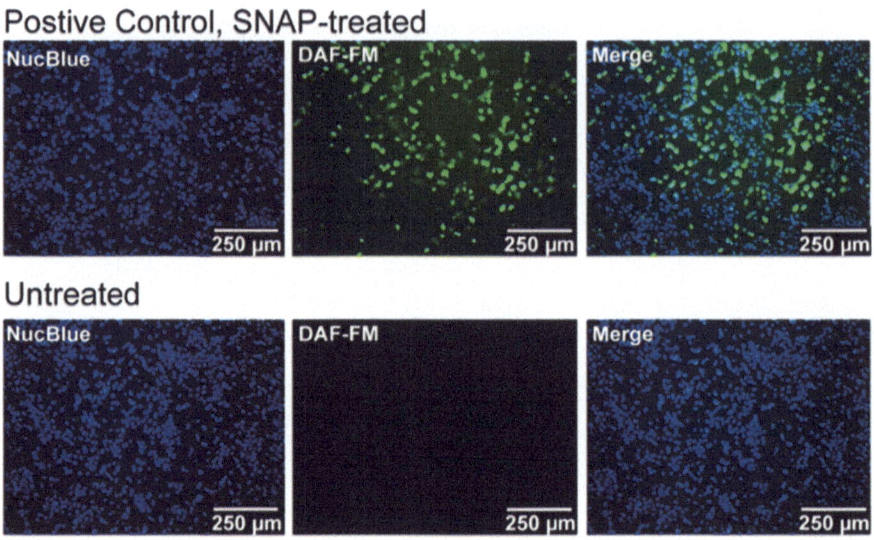

Fig. 4.12 DAF-FM detection of NO in HUVECs with SNAP-treatment for positive control (upper panel) and in untreated cells (lower panel)

Summary

DAF-FM is a fluorescent probe used to specifically detect NO. To use it for this purpose is correct, as it nearly only reacts with NO to form the fluorescent product. This makes the probe work effectively as a quantification tool for NO located within cells. To precisely determine the quantity of NO specifically generated from nitric oxide synthase (NOS) in cells, NOS inhibitors, such as L-N^{ω}-nitroarginine methyl ester (L-NAME) and L-N^{ω}-nitroarginine (L-NNA) can be used. Unfortunately, due to relatively fast signal decay, it is difficult to use for measuring concentration over time and should only be used to time point concentration measurement [40]. Alternatively, cellular-generated NO can be reacted with Iron-*N*-methyl-D-glucamine dithiocarbamate (iron-MGD) to generate a stable NO-iron-MGD complex, which can be quantified using electron paramagnetic resonance (EPR) [11].

10-Acetyl-3,7-Dihyrdroxyphenoxazine (Amplex Red) for Detection of H_2O_2

How It Works

Amplex red reagent is a highly sensitive, nonfluorescent colorless compound used as a stable probe for detecting H_2O_2 generation [42]. It is available from Thermo Fisher Scientific (Cat A1222). Amplex red is a cell impermeable probe, used to

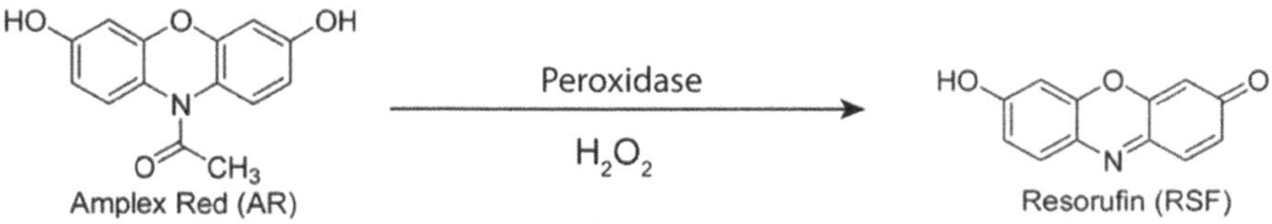

Fig. 4.13 Mechanism of oxidation of amplex red in the presence of H_2O_2 and peroxidase

detect ROS generation in either cells or extracellular environment. As H_2O_2 is freely diffusible, it can be used as a direct indicator of cellular ROS. Amplex red reacts with H_2O_2 at 1:1 molar ratio in the presence of horseradish peroxidase (HRP) [43]. It is oxidized by HRP and H_2O_2 into a fluorescent product called resorufin (Fig. 4.13) [44] which has excitation and emission wavelengths of 560 nm and 590 nm, respectively.

Methodology

The major site of ROS generation in mitochondria are NADH dehydrogenase (complex I) and ubiquinone-cytochrome (complex III) [45, 46]. The leakage of electrons from complex III results in reduction of oxygen to superoxide. In the case of complex I, reverse flow of electrons from complex II to complex I also contributes to increased ROS production [47]. ROS generation by cardiac mitochondria was detected by amplex red using a fluorescence spectrophotometer (Hitachi F-2710) [48, 49]. Briefly, 5 µg of horseradish peroxidase (Sigma-Aldrich) was added to the ROS buffer [mMol/L: 20 Tris-HCl, 250 sucrose, 1 EGTA-Na_4, 1 EDTA-Na_2, pH 7.4 at 37 °C] and the baseline fluorescence was measured at excitation wavelength of 560 nm and emission wavelength of 590 nm. After 60 s, 10 µmol/L amplex red was added followed by 25 µg of mitochondria (120 s). This was followed by addition of 3 mmol/L succinate at 210 s (Sigma-Aldrich) to activate mitochondria. Fluorescence was monitored continuously for 45 min at 5 s resolution.

Results

ROS production in the cardiac mitochondria was measured in the presence and absence of IAA-94 (indanyloxyacetic acid). The IAA-94 is a blocker of chloride channels, specifically chloride intracellular channel (CLIC) proteins [50–55]. In a

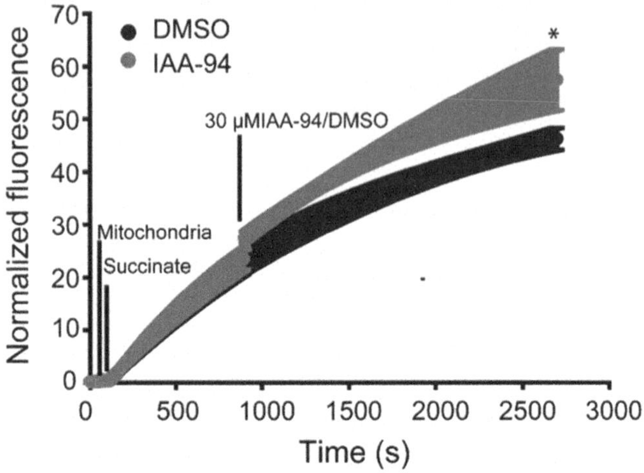

Fig. 4.14 Pharmacological inhibition of chloride channels using IAA-94 in cardiac mitochondria increases ROS generation. Representative traces showing the ROS production in cardiac mitochondria isolated from 3-month-old CD1 mice in the presence and absence of IAA-94. Blocking chloride channels with IAA-94 significantly increased ($p \leq 0.05$, $n = 3$) ROS production upon activation of mitochondrial electron transport chain by complex III substrate, succinate

recent study it was shown that two of the paralogs of mammalian CLIC namely, CLIC4 and CLIC5, are present in the cardiac mitochondria isolated from *R. Norvegicus* [49]. As shown in Fig. 4.14, there was an increased production of ROS in the presence of IAA-94 indicating that chloride channels modulate the ROS production by the cardiac mitochondria. These results are in agreement with the earlier study wherein it was shown that absence of CLIC5 from cardiac mitochondria increases ROS in *clic5*$^{-/-}$ mice [49].

Assessment

Pros It is a highly sensitive method for the detection of H_2O_2 produced with a detection limit of 5 pmol of H_2O_2. The detection range is linear as the stoichiometry of amplex red to H_2O_2 is 1:1. The assay procedure is very simple and the product formed, resorufin, is stable for quite some time [34].

Cons The amplex red dye is unstable and the assay should be performed in the dark. It can get auto-oxidized at higher concentrations and can produce reactive oxygen species. It is cell impermeable and cannot be used to detect intracellular H_2O_2 [34].

Summary

Amplex red is a very sensitive method for detecting ROS in organelles as well as extracellular ROS, which is freely diffusible. It may also be used to analyze permeabilized cells and organelles.

Conclusions and Future Perspectives

Cellular ROS/RNS generation under healthy or normal conditions are constantly modulated by cellular defenses or regulatory systems, namely catalase, superoxide dismutase, or nitric oxide synthase. However, under pathophysiological or disease conditions, these cellular defenses or regulatory systems are perturbed leading to uncontrolled ROS and RNS formation. The overproduction of ROS/RNS is the major cause of many diseases, such as cancer, neurodegenerative diseases, and cardiovascular diseases. Thus, the precise determination of the level and location of ROS/RNS generation in cells could provide suitable treatments for these diseases. In this chapter, we have provided comprehensive information on each individual fluorescent probe in the detection of cellular ROS and RNS generation. These include the basic chemistry behind each probe; the proper way in using them for ROS/RNS measurement; their limitations; and providing alternative methods for each measurement. Even though these commercially available fluorescent probes have stated their specificity toward certain oxygen/nitrogen-free radicals; the chemistry tells the different story. Thus, the use of these fluorescent probes requires careful consideration and proper methodological design. For example, DHE specifically detects superoxide in cells, but the chemistry behind the oxidation of DHE shows that DHE reacts not only with superoxide to form 2-hydroxyethidium but also with other oxidants to form ethidium. Both 2-hydroxyethidium and ethidium exhibit similar fluorescent properties; thus they are unable to be differentiated by fluorescent microscopy. To overcome this potential problem, the use of proper control experiments is required. An alternative method is to separate and quantify these two fluorescent compounds using HPLC. Therefore, we hope that the comprehensive review of these commercially available fluorescent probes will provide general guidelines in the selection and usage of these probes in the measurement of ROS/RNS generation in cells.

Acknowledgement I would like to acknowledge funding from National Institutes of Health (NIH) R01 Grant HL136232 to MK and HL133050 to HS.

References

1. Brieger K, et al. Reactive oxygen species: from health to disease. Swiss Med Wkly. 2012;142:w13659.
2. Valko M, et al. Free radicals and antioxidants in normal physiological functions and human disease. Int J Biochem Cell Biol. 2007;39(1):44–84.
3. Kovacic P, Jacintho JD. Mechanisms of carcinogenesis: focus on oxidative stress and electron transfer. Curr Med Chem. 2001;8(7):773–96.
4. Ridnour LA, et al. Nitric oxide regulates angiogenesis through a functional switch involving thrombospondin-1. Proc Natl Acad Sci U S A. 2005;102(37):13147–52.
5. Valko M, et al. Oxygen free radical generating mechanisms in the colon: do the semiquinones of vitamin K play a role in the aetiology of colon cancer? Biochim Biophys Acta. 2001;1527(3):161–6.
6. Dalle-Donne I, et al. Biomarkers of oxidative damage in human disease. Clin Chem. 2006;52(4):601–23.
7. Dhalla NS, Temsah RM, Netticadan T. Role of oxidative stress in cardiovascular diseases. J Hypertens. 2000;18(6):655–73.
8. Jenner P. Oxidative stress in Parkinson's disease. Ann Neurol. 2003;53(Suppl 3):S26–36; discussion S36–8.
9. Sayre LM, Smith MA, Perry G. Chemistry and biochemistry of oxidative stress in neurodegenerative disease. Curr Med Chem. 2001;8(7):721–38.
10. Skulachev VP. Mitochondria-targeted antioxidants as promising drugs for treatment of age-related brain diseases. J Alzheimers Dis. 2012;28(2):283–9.
11. Chen CA, et al. S-glutathionylation uncouples eNOS and regulates its cellular and vascular function. Nature. 2010;468(7327):1115–8.
12. Lushchak VI. Glutathione homeostasis and functions: potential targets for medical interventions. J Amino Acids. 2012;2012:736837.
13. McCord JM, Fridovich I. Superoxide dismutase. An enzymic function for erythrocuprein (hemocuprein). J Biol Chem. 1969;244(22):6049–55.
14. Commoner B, Townsend J, Pake GE. Free radicals in biological materials. Nature. 1954;174(4432):689–91.
15. Pastor N, et al. A detailed interpretation of OH radical footprints in a TBP-DNA complex reveals the role of dynamics in the mechanism of sequence-specific binding. J Mol Biol. 2000;304(1):55–68.
16. Radi R. Peroxynitrite, a stealthy biological oxidant. J Biol Chem. 2013;288(37):26464–72.
17. Ferrari CK, et al. Oxidative and nitrosative stress on phagocytes' function: from effective defense to immunity evasion mechanisms. Arch Immunol Ther Exp. 2011;59(6):441–8.
18. Bild W, et al. The interdependence of the reactive species of oxygen, nitrogen, and carbon. J Physiol Biochem. 2013;69(1):147–54.
19. Ghafourifar P, Cadenas E. Mitochondrial nitric oxide synthase. Trends Pharmacol Sci. 2005;26(4):190–5.
20. Bergendi L, et al. Chemistry, physiology and pathology of free radicals. Life Sci. 1999;65(18–19):1865–74.
21. Carr AC, McCall MR, Frei B. Oxidation of LDL by myeloperoxidase and reactive nitrogen species: reaction pathways and antioxidant protection. Arterioscler Thromb Vasc Biol. 2000;20(7):1716–23.
22. Martinez MC, Andriantsitohaina R. Reactive nitrogen species: molecular mechanisms and potential significance in health and disease. Antioxid Redox Signal. 2009;11(3):669–702.
23. Bayr H. Reactive oxygen species. Crit Care Med. 2005;33(12):S498–501.
24. Pizzino G, et al. Oxidative stress: harms and benefits for human health. Oxidative Med Cell Longev. 2017;2017:8416763.
25. National Center for Health Statistics. Health, United States, 2016: with chartbook on long-term. Trends in Health; 2017.

26. Mittler R. Oxidative stress, antioxidants and stress tolerance. Trends Plant Sci. 2002;7(9):405–10.
27. Held, P. (2015). "An introduction to reactive oxygen species." Acessed Sept 2018, from https://www.biotek.es/assets/tech_resources/ROS%20White%20Paper_2015.pdf.
28. Thor H, et al. The metabolism of menadione (2-methyl-1,4-naphthoquinone) by isolated hepatocytes. A study of the implications of oxidative stress in intact cells. J Biol Chem. 1982;257(20):12419–25.
29. Faulkner KM, Liochev SI, Fridovich I. Stable Mn(III) porphyrins mimic superoxide dismutase in vitro and substitute for it in vivo. J Biol Chem. 1994;269(38):23471–6.
30. Probes M. MitoSOX™ Red mitochondrial superoxide indicator, for live-cell imaging (M36008). Eugene, OR; 2005.
31. Loor G, et al. Menadione triggers cell death through ROS-dependent mechanisms involving PARP activation without requiring apoptosis. Free Radic Biol Med. 2010;49(12):1925–36.
32. Barajas-Espinosa A, et al. Redox activation of DUSP4 by N-acetylcysteine protects endothelial cells from Cd(2+)-induced apoptosis. Free Radic Biol Med. 2014;74:188–99.
33. Barajas-Espinosa A, et al. Modulation of p38 kinase by DUSP4 is important in regulating cardiovascular function under oxidative stress. Free Radic Biol Med. 2015;89:170–81.
34. Dikalov S, Griendling KK, Harrison DG. Measurement of reactive oxygen species in cardiovascular studies. Hypertension. 2007;49(4):717–27.
35. Eruslanov E, Kusmartsev S. Identification of ROS using oxidized DCFDA and flow-cytometry. In: Armstrong D, editor. Advanced protocols in oxidative stress II. Totowa, NJ: Humana Press; 2010. p. 57–72.
36. Aebi H. Catalase in vitro. Methods Enzymol. 1984;105:121–6.
37. Kondoh M, et al. Hypoxia-induced reactive oxygen species cause chromosomal abnormalities in endothelial cells in the tumor microenvironment. PLoS One. 2013;8(11).
38. Figueroa D, Asaduzzaman M, Young F. Real time monitoring and quantification of reactive oxygen species in breast cancer cell line MCF-7 by 2′,7′–dichlorofluorescin diacetate (DCFDA) assay. J Pharmacol Toxicol Methods. 2018;94:26–33.
39. Probes M. Nitric oxide indicators: DAF-FM and DAF-FM diacetate. Eugene, OR; 2001.
40. Namin SM, et al. Kinetic analysis of DAF-FM activation by NO: toward calibration of a NO-sensitive fluorescent dye. Nitric Oxide. 2013;28:39–46.
41. Zhang Y, et al. The nitric oxide donor S-nitroso-N-acetylpenicillamine (SNAP) increases free radical generation and degrades left ventricular function after myocardial ischemia-reperfusion. Resuscitation. 2003;59(3):345–52.
42. Zhao B, Summers FA, Mason RP. Photooxidation of Amplex Red to resorufin: implications of exposing the Amplex Red assay to light. Free Radic Biol Med. 2012;53(5):1080–7.
43. Gorris HH, Walt DR. Mechanistic aspects of horseradish peroxidase elucidated through single-molecule studies. J Am Chem Soc. 2009;131(17):6277–82.
44. Mishin V, et al. Application of the Amplex red/horseradish peroxidase assay to measure hydrogen peroxide generation by recombinant microsomal enzymes. Free Radic Biol Med. 2010;48(11):1485–91.
45. Murphy MP. How mitochondria produce reactive oxygen species. Biochem J. 2009;417(1):1–13.
46. Zorov DB, Juhaszova M, Sollott SJ. Mitochondrial reactive oxygen species (ROS) and ROS-induced ROS release. Physiol Rev. 2014;94(3):909–50.
47. Kurutas EB. The importance of antioxidants which play the role in cellular response against oxidative/nitrosative stress: current state. Nutr J. 2016;15(1):71.
48. Singh H, et al. Visualization and quantification of cardiac mitochondrial protein clusters with STED microscopy. Mitochondrion. 2012;12(2):230–6.
49. Ponnalagu D, et al. Molecular identity of cardiac mitochondrial chloride intracellular channel proteins. Mitochondrion. 2016;27:6–14.
50. Landry DW, et al. Epithelial chloride channel. Development of inhibitory ligands. J Gen Physiol. 1987;90(6):779–98.

51. Landry DW, et al. Purification and reconstitution of chloride channels from kidney and trachea. Science. 1989;244(4911):1469–72.
52. Ponnalagu D, Singh H. Anion channels of mitochondria. Handb Exp Pharmacol. 2017;240:71–101.
53. Jiang L, et al. Intracellular chloride channel protein CLIC1 regulates macrophage function through modulation of phagosomal acidification. J Cell Sci. 2012;125(Pt 22):5479–88.
54. Gururaja Rao S, et al. Three decades of chloride intracellular channel proteins: from organelle to organ physiology. Curr Protoc Pharmacol. 2018;80(1):11.21.1–11.21.17.
55. Singh H. Two decades with dimorphic chloride intracellular channels (CLICs). FEBS Lett. 2010;584(10):2112–21.

Chapter 5
Oxidative Lipidomics: Analysis of Oxidized Lipids and Lipid Peroxidation in Biological Systems with Relevance to Health and Disease

Surya T. Kodali, Philip Kauffman, Sainath R. Kotha, Anita Yenigalla, Rengasayee Veeraraghavan, Sonal R. Pannu, Thomas J. Hund, Abhay R. Satoskar, Jodi C. McDaniel, Rao K. Maddipati, and Narasimham L. Parinandi

Abstract Oxidation of polyunsaturated (PUFA) lipids, catalyzed by free radicals either by non-enzymatic or enzymatic mechanisms, is highly critical in living cells during toxicity, pathophysiological events and disease states, radiation exposure, xenobiotic and drug metabolism, and aging. However, the methods of analysis to characterize the structures of different peroxidized lipids in cells (including PUFA, either free or esterified in the cellular membrane PLs) and the oxidative lipid metabolites, and determination of the extent of lipid peroxidation in the biological systems, are challenging and tedious due to the choice/availability of a suitable

---

S. T. Kodali · P. Kauffman · S. R. Kotha · A. Yenigalla · S. R. Pannu
Division of Pulmonary, Critical Care, and Sleep Medicine, Department of Medicine, The Ohio State University, Columbus, OH, USA

R. Veeraraghavan · T. J. Hund
Department of Biomedical Engineering, College of Engineering, The Ohio State University, Columbus, OH, USA

A. R. Satoskar
Department of Pathology, College of Medicine, The Ohio State University, Columbus, OH, USA

J. C. McDaniel
College of Nursing, The Ohio State University, Columbus, OH, USA

R. K. Maddipati
Department of Pathology, Wayne State University, Detroit, MI, USA

N. L. Parinandi (✉)
Division of Pulmonary, Critical Care, Sleep Medicine, Lipid Signaling, Lipidomics, and Vasculotoxicity Laboratory, Dorothy M. Davis Heart & Lung Research Institute, The Ohio State University College of Medicine, Columbus, OH, USA
e-mail: narasimham.parinandi@osumc.edu

L. J. Berliner, N. L. Parinandi (eds.), *Measuring Oxidants and Oxidative Stress in Biological Systems*, Biological Magnetic Resonance 34,
https://doi.org/10.1007/978-3-030-47318-1_5

analytical method and also the complexity of different peroxidation products of lipids formed in a temporal scale which is system-dependent. Therefore, the current chapter discusses various available analytical methods/techniques to analyze the oxidized lipids and determine lipid peroxidation in the biological systems with an emphasis on health and disease. Finally, the importance of oxidative lipidomics has been discussed as a rapidly emerging discipline of measuring oxidative stress.

Keywords Analysis of lipid peroxides · Lipid peroxidation assay · Determination of oxidized lipids · Oxidative lipidomics · Polyunsaturated fatty acid peroxidation analysis · PUFA hydroperoxide determination

Abbreviations

4-HNE	4-hydroxynonenal
BD-AA	BODIPY-conjugated arachidonic acid
BD-IAM	BODIPY-iodoacetamide
BHT	Butylated hydroxytoluene
BODIPY	Boron-dipyrromethene
COX	Cyclooxygenase
CVD	Cardiovascular/cerebrovascular disease
DMPO	5,5-Dimethyl-pyrroline-1-oxide
DNPH	2,4-dinitrophenylhydrazine
DTC	Differential thermal calorimetry
ELISA	Enzyme-linked immunosorbent assay
EPR	Electron paramagnetic resonance
ESI	Electrospray ionization
FOX	Ferrous oxidation-xylenol orange
GC-MS	Gas chromatography-mass spectrometry
HDL	High-density lipoprotein
HETE	Hydroxyeicosatetraenoic acid
HPLC	High-performance liquid chromatography
LDL	Low-density lipoprotein
LOX	Lipoxygenase
MALDI	Matrix-assisted laser desorption/ionization
MDA	Malonaldehyde
MDMS	Multi-dimensional mass spectrometry
NMR	Nuclear magnetic resonance
PD	Parkinson's disease
POBN	α-(4-pyridyl-1-oxide)-*N*-tert-butylnitrone

PUFA	Polyunsaturated fatty acids
RNS	Reactive nitrogen species
ROS	Reactive oxygen species
SDS-PAGE	Sodium dodecyl sulfate-polyacrylamide gel electrophoresis
SOD	Superoxide dismutase
TBA	Thiobarbituric acid
TBARS	Thiobarbituric acid-reactive substances
TLC	Thin-layer chromatography
TXB2	Thromboxane B2

Introduction

Lipids constitute the predominant class of macromolecules that are critical for both structure and function along with carbohydrates and proteins in the biological systems. Phospholipids (PLs), an important class of lipids, are the chief integral molecules of the cellular/biological membranes and bioactive signal molecules. PLs contain both saturated and unsaturated fatty acids esterified to their glycerol backbones which contribute to the fluidity of the biological membranes. Triglycerides are mainly storage lipids, which contain esterified fatty acids in the glycerol backbone. The unsaturated fatty acids, especially the polyunsaturated ones (PUFAs) (Fig. 5.1), are vulnerable to oxidative attack by highly reactive species [oxygen radicals, organic radicals, reactive oxygen species (ROS), and reactive nitrogen species (RNS)] that cause peroxidation of PUFA, which leads to the deterioration of the lipids. Since the dawn of civilization, it has been known that fats (rich in PUFA) become rancid. Rancidity of fats is defined as the "Deterioration or Degradation of Fats," which renders the fats unpalatable and unhealthy due to the autoxidation of PUFA in the fats. This has been known in the food industry, and it is a challenge to counteract/prevent autoxidation of fats leading to rancidity. Oxygen (which is eight times more soluble in fats/oils) and water are the culprits causing the rancidity of fats through autoxidation. As the stability of PUFAs from oxidative deterioration is highly important, the peroxidation of PUFA (commonly called "lipid peroxidation") is an important process of focus in agriculture (dairy, meat, food, and oil industry), nutrition, food technology, toxicology, pharmacology, aging, health, and disease. In order to determine the oxidative/peroxidative damage or deterioration of PUFA-containing lipids in the biological systems, a suitable method of analysis is fundamental. Hence, the current chapter discusses (*i*) process and mechanisms of lipid peroxidation, (*ii*) impact of oxidized lipids and lipid peroxidation on health and disease; and (*iii*) several available methods of determination/analysis of oxidized lipids and lipid peroxidation.

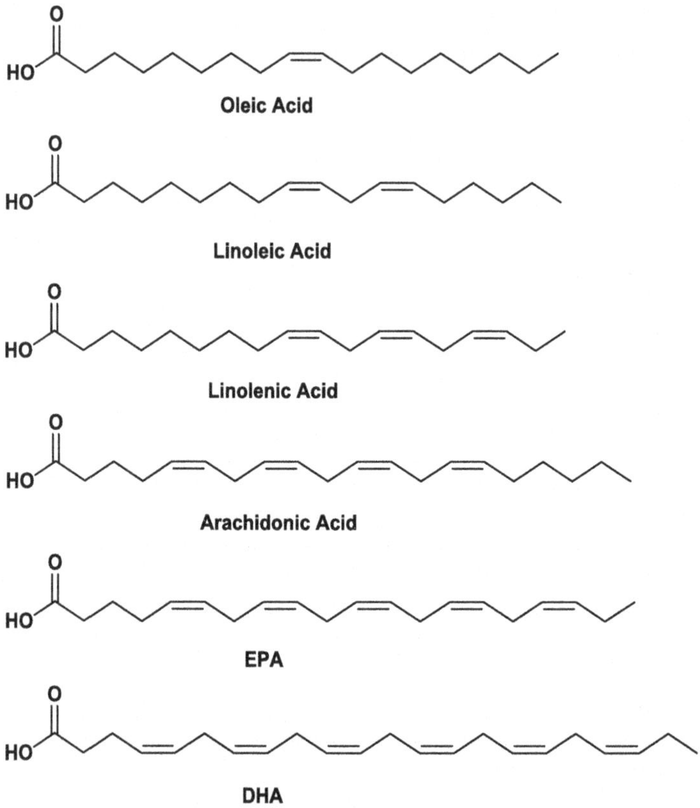

Fig. 5.1 Structures of polyunsaturated fatty acids (PUFAs). These are selected unsaturated fatty acids. Oleic acid (C18:1ω9; (9Z)-Octadec-9-enoic acid) is a ω9-monounsaturated fatty acid. Linoleic acid (C18:2ω6; (9Z,12Z)-octadeca-9,12-dienoic acid is a ω6-PUFA. α-Linolenic acid (C18:3ω3; (9Z,12Z,15Z)-octadeca-9,12,15-trienoic acid) is a ω3-PUFA. Arachidonic acid (C20:4ω6; (5Z,8Z,11Z,14Z)-5,8,11,14-Eicosatetraenoic acid) is a ω6-PUFA. Eicosapentaenoic acid (C20:5ω3; (5Z,8Z,11Z,14Z,17Z)-5,8,11,14,17-eicosapentaenoic acid) is a ω3-PUFA. Docosahexaenoic acid (C22:6ω3; (4Z,7Z,10Z,13Z,16Z,19Z)-docosa-4,7,10,13,16,19-hexaenoic acid) is a ω3-PUFA. All double bonds in these unsaturated fatty acids are in *cis* configuration. These fatty acids are either free or esterified in glycerides, lipoproteins, and phospholipids (PLs) of membranes) contributing to the structure and function of the cell and the whole system in general. The methylene-interrupted double bonds of PUFA are the targets of free radical-mediated oxidative attack

Process and Mechanisms of Peroxidation and Oxidative Deterioration of Lipids

As stated by Pryor [1], free radical biology (role/participation of free radicals in many biochemical reactions) was enigmatic and considered as an arcane science for a considerable time in the scientific community. However, three prominent discoveries, mainly (*i*) the function and role of superoxide dismutase (SOD) by

McCord and Fridovich [2], (*ii*) the biosynthesis of prostaglandins from arachidonic acid [3], and (*iii*) the environmental toxins-mediated toxicity through free radical reactions in the biological systems have transformed the discipline of biological chemistry and have created a new field of science called "Free Radical Biology." According to Bolton [4], a free radical is any molecule that has an odd number of electrons. Free radicals are involved in the normal cellular metabolism, pathological states, and aging [5]. The free radicals are formed during the: (*i*) unimolecular hemolysis of molecules, (*ii*) radiolysis, (*iii*) photolysis, (*iv*) one-electron transfer from transition metals to organic species, and (*v*) enzymatic generation in biological systems (xanthine oxidase, aldehyde oxidase, dihydroorotic acid dehydrogenase, galactose oxidase, flavin dehydrogenase, indoleamine dioxygenase, diamine oxidase, ribulose-1,5-bisphosphate carboxylase [6–9]. The autoxidation of molecules including the hydroquinones, flavins, catecholamines, thiols, tetrahydropterins, ferrodoxins, and hemoglobin is also known to form free radicals [7, 10, 11]. In plants, the chloroplasts are shown to generate free radicals (e.g. superoxide) during photosynthesis [7]. Lipid oxygenases such as the cyclooxygenases (COXs) and lipoxygenases (LOXs) form PUFA peroxides from PUFAs (mainly arachidonic acid) through free radical-mediated catalysis [12]. It is established that the oxygen free radicals and ROS are produced during the bacterial infection, inflammation, and phagocytosis in neutrophils, monocytes, and macrophages [13]. Xenobiotic (toxic) compounds such as herbicides (e.g. paraquat), carbon tetrachloride, bromotrichloromethane, drugs (e.g. adriamycin and daunomycin), nitrogen oxides, cigarette smoke, ozone, sulfur dioxide, sulfite, and transition metals induce or mediate the formation of the reactive free radical intermediates in the biological systems [1, 14–18]. Also, in recent times, free radical reactions in the biological systems have been established in several metabolic pathways such as the normal metabolism (mitochondrial electron transport and respiration, photosynthesis, phagocytosis, arachidonic acid cascade, and eicosanoid biosynthesis), xenobiotic metabolism (toxicity of ozone, drugs, herbicides, environmental contaminants, transition metals), pathological states (cancer, cardiovascular and cerebrovascular diseases, and ischemia-reperfusion damage of myocardium), and aging [1, 5, 19] (Fig. 5.2).

As the biological systems are rich in PUFA (mostly in the cellular membrane PLs), surrounded by an oxygen-rich and transition metal-containing environment (usually iron, Fe), membrane PUFA are vulnerable to oxidative attack by free radicals; such reactions are encountered in numerous pathophysiological states leading to cell membrane alterations and deterioration [20]. The living cell has a PL (with esterified PUFA) bilayer-containing outer plasma membrane and membrane-bound intracellular organelles such as the Golgi body, lysosomes, mitochondria, endoplasmic reticulum, nucleus, and chloroplasts (Fig. 5.3). The cellular membrane bilayer PLs, either hydrophobic or amphipathic, support the formation of radical centers in a nonaqueous environment, while the amphipathic lipids are considered as the most critical components of cellular membranes [21]. PUFAs have an intrinsic characteristic to undergo oxidative attack (autoxidation) and deterioration caused by free radicals, especially in the presence of oxygen [20]. The following is the most

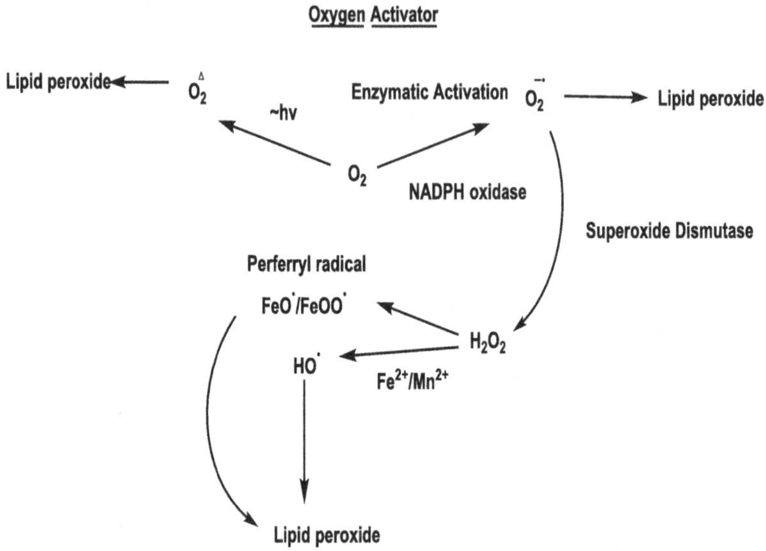

Fig. 5.2 Oxygen activation and formation of reactive oxygen species (ROS) and other reactive intermediates. Ground-state molecular oxygen (O_2) is activated by different mechanisms. Molecular oxygen is excited by high-energy radiation into singlet oxygen, which is a strong oxidant and causes lipid peroxidation of PUFA. Also, molecular oxygen is activated by different enzymes (oxygen-activating enzymes, e.g., NADPH oxidase) into superoxide ($O_2^{-\cdot}$) anion that is converted into hydrogen peroxide by the action of superoxide dismutase (SOD), which forms the most potent hydroxyl radical (OH^-) through reaction with iron (Fe). Hydroxyl radical causes peroxidation of PUFA. Oxygen reactive intermediates of iron, the perferryl species, are potent oxidants capable of causing peroxidation of PUFA

generalized and widely acceptable sequence of reactions during the course of peroxidation of PUFAs:

$$\text{PUFA}-\text{H}+\text{R}^{\cdot}\rightarrow\text{PUFA}^{\cdot}+\text{RH} \tag{5.1}$$

$$\text{PUFA}^{\cdot}+\text{O}_2\rightarrow\text{PUFA}-\text{O}-\text{O}^{\cdot} \tag{5.2}$$

$$\text{PUFA}-\text{O}-\text{O}^{\cdot}+\text{PUFA}\rightarrow\text{PUFA}-\text{O}-\text{O}-\text{H}+\text{PUFA}^{\cdot} \tag{5.3}$$

In the above sequence of reactions, an initiation of free radical ($R^{\cdot}$) attack on the PUFA leads to the formation of lipid (PUFA) hydroperoxides: (5.1) $R^{\cdot}$ abstracts hydrogen from the PUFA at the methylene-interrupted double bond; (5.2) a free radical of PUFA (PUFA·) is formed, following the reaction with molecular oxygen (O_2) that leads to the formation of a PUFA peroxy radical (PUFA–O–O˙); and (5.3) this PUFA peroxy radical reacts with another PUFA molecule to form a lipidfree radical (PUFA·) and the PUFA hydroperoxide (Fig. 5.4).

Oxidative breakdown/deterioration of PUFA leads to the formation of carbonyls including the malonaldehyde (MDA) [22] through a sequence of reactions as out-

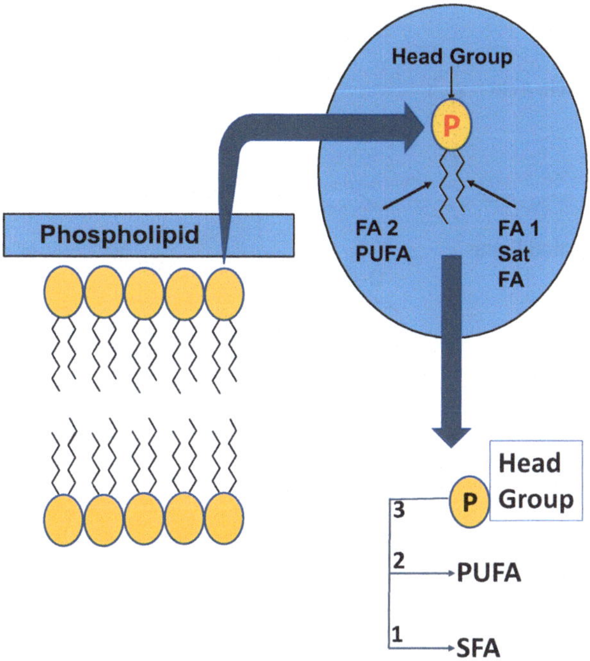

Fig. 5.3 Structure of membrane phospholipid (PL). PLs are integral components of the cellular membrane bilayers. The typical PL has the glycerol backbone with the esterified saturated fatty acid at *sn*-1 and the unsaturated fatty acid at the *sn*-2 position. The third carbon contains the phosphorus and head group of the PL. The PUFA ester at the *sn*-2 position is vulnerable to the oxidative attack by the reactive radicals, leading to formation of the PL-esterified PUFA lipid peroxidation products

lined here: (*i*) PUFA undergoes hydrogen abstraction through the reaction with the radical ($R^{\cdot}$) at the methylene-interrupted double bond; (*ii*) rearrangement of double bonds; (*iii*) oxygen uptake; (*iv*) formation of 5-membered ring; and finally the (*v*) bond scission leading to the formation of MDA. The conjugated diene arrangement of double bonds occurs in the PUFA (in reaction-*ii*) that is responsible for the increased absorption of UV light at 232–233 nm by the PUFA undergoing peroxidation—this conjugated diene rearrangement of PUFA double bonds is a characteristic feature of the peroxidizing lipids that can be used in the analytical determination of peroxidation of PUFA. On the other hand, lipid peroxidation causes extensive damage to the fine structure of the cellular membranes. The degradation/decomposition products arising from the PUFA peroxidation of cellular membranes (carbonyls) can crosslink with the proteins, lipids, and nucleic acids and form adducts. Thus, the scission, addition, and crosslinking reactions are critical in the free radical-mediated cellular membrane damage, especially in the exposure to radiation, xenobiotics, and in many disease states and aging (Figs. 5.5 and 5.6). It has been emphasized that the rigidity of cellular membranes increases with the advancement of age, probably due to the free radical-induced peroxidation of membrane PL PUFAs [23]. The

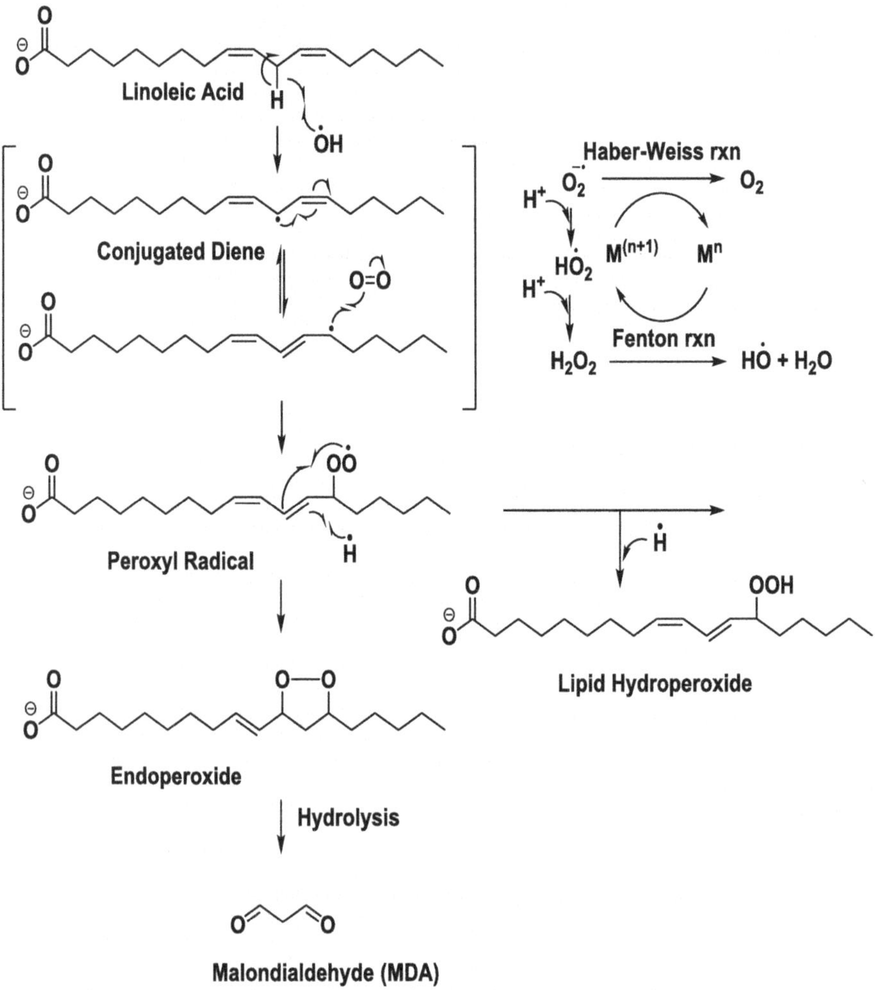

Fig. 5.4 Mechanism of lipid peroxidation. Oxygen free radicals (˙OH) initiate free radical attack at the methylene-interrupted double bonds of the PUFA followed by the conjugated diene rearrangement of the double bond, molecular oxygen insertion, formation of PUFA-OOH (lipid hydroperoxide), lipid peroxy radical, and further chain reaction leading to enhancement of lipid peroxidation. The PUFA-OO˙ (lipid peroxy radical) is converted into an endoperoxide that undergoes hydrolysis leading to the formation of malondialdehyde (MDA), the lipoperoxidative aldehyde. Redox-active transition metals (e.g. Fe) play a major role (in Haber-Weiss and Fenton reactions) here both in the formation of oxygen radicals and ROS and also scission of hydrogen peroxide and the PUFA hydroperoxide in to different reactive oxidized lipid species

crosslinking of aldehydes (electrophilic carbonyls) with the most crucial cellular biomolecules (peptides, proteins, and nucleic acids) could lead to both structural and functional modifications/alterations in the biological systems. One of the important reactive products (aldehydes) of lipid peroxidation, MDA (CHO–CH_2–CHO),

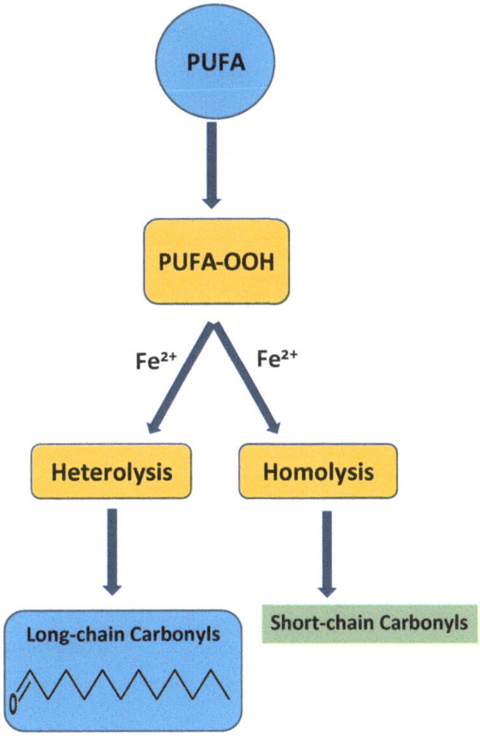

Fig. 5.5 Scission of polyunsaturated fatty acid hydroperoxides (PUFA-OOH) into carbonyls. Transition metals (e.g. iron, Fe^{2+}) cause the scission of pre-formed PUFA hydroperoxide (PUFA-OOH) in to either long-chain or short-chain carbonyls by either heterolysis or hemolysis, respectively

has been demonstrated to form the Schiff base adducts with amines of proteins, PLs, and nucleic acids in the biological systems [24]. These Schiff base adducts (biopolymers) attain large sizes with unusual chemical bonds, accumulate in lysosomes, and exhibit fluorescence at 360–380 nm excitation maxima and 440–470 nm emission maxima; these fluorescent Schiff base adducts originating from lipid peroxidation are called the "aging pigments" or "lipofuscin" or "ceroid," whose accumulation is correlated with the advancement of age of the organism [24, 25]. Thus, the free radical-mediated oxidative deterioration of membrane lipids has been shown to play a crucial role in the pathological state of cellular membranes, crosslinking and inactivation of enzymes, alterations in the function(s) of the genetic material (DNA), and ultimately changes in cellular metabolism. Therefore, the nature (molecular structure) and extent (levels/concentration) of formation of peroxidized PUFA, carbonyls (aldehydes such as MDA and others), crosslinking and formation of fluorescent Schiff base adducts (lipofuscin and other fluorescent pigments) could be suitable/appropriate indices to determine the extent/degree of oxidative damage of cellular membrane lipids (peroxidation) in the biological systems. During the peroxidation of PUFA, in addition to the formation of highly reactive electrophilic carbonyls (aldehydes), certain volatile hydrocarbons including ethane and pentane are produced by the biological systems and it has been demonstrated that the nature and concentrations of those volatile hydrocarbons serve as a reliable index of

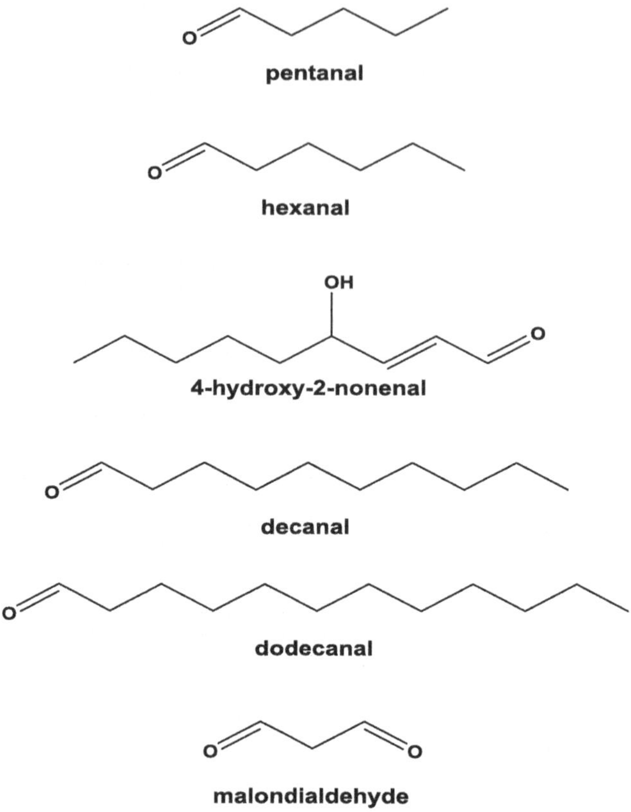

Fig. 5.6 Structures of lipid peroxidation-generated carbonyls (aldehydes). Structures of some selected aldehydes (carbonyls) formed during peroxidation of the PUFA. Some of these carbonyls are strong electrophiles and react with proteins, PLs (e.g. phosphatidylethanolamine, PE), and nucleic acids to form Shciff base adducts (reacting with the nucleophilic $-NH_2$) and Michael addition products (reacting with $-NH_2$ and –SH). The protein crosslinking (adducts) arises from lipid peroxidation upon reaction of these carbonyls with the $-NH_2$ and –SH groups

noninvasive analysis of in vivo lipid peroxidation [26]. Loss of PUFAs and alterations in the cellular membrane PLs can also serve as dependable indices of lipid peroxidation of membranes in biological systems.

Impact of Oxidized Lipids and Lipid Peroxidation on Health and Disease

Cancer, inflammation, cardiovascular and cerebrovascular diseases (CVDs), and neurodegenerative diseases are pathological states mediated by ROS and RNS [27] (Fig. 5.7). ROS and RNS (biological oxidants) have been shown to cause redox

Fig. 5.7 Lipid peroxidation association with diseases. Association of lipid peroxidation (oxidized lipids) with diseases, drug toxicity, environmental toxicity, hyperoxic damage, radiation sickness, and aging

dysregulation altering the structure, activity, and physical properties of cells. In addition to their reactive nature, ROS is the most common oxidant available in cells, which include the oxygen radicals (superoxide, hydroxyl radical, peroxy radicals, etc.), lipid peroxides and hydroperoxides, and lipid radicals [28]. Lipid peroxides can cause cytotoxicity through different mechanisms. Lipid peroxides can impair the structure and function of cell membranes, proteins, and DNA [28]. Cellular membrane PLs are affected the most by peroxidative alteration/degradation due to the presence of esterified PUFA in the *SN-2* position of the glycerol backbone that is the substrate for free radical-initiated and enzymatic oxidation [29–31]. This could also lead to the formation of membrane PL-containing oxidized PUFA still esterified in the PL, leading to the formation of intact oxidized PL. At times, the oxidized PL may have a PUFA hydroperoxide or a PUFA radical or carbonyl of different chain length(s) still esterified in the *Sn-2* position. Each species of this oxidized membrane PL has distinct/unique physiological, biochemical, biophysical, molecular, and toxicological actions, and therefore the nature and extent of formation of these different types of oxidized membrane PLs should be analyzed to establish their specific structural features and quantities (concentrations) in the cells during oxidative degradation of cellular membrane lipids. Although undesired lipid peroxidation may be detrimental for cell survival, it is an important process in the biosynthesis of regulatory molecules such as leukotrienes and prostaglandins (eicosanoids – the bioactive lipid mediators formed from arachidonic acid of PLs upon the actions of cyclooxygenases, COXs and lipoxygenases, LOXs), which are crucial for inflammation [32]. Enzymes such as COXs, LOXs, and cytochrome p-450 epoxygenases can oxygenate free and esterified PUFAs to produce lipid peroxy radicals, lipid hydroperoxides, and lipid endoperoxides [28]. The resultant lipid hydroperoxides can then be decomposed by reacting with either reduced iron (Fe^{2+}) to form alkoxy radicals or oxidized iron (Fe^{3+}) to generate peroxy radicals. Both radicals can further propagate radical chains until terminated by an antioxidant, but result in the formation of cytotoxic aldehydes and alkanes from the breakdown of lipid hydroperoxides [33]. Thus, the determination or analysis of eicosanoids (COX and LOX metabolites of arachidonic acid), PUFA radicals, peroxides, and epoxides will serve as suitable indices (specific biomarkers) of oxidative modification of cellular membrane lipids in both normal and pathophysiological states. PUFAs are particularly susceptible to the free radical-catalyzed oxidation initiated at the methylene-interrupted double bond (Fig. 5.8). The sp^2 hybridization causes the C–H bond to be weak allowing the abstraction of a hydrogen atom [34]. The lipid radical, produced from oxidation of a PUFA, can then be oxidized by molecular oxygen to form a lipid peroxyl radical, which can then react with a hydrogen atom from another lipid to form a lipid hydroperoxide [30]. The lipid radicals either propagate further generation of ROS or terminate creating lipid hydroperoxides and aldehyde end products (electrophilic carbonyls). In the presence of antioxidants, such as α-tocopherol (vitamin-E) or ascorbic acid (vitamin-C), the lipid peroxy radical will be unable to abstract a hydrogen atom from another PUFA and terminates the radical cycle [35]. The products of lipid radicals can form bioactive aldehydes with greater stability over their parent molecules. These aldehyde products are the main cause of damage by the lipid peroxides [36]. Many analytical methods in detecting/determining the oxidative stress are centered around detection and measurement of the aldehyde products (highly reactive electrophilic carbonyls) of lipid peroxidation (Fig. 5.9).

Adverse Actions of Lipid Peroxidation

Lipid peroxidation hampers cell survival through alterations in lipid–lipid interactions, membrane permeability, ion transport/gradients, and membrane fluidity [37]. Increased concentrations of oxidized lipids in liposomal membranes have been

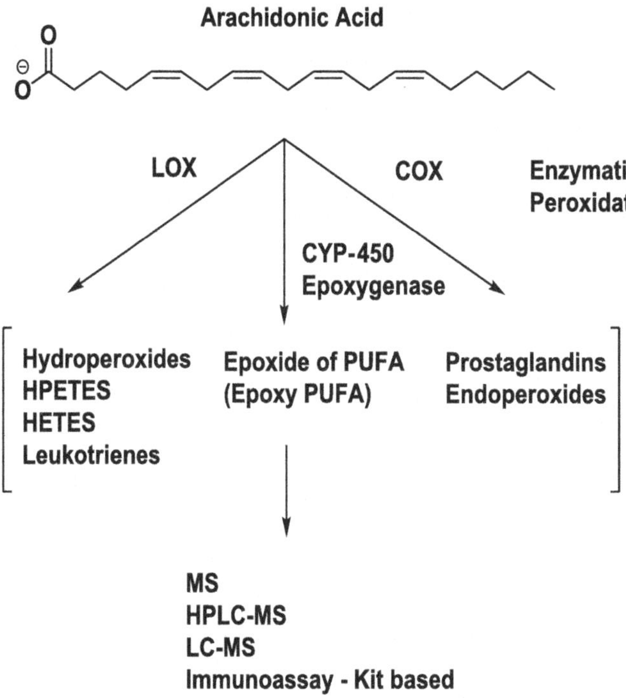

Fig. 5.8 Cytochrome-P450 epoxygenase (CYP-450 Epoxygenase)-, cyclooxygenase (COX), and lipoxygenase (LOX-)-catalyzed formation of eicosanoids from arachidonic acid. Enzymatic oxidation of arachidonic acid by the CYP-450 epoxygenase, COXs, and LOXs into epoxy-arachidonic acid metabolites and eicosanoids (prostaglandins, lipid hydroperoxides, and leukotrienes)

shown to alter membrane fluidity and slow lipid movement across the membrane [28]. PUFA oxidation by the 5-LOX requires prior hydrolysis at the *Sn-2* position and release of free PUFA from the parent membrane PL catalyzed by the phospholipase A_2, which also generates lysophospholipids. These lysophospholipids are readily soluble in cytosol and the oxidized PLs can increase membrane permeability allowing passage for unwanted biomolecules [28]. The products of lipid peroxidation can contribute to cytotoxicity. Ferrous iron (Fe^{2+}) can react with lipid peroxides to form radicals which further propagate new peroxidation reactions. 4-Hydroxynonenal (4-HNE), malonaldehyde (MDA), and acrolein (ACR) are highly reactive and toxic lipid peroxidation degradation products [38] and are generally produced from oxidation of the ω-6 PUFAs [39]. MDA can react with primary amines of proteins or DNA to form crosslinks and may also form 1,4-dihydropyridine adducts with primary amines [28]. MDA is a Michael acceptor and forms covalent adducts with side chains of nucleophilic amino acids [28]. Similar to MDA, 4-HNE forms Schiff base adducts with primary amines and cyclization products [40]. The molecular changes caused by the reactions of products of lipid peroxidation alter the structure and

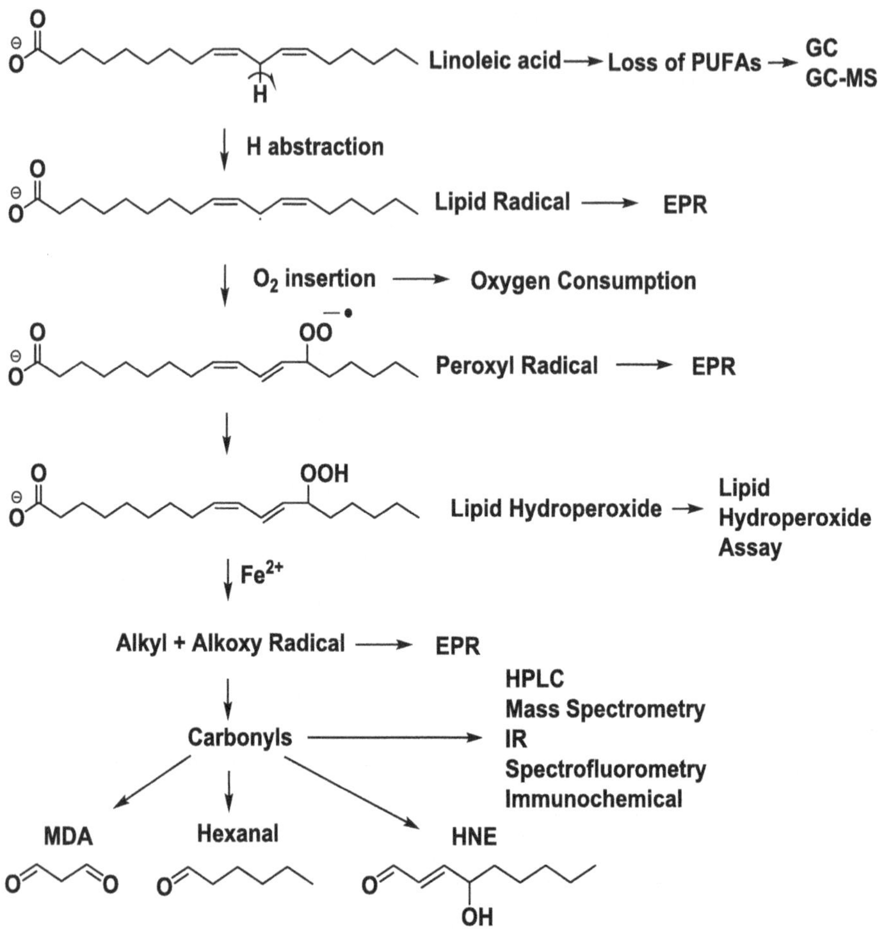

Fig. 5.9 Formation of secondary lipid peroxidation products from linoleic acid and different targets for analysis. Peroxidation of PUFA leads to the formation of lipid peroxidation products such as PUFA hydroperoxides, lipid alkyl and alkoxy radicals, and carbonyls including malondialdehyde, hexanal, and 4-hydroxy-2-nonenal (4-HNE). The target molecules for analysis have been identified with specific analytical methods. Following H abstraction by a radical species, the resulting conjugated diene rearrangement of the double bond is analyzed/quantified by the UV spectrophotometry. PUFAs are quantified by the gas chromatography (GC) and gas chromatography-mass spectrometry (GC-MS). Lipid radicals are analyzed by the electron paramagnetic resonance (EPR) spectrometry. Oxygen consumed during PUFA peroxidation is measured by polarography. Lipid hydroperoxides and carbonyls are analyzed/determined by spectrophotometry, high-performance liquid chromatography (HPLC), MS, infrared spectrometry, spectrofuorometry, and immunochemical methods

function of proteins and DNA and are responsible for the cytotoxicity induced by these molecules.

Analysis of Indices of Lipid Peroxidation

Lipid peroxidation occurs through either enzymatic or nonenzymatic catalysis producing unique products along the way. Each of these products is crucial to the understanding and analysis of lipid peroxidation, and each product is itself a source of concern for cell damage. Hydroxyl radicals are a common initiator of lipid peroxidation and are produced frequently by the cell during the Fenton reaction and by other metals such as Cu, Co, Ni, and V [29]. The core dilemma arises from the ambiguity of these radicals attacking any nearby biomolecules as well as the imbalance between the production of radicals and antioxidants, which all increase the likelihood of lipid peroxidation [29]. Analysis of the products of lipid peroxidation is necessary to determine the extent of lipid peroxidation and the potential cytotoxicity in a biological system. Primary oxidation products, such as lipid peroxides, lipid hydroperoxides, and conjugated dienes are central to lipid peroxidation analysis, which focuses on the physical and chemical alterations/modifications/loss of the PUFAs, PLs, and lipoproteins [41]. In addition to the oxidized PUFAs, the formation of secondary products including malondialdehyde (MDA), F_2-isoprostanes, 4-hydroxyalkenals [42], and 4-hydroxy-2E-nonenal can provide reliable indices of lipid peroxidation in the biological systems [43]. In order to conduct analysis of the oxidized lipids and lipid peroxidation, often the investigator chooses a suitable, reliable, and reproducible method of analysis depending on the choice of the analyte (e.g. oxygen consumption, PUFA or peroxidized PUFA, PL modifications, PUFA conjugated dienes and hydroperoxides, reactive carbonyls and aldehydes, isoprostanes, Schiff base fluorescent adducts, and volatile hydrocarbons, to name a few). These analytical methods require careful and exhaustive extraction and derivatization procedures with unique standards coupled with the use of techniques including polarography, UV-visible spectrophotometry, fluorescence spectrometry, column, chromatography, thin-layer chromatography (TLC), high-performance liquid chromatography (HPLC), gas chromatography-mass spectrometry (GC-MS), liquid chromatography-mass spectrometry (LC-MS), nuclear magnetic resonance (NMR), electron spin resonance (EPR), differential thermal calorimetry (DTC), immunoassay, SDS-PAGE, Western blotting, and fluorescence microscopy (Fig. 5.10).

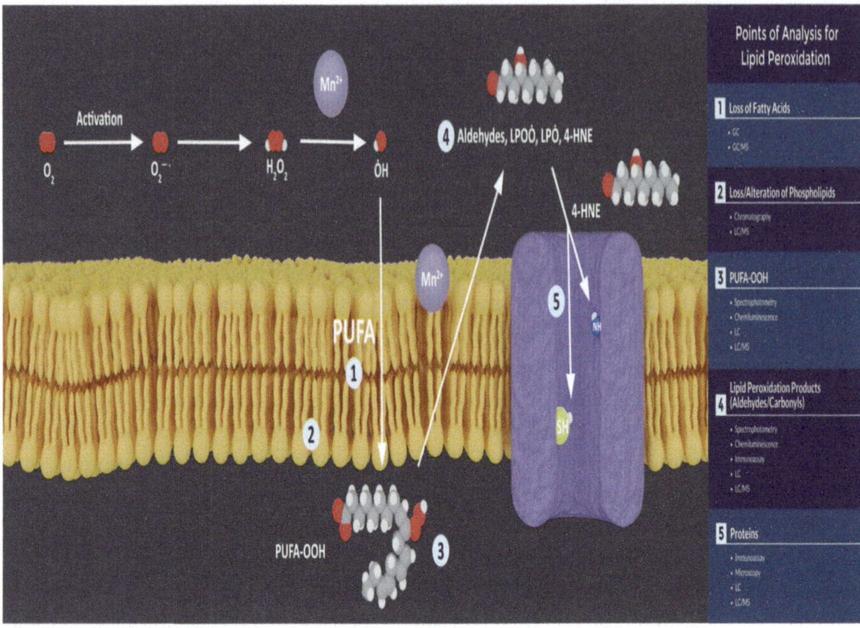

Fig. 5.10 Oxygen radical-induced membrane lipid peroxidation and analysis of products oxidized lipids by different analytical techniques. Oxygen radicals and reactive oxygen species peroxidize the PL-esterified PUFAs of the cellular membranes. The oxidized PUFAs and their secondary oxidized products are measured by different analytical methods

Classical and Traditional Methods of Analysis of Lipid Peroxidation

Rancidity causes deterioration of fats making them unpalatable and unhealthy, and therefore the fat chemists have focused on the suitable analysis of the oxidative deterioration/degradation of fats and oils (rancidity). The early classical tests and analytical methods of oxidative degradation of fats include (*i*) determination of the acid value (AV), (*ii*) determination of the peroxide value (PV), (*iii*) thiobarbituric acid (TBA) test, (*iv*) Kreis test, (*v*) analysis of total volatile carbonyl compounds, and (*vi*) determination of oxiranes [44, 45]. These tests are simple and easily applicable to assess the oxidative deterioration and stability of fats in the food industry and they have been introduced and widely used in fields including nutrition, pharmacology, toxicology, health, and disease. Most of these methods have been transformed into the commercial kits that are used widely to determine the lipid peroxidation in many systems.

UV-Visible and Fluorescence Spectrophotometry of Lipid Peroxidation

The standard, feasible, and widely practiced approaches to analyze lipid peroxidation are the UV-visible spectrophotometric and fluorometric methods termed "batch" assays, as they do not involve any electrophoretic or chromatographic separation [43]. The high reactivity of the aldehyde functional groups, formed from lipid peroxidation, can be exploited using the nucleophilic reactive reagents that can create photosensitive compounds. In addition to the aldehyde reactivity, the increased acidity of the C-H in the methylene H atoms in aqueous solutions can also be utilized to create fluorescent compounds [43]. Analysis of lipid peroxidation utilizing TBA offers simple and sensitive method of determination of the oxidized lipids. Malondialdehyde or malonaldehyde (MDA) formed from lipid peroxidation reacts with 2-thiobarbituric acid (TBA) to form light red- to pink-colored TBA-reactive substances (TBARS). One molecule of MDA reacts with two molecules of TBA constituting a condensation product [46] with a distinct red color. Resultant TBARS ($MDA\text{-}TBA_2$ complex) can be quantified spectrophotometrically by measuring absorption of the complex at 532 nm and fluorometrically by measuring the fluorescence excitation and emission at 515 nm and 553 nm, respectively [43]. Despite the TBARS assay being a commonly used lipid peroxidation determination method, it lacks the chemical specificity to accurately confirm lipid peroxidation as TBA is known react with aldehydes of non-lipoperoxidative origin in the biological systems [46, 47]. It has also been reported that MDA-like molecules with TBA reactivity can be developed from the nonlipid molecules such as proteins, carbohydrates, and DNA. Due to these limitations, the TBA method should only be used amidst other assays or analytical methods while determining the extent of lipid peroxidation in biological systems [47]. However, there have been modifications to determine more specifically and accurately the lipid peroxidation-derived aldehydes such as MDA and 4-HNE by adopting the 2,4-dinitrophenylhydrazine derivatization (DNPH) and quantifying the aldehydes/carbonyls by HPLC [48, 49].

Conjugated dienes can reveal the molecular reorganization of PUFAs during the early stages of lipid peroxidation [34]. While analyzing the secondary oxidation products, specifically aldehydes, the α- and β-unsaturated aldehydes will contain conjugated double bonds; however, aldehydes containing hydroperoxide groups can also be observed. This reveals that both conjugated diene and hydroperoxy groups are not limited to primary oxidation products [50]. The conjugated diene hydroperoxides formed from the oxidation of PUFAs can be analyzed by the spectrophotometric methods since the conjugated dienes exhibit maximum absorbance of the UV radiation [51] at 232–234 nm [36]. The disadvantage of this method is that while analyzing serum lipids for determination of lipid peroxidation, the sample must be thoroughly diluted and also albumin and other molecules present in the serum also absorb UV radiation, leading to interferences [43]. In order to avoid that, lipids can be extracted by Folch extraction (2:1 chloroform-methanol, vol/vol) and then are reconstituted in cyclohexane (which has no UV absorbance at 232–234 nm) and subjected to the measurement of absorbance of conjugated dienes in the peroxidized PUFA [52, 53]. Sometimes, while analyzing the complex lipids, it is advisable to obtain a clear differential conjugated diene spectrum by measuring absorbance of the oxidized PUFA (lipid) against the un-oxidized PUFA (lipid) at 232 nm. The conjugated diene assay of lipid peroxidation can be conveniently applied to free and esterified fatty acids, triglycerides, and PLs [52, 53].

Determination of Lipid Peroxides by Spectrophotometric Iodometry and Fox Assay

Lipid hydroperoxides are known to react with iodide (KI), oxidize it, and release stoichiometric iodine (I_2), which reacts with iodide (I^-) to form triiodide (I_3^-) that can be determined spectrophotometrically at 353 nm as an index of lipid peroxidation [52]. We have utilized this iodometric-spectrophotometric assay to determine lipid peroxidation in the myocardial cardiolipin [52]. Ferrous oxidation-xylenol orange (FOX) assay can be used to detect lipid peroxidation and when combined with HPLC separation, it can determine different classes of lipid hydroperoxides. There are currently two commonly used FOX assays: FOX 1 and FOX 2 assays. FOX 1 is only capable of determining low levels of hydrogen peroxide in aqueous buffers; whereas, FOX 2 can specifically determine lipid hydroperoxides in low concentrations despite the presence of high background levels of nonperoxidized fatty acids. This assay relies on the oxidation of reagent iron (II) to iron (III) by the lipid hydroperoxide, then reacting the xylenol orange reagent with iron (III) to produce a blue–purple complex with an absorbance maximum at 560 nm. The FOX 2 assay also employs butylated hydroxytoluene (BHT), an antioxidant that inhibits/blocks further propagation of hydroperoxides, for a more accurate assessment of lipid hydroperoxides. This assay is convenient, inexpensive, and precise with broad applicability [36].

Determination of PUFAs and Lipid Hydroperoxides by Gas Chromatography (GC) and High-Performance Liquid Chromatography (HPLC)

Lipid peroxidation can be determined by (*i*) loss of PUFA and (*ii*) formation of the lipid hydroperoxides. Changes in fatty acid composition and loss of PUFA in simple to complex lipids such as the membrane PLs are determined by GC [53] following hydrolysis of PLs and converting the fatty acids into methyl esters. Loss of different PL classes can also be determined (as changes in lipid phosphorus) in the biological systems (cellular membranes) as an index of lipid peroxidation [53]. The fatty acid composition, changes in PUFA composition, and loss of PLs can be determined as indices of lipid peroxidation in the biological membranes by GC and HPLC [53]. HPLC has been used in characterization of the lipoxygenase-catalyzed synthesis of phosphatidylinositol peroxides [54].

Mass Spectrometry (MS) of Oxidized Lipids

The analysis of specific oxidized lipids is increasingly challenging and tedious as the number of chemical structures and differences among them increase [55]. This is where MS takes the spotlight as it can distinguish up to 50 different lipid classes and thousands of different lipid species with great accuracy and precision [42]. In MS, ionized analytes are measured for their mass to charge ratio (*m*/*z*) and as the lipids undergo oxidative modifications, this ratio changes [56]. The benefit of MS, often couples with the liquid chromatography (LC), is that there is no need for chemical manipulation of the analytes to release the PUFAs necessary for analysis or even the need for derivatization to produce the volatile species as required for the GC columns [56]. Earlier, the electrospray ionization (ESI) and matrix-assisted laser desorption/ionization (MALDI) were effective low energy ionization techniques used for their simplicity and minimal artifacts allowing for nominal loss in structure of PLs; however, the concerning detriment of this method is that there is currently no way to differentiate the positional distribution of the oxygenated groups without hydrolyzing the sample and analyzing the individual fatty acids [55]. The ability of ESI and MALDI to analyze simultaneously multiple species among the lipid classes has given rise to the shotgun lipidomics [55]. The shotgun lipidomics is a direct infusion-based lipid analysis technique that can be approached using the tandem MS-based, high mass accuracy MS-based, or multi-dimensional MS-based (MDMS-SL) methods [42]. Tandem MS-based methods provide structural information on the sample being analyzed by fragmenting ions. Here, two analyzers are connected by a collision cell where the first analyzer contains an ion which is broken down by an inert gas in the collision cell to produce fragmented ions which are then examined in the second analyzer [55]. The most common analyzer is the quadrupole mass analyzer which only allows ions of specific m/z values to pass through the detector [55]. Modern methods may contain many more analyzers and fragmentation levels to increase the accuracy of the measurement [55].

MDMS-SL method of the shotgun lipidomics overcomes most of the issues of the other techniques through its ability to examine the unique chemical differences among lipid classes and to utilize very low-abundance values [42]. This allows for the incredible scope of lipid class analysis along with the high precision and accuracy of the technique [42]. Using the shotgun lipidomics, oxidized fatty acids, secondary aldehyde products, and affected/altered proteins can be measured to determine the extent of lipid peroxidation in a biological system [42]. 4-Hydroxy-2E-nonenal (4-HNE), the most common among the class of 4-hydroxyalkenals, is a highly reactive aldehyde product formed during the oxidation of the n-6 PUFAs such as arachidonic and linoleic acid as well as arising from the15-lipoxygenase metabolites. The hydroxyalkenals originate from the enzymatic and nonenzymatic pathways forming a family of hydroxyalkenal products, each of which exhibiting specific pathways of oxidation. 4-Hydroxyalkenals are very unstable, nonionizable, and present in low abundance, but can be quantified using various MDMS-SL methods [42]. Thus, MS has been a very important and useful technique to analyze the lipid peroxidation products at the structural and quantification levels, and this analytical technique has opened up entirely the new discipline of oxidative lipidomics.

Electron Paramagnetic Resonance (EPR) Spectroscopy of Oxidized Lipids

The EPR spectroscopy is an invaluable analytical technique in detecting and characterizing the lipid radicals. This method uses a diamagnetic compound (spin trap) that reacts with a radical species to produce a more stable free radical product (spin adduct) that can be analyzed by the EPR spectroscopy. The EPR spectrum provides the g factor and hyperfine splitting patterns that allow to determine the radical species [57, 58]. The g value provides information about the orbital spin and angular momentum of an electron, whereas the hyperfine splitting patterns can determine the distance of an electron from the nuclei and identify the specific nuclei and its abundance [57]. The EPR spin trapping has the potential in the biological systems for detection and analysis of hydroxyl and superoxide radicals, free radical metabolites, and most relevantly the peroxidized lipids. The EPR spin trapping is effective because of the specific spin trap used to extend the lifespan of the resultant spin adduct. 5,5-Dimethyl-pyrroline-1-oxide (DMPO) is commonly used as a spin trap because it reacts with most radicals and forms a spin adduct with a relatively long lifespan. This long lifespan allows for the accumulation of the spin adduct to a concentration detectable by the continuous wave EPR. Although DMPO is widely used, α-(4-pyridyl-1-oxide)-*N*-tert-butylnitrone (POBN) is used instead because of its ability to trap the carbon-centered radicals and is necessary for spin trapping lipid radicals. The carbon-centered adducts of POBN can persist for hours and even days in standard neutral solutions, allowing them to be more effective than the DMPO spin adducts in detection and analysis of the carbo-centered radicals. However, the hyperfine splitting patterns are very similar to one another making it difficult to differentiate the radicals [58].

Nuclear Magnetic Resonance (NMR) Spectroscopy of Oxidized Lipids

High-resolution NMR spectroscopy techniques including the ^{1}H-,^{13}C-, and ^{31}P-NMR spectroscopy have been extensively used in the field of lipidomics to measure both primary and secondary oxidation products of lipid peroxidation [43, 50].

Proton NMR spectroscopy analyzes the primary oxidation products of lipid peroxidation by identifying the amount of bis-allylic hydrogen atoms that arise from oxidation of the PUFAs. The proton NMR spectroscopy is advantageous in the sense that there are no modifications required to prepare the sample for analysis and therefore, the original integrity of the sample still can be retained intact during the NMR analysis [51]. When analyzing the lipid hydroperoxides by the proton NMR spectroscopy, it is important to see a significant difference in peak area of the olefinic and aliphatic protons as they serve as good indicators of lipid oxidation. The ratios of aliphatic to olefinic protons and aliphatic to diallylmethylene protons are shown to gradually increase during the oxidation of lipids [43]. The resonances in the 10.5–8 ppm range can usually be attributed to the oxidation of PUFAs resulting in the formation of lipid hydroperoxides and subsequent shorter chain saturated and unsaturated aldehydes [50]. Hence, the NMR spectroscopy offers a complimentary technique to the MS for the structural elucidation of the oxidized lipids.

Fluorescence Microscopy

During the course of lipid peroxidation, extensive oxidative modifications of proteins may take place but are not recognized/analyzed easily as compared to the oxidatively modified lipid species. These oxidatively modified proteins may be trapped inside certain intracellular organelles and regions of the cell that are not as easily accessible. Therefore, the fluorescent dyes using boron-dipyrromethene (BODIPY: *4,4-difluoro-4-bora-3a,4a-diaza-s-indacene*) can be paired with certain organelle probes allowing for clear identification/localization of the biological processes and signal pathways involved in the lipid peroxidation. The analytical advantages of BODIPY are based on its high yield and narrow range of the excitation and emission wavelengths, allowing the probe to work smoothly with other fluorophores and probes. Modifications, consequential to oxidative damage within the organelles and cellular membranes, can then be observed/analyzed by the fluorescence microscopy [59]. Fluorescent microscopic methods use both positive and negative labeling. Positive labeling is the pre-loading of cells with the BODIPY-conjugated arachidonic acid (BD-AA) followed by the treatment of cells with a pro-oxidant. Thus, oxidative stress can damage the intracellular organelles including mitochondria, endoplasmic reticulum, and the autophagy machinery leading to apoptosis. Along this approach, the lipid peroxidation products will be generated intracellularly and develop multiple fluorescently labeled reactive species. BD-AA is uncharged and hydrophobic allowing it to lodge primarily within the cellular membranes where lipid peroxidation mostly takes place. In the negative labeling conditions, the cells are first exposed to the pro-oxidant to induce oxidative damage leading to the protein damage and alterations, which can be observed by post-labeling with the BODIPY-iodoacetamide (BD-IAM). This will then decrease the labeling in cells without the effects of oxidative stress [59]. Analysis of the products and metabolites of lipid peroxidation may show general outcome of the overall extent of lipid peroxidation, but the mechanism by which lipid peroxidation may occur or be impacted by certain antioxidants, cannot be determined with those methods. Here, lipid peroxidation and radical formation can be observed in the mitochondria with the use of fluorescent dye, specifically the fluorogenic dyes that behave as antioxidants scavenging the peroxy radicals and provide useful data pertinent to this scenario [60].

Western Blotting and Immunoassays for Analysis of Lipid Peroxidation

Often, lipid peroxidation causes nonenzymatic and posttranslational protein modifications in cells, which can be identified and analyzed. The cellular proteins are modified by molecules arising from free radical damage that causes alterations in the structure and folding of proteins. These protein alterations are caused by lipid peroxidation secondary products, especially lipoperoxidative aldehyde products. The resultant protein alterations can be analyzed and quantified by Western blotting and immunoassays [39]. 4-Hydroxynonenal (4-HNE), a reactive electrophilic carbonyl product of lipid peroxidation, covalently binds to proteins and forms relatively stable protein adducts, which can be analyzed and quantified by an enzyme-linked immunosorbent assay (ELISA) technique [38]. The ELISA technique employs antibodies to identify specific proteins and analyze changes through the absorbance spectra. Earlier, ELISA was used for analysis of cell lysates, but recently the use of this method has been extended to analyze the human plasma and serum samples [38]. This method requires the use of a specific monoclonal antibody that can distinguish the HNE-histidine epitope [38]. Since the ELISA method requires only a few micrograms of protein, it is highly convenient for the clinical applications wherein the samples are generally small [38].

Chemiluminescence Analysis of Lipid Peroxidation

The chemiluminescence analytical methods are based on specific chemical reactions that excite the molecules (excited state) in the sample to be analyzed and then bring them down to the electronic ground state, resulting in the emission of light [30]. The singlet molecular oxygen and triplet carbonyl compounds are important chemiluminescent species of the lipid peroxides (oxidized lipids) formed from the oxygen radical attack on the PUFAs [30, 61]. Singlet oxygen produces dimol red emission detectable between 634 and 703 nm wavelengths, whereas the triplet carbonyls produce a blue-green phosphorescence detectable between 450 and 550 nm wavelengths [61]. These compounds emit a light intensity which is directly proportional to the concentration of the sample [30, 62]. Chemiluminescence assays have a higher detectability limit in comparison to the other absorption techniques including fluorometry and spectrophotometry due to low background light, less heat, and low light scattering of the source and detector [30].

Commercial Kits for Lipid Peroxidation Assay

The spectrophotometric (often called colorimetric) and fluorometric techniques are the most widely used consumer kits and assays. Manufacturers and commercial vendors including the ABCAM, Cayman Chemical, and Sigma Aldrich sell assay kits to determine the levels of TBARS, MDA, hydroxyalkenals, and 4-HNE products of lipid peroxidation. Cayman offers a kit that measures lipid hydroperoxides in biological samples, which requires first lipid extraction from the sample into chloroform, eliminating interference and then performing redox reactions with ferrous iron and then the analyte can be measured spectrophotometrically. Cell Biolabs offers multiple oxidized low-density lipoprotein (LDL), oxidized high-density lipoprotein (HDL), and isoprostane ELISA kits as well as specific HNE antibodies that can all be used to measure the extent/levels of lipid peroxidation. It is apparent that the commercial kits are convenient and less tedious to perform the analysis of lipid peroxidation in the biological samples.

Techniques for Clinical Analysis of Lipid Peroxidation

Elevated levels of lipid peroxidation are akin to development of a variety of pathological/disease conditions, wherein the detection/analysis of the lipid peroxidation products can be suitable [63]. The clinical analysis of lipid peroxidation consists of measuring the lipid peroxidation products in the biological samples including plasma, tissues, urine, and breath [64], but analysis of the lipid peroxidation products is more reliable and convenient in comparison to measurement of ROS and other oxidants [65]. There has been an improvement in the analysis of lipid peroxidation in the tissue specific to experiments. The lipid peroxidation products such as MDA and F_2-isoprostanes have been the most commonly used biomarkers of oxidized lipids in the clinical setting while examining the breath for these secondary lipid oxidation products [64, 66]. Choice of MDA as a lipid peroxidation biomarker has been questioned for its specificity issues, whereas, 15-F_2-isoprostane and 15-F_2-IsoPM are considered as the benchmark of in vivo analysis of lipid peroxidation and are measured using the GC/NICI MS [64, 66]. Thus, GC-MS can be applied to analyze/determine lipid peroxidation in vivo in the clinical samples.

Oxidative Lipidomics and Bioinformatics in Lipidomics

Oxidative lipidomics is a sub-omics in lipidomics that focuses on the structure, function, and quantification of the oxidized lipids. This involves lipid extraction, LC-MS/MS analysis, data analysis, and integration of this data into a biological system or mechanism [67]. Advancements in computer modeling and machine learning have offered simple tools for the identification and quantification of lipids in high throughput. Utilizing the fragment dissociation and MSn analysis, lipid analysis software can determine the structures of unknown lipids accurately and instantaneously. There are several databases with large pools of training data sets that will only expand in the future, and these data sets can be used to improve prediction models. LIPID MAPS, LipidHome, and SwissLipids are the most commonly used databases for MS analysis. ALEX123 is a database that contains fragmentation patterns and a software tool that allows the automatic analysis of MSn spectra. Although the fragmentation analysis software appears promising, there are still unresolved issues limiting the potential of this method, mainly cases where there are inadequate fragments or situations with the fragments may not be specific to a certain lipid, and an accurate identification thus is not possible. Predictive models have also used in other important omics fields such as the proteomics in collaboration with lipidomics as they are tightly interrelated. Skyline is a software for proteomics that can be applied to the analysis of complex lipids [68]. Overall, the bioinformatics tool offers great advantage in the analysis and modeling of the oxidized lipids in the rapidly growing field of oxidative lipidomics.

Clinical and Translational Lipidomics

Lipid peroxidation tremendously influences the inflammation-based pathologies as the n-3 and n-6 metabolites and eicosanoids are required to generate bioactive pro- and anti-inflammatory mediators such as the eicosanoids and the metabolites of docosahexaenoic (DHA) and eicosapentaenoic acid (EPA) metabolites. The use and application of translational lipidomics has been shown in prediction of pre-eclampsia, a condition where there is increased blood pressure and risk for kidney failure post pregnancy, by using the ratio of prostacyclin/thromboxane-B2 (TXB2) [69, 70]. During the course of lipid peroxidation, the levels of prostacyclin is low altering the ratio [69]. The isoprostanes, generated through the peroxidation of arachidonic acid, have also been considered as an important clinical marker of pre-eclampsia wherein the high levels isoprostanes in urine of pregnant women show a five-fold risk [69]. Lipids are integral to the signaling and proliferation of tumor growth and metabolism. Several cancer cells exhibit elevated levels of ROS leading to a greater susceptibility to lipid peroxidation [71]. Studies have shown an increase in the lipid peroxidation products such as MDA and HNE in the colon adenocarcinoma tissue [71]. However, the extent of lipid peroxidation and formation of its products are still debated as the cases are different. In vivo studies have shown a decrease in the levels of HNE in the tissue biopsies of colon canner [71]. Elevated levels of saturated PLs have shown a lower susceptibility to lipid peroxidation in the cancer cells [71, 72]. Antitumor drugs including cisplatin, mitomycin C, vinblastine, doxorubicin, and camptothecin generate ROS which can lead to cause enhanced lipid peroxidation in the cancer cells [71]. The analysis of the lipid peroxidation products can provide insights into the mechanisms of action of anti-cancer drugs as well as the analysis of the potential effectiveness of certain anti-cancer drugs. Brain consists of 60% lipid and thus changes in the lipid structure and composition of brain can severely impact the function of the neurological tissue and contribute to the neurodegenerative diseases. Studies have revealed the presence of high levels of F_2-isoprostanes and HNE in the biological fluids of Alzheimer's patients [31]. The formation of these lipid peroxidation products is caused by the aggregation of amyloid ß peptide (1–42) in the extracellular plaques inserted into the lipid bilayer, leading to the formation of hydrogen peroxide and peroxy radicals through the Fenton reaction, and finally resulting in the formation of oxidized lipids [31]. Parkinson's disease (PD) has also been shown exhibiting presence of HNE as well as elevated levels of F_2-isoprostanes and hydroxyeicosatetraenoic acids (HETEs) in the plasma of PD patients in the early stages [31]. These studies support the utilization of oxidative lipidomics (analysis of lipid peroxidation) in the clinical and translational settings to evaluate the use of oxidized lipids as biomarkers of oxidative stress in disease states.

Analysis of Lipid Peroxidation in Nutrition and Preventive Medicine

Vitamins including the vitamin C (ascorbic acid) and vitamin E (tocopherols) have been considered as antioxidants in terminating the radical propagation and preventing/attenuating the oxidation of lipids, regardless of certain inconsistencies associated with the treatments/actions of these vitamins [35, 70]. Patients with AD or mild cognitive impairments receiving the vitamin E as treatments have exhibited no significant improvement of the symptoms [31]. However, in a group of 3000 elderly woman, individuals with higher intake of vitamin E-containing foods have significantly less cognitive decline [73]. Similarly, Parkinson's patients have shown inconsistent results with vitamin E treatments [31]. Subjects with pre-eclampsia have shown to have lower levels of vitamins C and E, but supplementation with these vitamins have revealed different results [70]. Certain studies have shown that intake of vitamins promotes aging through the disruption of natural ROS production to counteract mutation [74]. Prevention of cognitive decline may be enforced through the regimental/controlled consumption of vitamins as opposed to the treatment of the afflicted patients. Low saturated fatty acid content has been shown to cause less extent of formation of the lipid peroxidation products in vivo [74]. Fish with high concentration of omega-3 PUFAs appear to be susceptible to oxidation; however, they have in place furan fatty acids acting as radical scavengers supporting the cardioprotective benefits offered by the fish diets [74, 75]. Thus far, the nutritional lipidomics appears to be a powerful analytical platform to evaluate the nutritional and disease preventive actions of lipid nutrition.

Analysis of Lipid Peroxidation in Agriculture and Food Industry

This is where the saga of chemistry and biochemistry of lipid peroxidation and analysis of lipids and oxidized lipids started first—in agriculture and food industry to prevent/stop rancidity of fats, protect fats, oils, and meat from oxidative deterioration and preserve the quality of foods, and to stop oxidative deterioration of oil seeds in storage. All the knowledge obtained earlier in this field, both mechanistic and analytical, has been applied to the biomedical fields. The oxidative deterioration of soybean oils caused by the linoleate LOX in the soybean seeds during storage has been an intense study for decades to stop the LOX-induced oxidative degradation of soybean oils. These early studies on soybean LOXs have paved the path to investigate the enzymology of mammalian LOXs and have led to several advancements in the field of LOXs in inflammation and disease. The most common issue in the meat industry is the oxidative deterioration of meat and meat products due to lipid peroxidation during processing and storage [76]. Oxidation (peroxidation) of lipids of the meat alters and jeopardizes the food quality and lowers the nutritional value resulting in an inferior product of unpalatability and unhealthy and may not be acceptable according to the guidelines and standards [76]. There are protocols and procedures in place for the live muscle to prevent the lipid oxidation, but these are lost during the mechanical conversion of meats that have higher lipid content and are thus vulnerable to the increased oxidative damage [77]. The most widely used biomarker of lipid peroxidation, MDA, is the most prominent secondary product generated during the peroxidation of lipids and often measured by the TBA assay [77]. Photooxidation is also a major inducer of lipid peroxidation since plants, oils, and fish are exposed to the solar radiation that causes the formation of singlet oxygen and peroxy radicals [75, 77]. The lipid peroxidation alters the flavor, aroma, color, texture, and shelf life of the foods [78]. The LC-MS analysis is often applied to analyze/quantify the oxidized lipids in the oils of soybean, corn, canola, and other plants which contain high amounts of linoleic and α-linolenic acids [79] in order to enhance the stability, shelf life, and quality of these oils. The use of natural and synthetic antioxidants prevents or lowers the peroxidation of lipids in foods and oils [76]. Butylated hydroxytoluene (BHT) is a common synthetic antioxidant used to scavenge lipid peroxy radicals [76]. The natural product phenolic compounds are natural antioxidants which trap and scavenge the free radicals and act as effective antioxidants [76]. These natural phenolic compounds have higher antioxidant activity as compared to the synthetic antioxidants such as BHT because of their vicinal OH groups which allow them to chelate (complex) transition metals [76]. Several spices including the oregano and rosemary (which contain the antioxidant polyphenols) have been well documented to preserve lipids from oxidative deterioration through the prevention/inhibition of lipid oxidation in poultry [76]. Several plant extracts, oils, herbs, and spices with the antioxidant actions have been explored but still not more studies need to be conducted to precisely evaluate their beneficial effects in nutrition and diet [76]. Thus, the analysis of lipid peroxidation and peroxidized lipids has the most important role and place in agriculture and food industry.

Conclusion

In this chapter, we have discussed different methods of analysis of lipid peroxidation, with their crucial roles, strengths, and weaknesses in the field of oxidative lipidomics. Although some investigators may consider certain analytical methods for accuracy to characterize lipid peroxides and determine lipid peroxidation in the biological systems, the right choice of a method depends on the desired/focused product of lipid peroxidation on a temporal scale of the lipid peroxidation cascade taking the sample of analysis into consideration (Fig. 5.11). It is extremely impor-

Index Component of Lipid Peroxidation Analyzed	Method of Analysis	Sensitivity
Aldehydes (MDA)	• Spectrophotometry • HPLC • MS	++
Carbonyls (HNE)	• Chemical Derivatized (Hydrazone) HPLC • LCMS • Immunodetection Assay & Western blotting	+++
Hydroperoxides	• Spectrophotometry / EPR	+
Oxygen Consumption (CN-insensitive)	• Clark Electrode • EPR	+
Fatty Acid Loss	• GC • GC-MS • Unsaturation Index	+++
Membrane Phospholipids (Alteration)	• HPLC • LC-MS • TLC / Spectrophotometry	++
Cellular Localization of Lipid Peroxides (Aging Pigments)	• Microscopy (Antibodies) • Immunofluorescence Microscopy	+++
Lipid Radical	• EPR • Chemiluminescence	+
Lipid Peroxide Addition	• Enzyme-based assay • Enzyme Immunoassay	++

Fig. 5.11 Sensitivity of methods of analysis of lipid peroxidation. The utilization of different analytical methods/techniques to analyze/determine lipid peroxidation (oxidized lipids and their secondary products) and their sensitivities are presented

tant to emphasize that no single method is adequate to confirm the occurrence, precisely characterize the oxidized lipid species, and determine the extent of lipid peroxidation. Multiple suitable analytical methods are highly recommended to acquire consistent, reproducible, and reliable data/results of oxidative lipidomics.

Acknowledgements Funding support of the NIH R01AG059981 is acknowledged.

Conflicts of Interest Statement The authors declare that they have NO affiliations with or involvement in any organization or entity with any financial interest or nonfinancial interest in the subject matter or materials discussed in this chapter.

References

1. Pryor WA. Free radical biology: xenobiotics, cancer, and aging. Ann N Y Acad Sci. 1982;393:1–22. https://doi.org/10.1111/j.1749-6632.1982.tb31228.x.
2. McCord, J.M. and Fridovich, I. Superoxide dismutated: a history. in Superoxide and superoxide dismutases. 1977; Ed, by A. Michelson, J.M. McCord, and I. Fridovich. Academic Press, New York, pp. 1–11.
3. Samuelson B, Ranwell PW, Paolette R. Advances in prostaglandin and thromboxane research, vol. VI. New York: Raven Press; 1980.
4. Bolton JR. In: Swartz HM, Bolton JR, Borg DC, editors. Biological applications of electron spin resonance. New York: Wiley-Interscience; 1972. p. 11–61.
5. Pryor WA. Free radical pathology. Chem Eng News. 1971;7:34–51.
6. Bhagwat AS, Sane PV. Evidence for the involvement of superoxide anions in the oxygenase reaction of ribulose-1,5-diphosphate carboxylase. Biochem Biophys Res Commun. 1978;84(4):865–73. https://doi.org/10.1016/0006-291x(78)91664-9.
7. Fridovich I. The biology of oxygen radicals. Science. 1978;201(4359):875–80. https://doi.org/10.1126/science.210504.
8. Pryor WA. The role of free radical reactions in biological systems. In: Pryor WA, editor. Free radicals in biology, vol. I. New York: Academic Press; 1976. p. 1–49.
9. Younes M, Weser U. Involvement of superoxide in the catalytic cycle of diamine oxidase. Biochim Biophys Acta. 1978;526(2):644–7. https://doi.org/10.1016/0005-2744(78)90156-0.
10. Cohen G, Heikkila RE. The generation of hydrogen peroxide, superoxide radical, and hydroxyl radical by 6-hydroxydopamine, dialuric acid, and related cytotoxic agents. J Biol Chem. 1974;249(8):2447–52.
11. Misra HP, Fridovich I. The generation of superoxide radical during the autoxidation of hemoglobin. J Biol Chem. 1972;247(21):6960–2.
12. Flower RJ. Biosynthesis of prostaglandins. In: Oxygen free radicals and tissue damage. Ciba Foundation Symp. 65. Amsterdam: Excerpta Medica; 1979. p. 123–42.
13. Leibovitz BE, Siegel BV. Aspects of free radical reactions in biological systems: aging. J Gerontol. 1980;35(1):45–56. https://doi.org/10.1093/geronj/35.1.45.
14. Albrecht DK, Kappus H, Remmer H. Lipid peroxidation and cell damage in isolated hepatocytes due to bromotrichloromethane. Toxicol Appl Pharmacol. 1978;46(2):499–505. https://doi.org/10.1016/0041-008x(78)90095-9.
15. Di Luzio NR. Antioxidants, lipid peroxidation and chemical-induced liver injury. Fed Proc. 1973;32(8):1875–81.
16. Henderson CA, Metz EN, Balcerzak SP, Sagone AL Jr. Adriamycin and daunomycin generate reactive oxygen compounds in erythrocytes. Blood. 1978;52(5):878–85.

17. Jose PJ, Slater TF. Increased concentrations of malonaldehyde in the livers of rats treated with carbon tetrachloride. Biochem J. 1972;128(4):141P. https://doi.org/10.1042/bj1280141pa.
18. Recknagel RO, Glende EA, Hruszkewycz AM. Chemical mechanisms in carbon tetrachloride toxicity. In: Pryor WA, editor. Free radicals in biology, vol. III. New York: Academic Press; 1977. p. 97–132.
19. Pryor WA. Methods for detecting free radicals and free radical-mediated pathology in environmental toxicology. In: Bhatnagar RS, editor. Molecular basis of environmental toxicity. Ann Arbor, MI: Ann Arbor Sci. Publ., Inc.; 1980. p. 3–36.
20. Slater TF. Free radical mechanisms in tissue injury. London: Pion Ltd.; 1972. p. 1–19.
21. Mead JF. Free radical mechanisms of lipid damage and consequences for cellular membranes. In: Pryor WA, editor. Free radicals in biology. New York: Academic Press; 1976. p. 51–68.
22. Dahle LK, Hill EG, Holman RT. The thiobarbituric acid reaction and the autoxidations of polyunsaturated fatty acid methyl esters. Arch Biochem Biophys. 1962;98:253–61. https://doi.org/10.1016/0003-9861(62)90181-9.
23. Nagy IZ. A membrane hypothesis of aging. J Theor Biol. 1978;75:189–95.
24. Tappel AL. Lipid peroxidation damage to cell components. Fed Proc. 1973;32(8):1870–4.
25. Shimasaki H, Nozawa T, Privett OS, Anderson WR. Detection of age-related fluorescent substances in rat tissues. Arch Biochem Biophys. 1977;183(2):443–51. https://doi.org/10.1016/0003-9861(77)90379-4.
26. Dougherty JJ, Hoekstra WG. Effects of vitamin E and selenium on copper-induced lipid peroxidation in vivo and on acute copper toxicity. Proc Soc Exp Biol Med. 1982;169(2):201–8. https://doi.org/10.3181/00379727-169-41332.
27. Weidinger A, Kozlov AV. Biological activities of reactive oxygen and nitrogen species: oxidative stress versus signal transduction. Biomolecules. 2015;5(2):472–84. https://doi.org/10.3390/biom5020472. Published 2015 Apr 15.
28. Gaschler MM, Stockwell BR. Lipid peroxidation in cell death. Biochem Biophys Res Commun. 2017;482(3):419–25. https://doi.org/10.1016/j.bbrc.2016.10.086.
29. Ayala A, Muñoz MF, Argüelles S. Lipid peroxidation: production, metabolism, and signaling mechanisms of malondialdehyde and 4-hydroxy-2-nonenal. Oxidative Med Cell Longev. 2014;2014:360438. https://doi.org/10.1155/2014/360438.
30. Catalá A. Five decades with polyunsaturated fatty acids: chemical synthesis, enzymatic formation, lipid peroxidation and its biological effects. J Lipids. 2013;2013:710290. https://doi.org/10.1155/2013/710290.
31. Shichiri M. The role of lipid peroxidation in neurological disorders. J Clin Biochem Nutr. 2014;54(3):151–60. https://doi.org/10.3164/jcbn.14-10.
32. Fuchs B. Mass spectrometry and inflammation--MS methods to study oxidation and enzyme-induced changes of phospholipids. Anal Bioanal Chem. 2014;406(5):1291–306. https://doi.org/10.1007/s00216-013-7534-5.
33. Repetto MG, Ferrarotti NF, Boveris A. The involvement of transition metal ions on iron-dependent lipid peroxidation. Arch Toxicol. 2010;84(4):255–62. https://doi.org/10.1007/s00204-009-0487-y.
34. El-Aal HHMA. Lipid peroxidation end-products as a key of oxidative stress: effect of antioxidant on their production and transfer of free radicals. In: Lipid peroxidation; 2012. p. 64–88. https://doi.org/10.5772/45944.
35. Shichiri M, Yoshida Y, Niki E. Unregulated lipid peroxidation in neurological dysfunction. In: Omega 3 fatty acids in brain and neurologic health; 2014. Elsevier Inc. https://doi.org/10.1016/B978-0-12-410527-0.00004-1.
36. Orefice I, Gerecht A, d'Ippolito G, Fontana A, Ianora A, Romano G. Determination of lipid hydroperoxides in marine diatoms by the FOX2 assay. Mar Drugs. 2015;13(9):5767–83. https://doi.org/10.3390/md13095767. Published 2015 Sep 11.
37. Romano R, Cristescu SM, Risby TH, Marczin N. Lipid peroxidation in cardiac surgery: towards consensus on biomonitoring, diagnostic tools and therapeutic implementation. J Breath Res. 2018;12(2):027109. https://doi.org/10.1088/1752-7163/aa9856. Published 2018 Feb 6.

38. Weber D, Milkovic L, Bennett SJ, Griffiths HR, Zarkovic N, Grune T. Redox biology measurement of HNE-protein adducts in human plasma and serum by ELISA—comparison of two primary antibodies. Redox Biol. 2013;1(1):226–33. https://doi.org/10.1016/j.redox.2013.01.012.
39. Spickett CM. The lipid peroxidation product 4-hydroxy-2-nonenal: advances in chemistry and analysis. Redox Biol. 2013;1(1):145–52. https://doi.org/10.1016/j.redox.2013.01.007. Published 2013 Jan 21.
40. Pizzimenti S, Ciamporcero E, Daga M, et al. Interaction of aldehydes derived from lipid peroxidation and membrane proteins. Front Physiol. 2013;4:242. https://doi.org/10.3389/fphys.2013.00242. Published 2013 Sep 4.
41. Davies SS, Guo L. Lipid peroxidation generates biologically active phospholipids including oxidatively N-modified phospholipids. Chem Phys Lipids. 2014;181:1–33. https://doi.org/10.1016/j.chemphyslip.2014.03.002.
42. Hu C, Wang M, Han X. Shotgun lipidomics in substantiating lipid peroxidation in redox biology: methods and applications. Redox Biol. 2017;12:946–55. https://doi.org/10.1016/j.redox.2017.04.030.
43. Tsikas D. Assessment of lipid peroxidation by measuring malondialdehyde (MDA) and relatives in biological samples: analytical and biological challenges. Anal Biochem. 2017;524:13–30. https://doi.org/10.1016/j.ab.2016.10.021.
44. Gray IJ. Measurement of lipid oxidation: a review. J Am Oil Chem Soc. 1978;55:539–46.
45. Kwon CW, Park K-M, Park JW, Lee JH, Choi SJ, Chang P-S. Rapid and sensitive determination of lipid oxidation using the reagent kit based on spectrophotometry (FOODLABfat System). J Chem. 2016;2016, Article ID 1468743, 6 pages. https://doi.org/10.1155/2016/1468743.
46. Domijan AM, Ralić J, Radić Brkanac S, Rumora L, Žanić-Grubišić T. Quantification of malondialdehyde by HPLC-FL—application to various biological samples. Biomed Chromatogr. 2015;29(1):41–6. https://doi.org/10.1002/bmc.3361.
47. Madany J. Serum malondialdehyde level and activity of total antioxidant status of dogs with age-related cataract. Pol J Vet Sci. 2016;19(2):429–31. https://doi.org/10.1515/pjvs-2016-0054.
48. Goldring C, Casini AF, Maellaro E, Del Bello B, Comporti M. Determination of 4-hydroxynonenal by high-performance liquid chromatography with electrochemical detection. Lipids. 1993;28(2):141–5. https://doi.org/10.1007/bf02535778.
49. Mendonça R, Gning O, Di Cesaré C, et al. Sensitive and selective quantification of free and total malondialdehyde in plasma using UHPLC-HRMS. J Lipid Res. 2017;58(9):1924–31. https://doi.org/10.1194/jlr.D076661.
50. Alexandri E, Ahmed R, Siddiqui H, Choudhary MI, Tsiafoulis CG, Gerothanassis IP. High resolution NMR spectroscopy as a structural and analytical tool for unsaturated lipids in solution. Molecules. 2017;22(10):1663. https://doi.org/10.3390/molecules22101663. Published 2017 Oct 5.
51. Tynkkynen T, Mursu J, Nurmi T, Tuppurainen K, Laatikainen R, Soininen P. NMR protocol for determination of oxidation susceptibility of serum lipids and application of the protocol to a chocolate study. Metabolomics. 2012;8(3):386–98. https://doi.org/10.1007/s11306-011-0323-2.
52. Parinandi NL, Weis BK, Schmid HH. Assay of cardiolipin peroxidation by high-performance liquid chromatography. Chem Phys Lipids. 1988;49(3):215–20. https://doi.org/10.1016/0009-3084(88)90009-6.
53. Parinandi NL, Weis BK, Natarajan V, Schmid HH. Peroxidative modification of phospholipids in myocardial membranes. Arch Biochem Biophys. 1990;280(1):45–52. https://doi.org/10.1016/0003-9861(90)90516-2.
54. O'Connor Butler ES, Mazerik JN, Cruff JP, et al. Lipoxygenase-catalyzed phospholipid peroxidation: preparation, purification, and characterization of phosphatidylinositol peroxides. Methods Mol Biol. 2010;610:387–401. https://doi.org/10.1007/978-1-60327-029-8_23.
55. Spickett CM, Pitt AR. Oxidative lipidomics coming of age: advances in analysis of oxidized phospholipids in physiology and pathology. Antioxid Redox Signal. 2015;22(18):1646–66. https://doi.org/10.1089/ars.2014.6098.

56. Verrastro I, Pasha S, Jensen KT, Pitt AR, Spickett CM. Mass spectrometry-based methods for identifying oxidized proteins in disease: advances and challenges. Biomolecules. 2015;5(2):378–411. https://doi.org/10.3390/biom5020378. Published 2015 Apr 14.
57. Cui L, Lahti PM, Decker EA. Evaluating electron paramagnetic resonance (EPR) to measure lipid oxidation lag phase for shelf—life determination of oils. J Am Oil Chem Soc. 2017;94:89–97. https://doi.org/10.1007/s11746-016-2927-1.
58. Venkataraman S, Schafer FQ, Buettner GR. Detection of lipid radicals using EPR. Antioxid Redox Signal. 2004;6(3):631–8. https://doi.org/10.1089/152308604773934396.
59. Cummins TD, Higdon AN, Kramer PA, et al. Utilization of fluorescent probes for the quantification and identification of subcellular proteomes and biological processes regulated by lipid peroxidation products. Free Radic Biol Med. 2013;59:56–68. https://doi.org/10.1016/j.freeradbiomed.2012.08.014.
60. Greene LE, Lincoln R, Cosa G. Spatio-temporal monitoring of lipid peroxyl radicals in live cell studies combining fluorogenic antioxidants and fluorescence microscopy methods. Free Radic Biol Med. 2018;128:124–36. https://doi.org/10.1016/j.freeradbiomed.2018.04.006.
61. Tímmins GS, dos Santos RE, Whitwood AC, et al. Lipid peroxidation-dependent chemiluminescence from the cyclization of alkylperoxyl radicals to dioxetane radical intermediates. Chem Res Toxicol. 1997;10(10):1090–6. https://doi.org/10.1021/tx970075p.
62. Mano CM, Prado FM, Massari J, et al. Excited singlet molecular $O_2(^1\Delta g)$ is generated enzymatically from excited carbonyls in the dark. Sci Rep. 2014;4:5938. https://doi.org/10.1038/srep05938. Published 2014 Aug 4.
63. van't Erve TJ, Kadiiska MB, London SJ, Mason RP. Classifying oxidative stress by F2-isoprostane levels across human diseases: a meta-analysis. Redox Biol. 2017;12:582–99. https://doi.org/10.1016/j.redox.2017.03.024.
64. Lee YY, Galano JM, Oger C, et al. Assessment of isoprostanes in human plasma: technical considerations and the use of mass spectrometry. Lipids. 2016;51(11):1217–29. https://doi.org/10.1007/s11745-016-4198-x.
65. Yoshida Y, Umeno A, Shichiri M. Lipid peroxidation biomarkers for evaluating oxidative stress and assessing antioxidant capacity in vivo. J Clin Biochem Nutr. 2013;52(1):9–16. https://doi.org/10.3164/jcbn.12-112.
66. Lee JD, Cai Q, Shu XO, Nechuta SJ. The role of biomarkers of oxidative stress in breast cancer risk and prognosis: a systematic review of the epidemiologic literature. J Womens Health (Larchmt). 2017;26(5):467–82. https://doi.org/10.1089/jwh.2016.5973.
67. Anthonymuthu TS, Kim-Campbell N, Bayır H. Oxidative lipidomics: applications in critical care. Curr Opin Crit Care. 2017;23(4):251–6. https://doi.org/10.1097/MCC.0000000000000419.
68. Pauling J, Klipp E. Computational lipidomics and lipid bioinformatics: filling in the blanks. J Integr Bioinform. 2016;13(1):299. https://doi.org/10.2390/biecoll-jib-2016-299. Published 2016 Dec 22.
69. Stephenson DJ, Hoeferlin LA, Chalfant CE. Lipidomics in translational research and the clinical significance of lipid-based biomarkers. Transl Res. 2017;189:13–29. https://doi.org/10.1016/j.trsl.2017.06.006.
70. Traber MG, Stevens JF. Vitamins C and E: beneficial effects from a mechanistic perspective. Free Radic Biol Med. 2011;51(5):1000–13. https://doi.org/10.1016/j.freeradbiomed.2011.05.017.
71. Barrera G. Oxidative stress and lipid peroxidation products in cancer progression and therapy. ISRN Oncol. 2012;2012:137289. https://doi.org/10.5402/2012/137289.
72. Beloribi-Djefaflia S, Vasseur S, Guillaumond F. Lipid metabolic reprogramming in cancer cells. Oncogenesis. 2016;5(1):e189. https://doi.org/10.1038/oncsis.2015.49. Published 2016 Jan 25.
73. Cervantes B, Ulatowski LM. Vitamin E and Alzheimer's disease-is it time for personalized medicine? Antioxidants (Basel). 2017;6(3):45. https://doi.org/10.3390/antiox6030045. Published 2017 Jun 24.

74. Gru'z P. Lipid peroxidation, diet, and the genotoxicology of aging. In: Molecular basis of nutrition and aging. Elsevier Inc.; 2016. p. 155–76. https://doi.org/10.1016/B978-0-12-801816-3.00012-1.
75. Ahmed N, Barrow CJ, Suphioglu C. Exploring the effects of omega-3 and omega-6 fatty acids on allergy using a HEK-blue cell line. Int J Mol Sci. 2016;17(2):220. https://doi.org/10.3390/ijms17020220. Published 2016 Feb 6.
76. Kumar Y, Yadav DN, Ahmad T, Narsaiah K. Recent trends in the use of natural antioxidants for meat and meat products. Compr Rev Food Sci Food Saf. 2015;14:796–812. . Institute of Food Technologists®. https://doi.org/10.1111/1541-4337.12156.
77. Reitznerová A, Šuleková M, Nagy J, et al. Lipid peroxidation process in meat and meat products: a comparison study of malondialdehyde determination between modified 2-thiobarbituric acid spectrophotometric method and reverse-phase high-performance liquid chromatography. Molecules. 2017;22(11):1988. https://doi.org/10.3390/molecules22111988. Published 2017 Nov 16.
78. Dridi W, Toutain J, Sommier A, et al. Characterization of lipid oxidation in plant oils by micro-calorimetry. Food Chem. 2016;197(Pt A):709–13. https://doi.org/10.1016/j.foodchem.2015.11.040.
79. Richardson CE, Hennebelle M, Otoki Y, Zamora D, Yang J, Hammock BD, Taha AY. Lipidomic quantitation of oxidized fatty acids in plant and algae oils. J Agric Food Chem. 2017;65(9):1941–51. https://doi.org/10.1021/acs.jafc.6b05559.

Part III
Clinically Related Models and Approaches

Chapter 6
Oxidant-Induced Models of Vascular Leak

Rengasayee Veeraraghavan, Narasimham L. Parinandi, and Thomas J. Hund

Abstract The vascular endothelium is an organ that regulates mass transfer between blood and various tissues around the body. Thus, vascular leak, the exchange of fluid between the vasculature and the interstitium of surrounding tissue, is an essential physiological process. The rate of vascular leak is determined by the tightness of the vascular barrier, which is composed of a monolayer of endothelial cells that form the inner lining of blood vessels. This in turn is regulated by redox-dependent modifications to the cytoskeletons of endothelial cells and the cell–cell junctions between them. Thus, understanding redox-dependent regulation of endothelial barrier function is of significant scientific interest. However, the impetus for this area of inquiry extends well beyond scientific curiosity: In several cardiovascular, renal,

R. Veeraraghavan (✉)
Department of Biomedical Engineering, The Ohio State University, Columbus, OH, USA

Bob and Corinne Frick Center for Heart Failure and Arrhythmia, The Ohio State University Wexner Medical Center, Columbus, OH, USA

Dorothy M. Davis Heart and Lung Research Institute, College of Medicine, The Ohio State University Wexner Medical Center, Columbus, OH, USA

Department of Physiology and Cell Biology, College of Medicine, The Ohio State University, Columbus, OH, USA
e-mail: veeraraghavan.12@osu.edu

N. L. Parinandi
Bob and Corinne Frick Center for Heart Failure and Arrhythmia, The Ohio State University Wexner Medical Center, Columbus, OH, USA

Dorothy M. Davis Heart and Lung Research Institute, College of Medicine, The Ohio State University Wexner Medical Center, Columbus, OH, USA
e-mail: Narasimham.parinandi@osumc.edu

T. J. Hund
Department of Biomedical Engineering, The Ohio State University, Columbus, OH, USA

Bob and Corinne Frick Center for Heart Failure and Arrhythmia, The Ohio State University Wexner Medical Center, Columbus, OH, USA

Dorothy M. Davis Heart and Lung Research Institute, College of Medicine, The Ohio State University Wexner Medical Center, Columbus, OH, USA

L. J. Berliner, N. L. Parinandi (eds.), *Measuring Oxidants and Oxidative Stress in Biological Systems*, Biological Magnetic Resonance 34,
https://doi.org/10.1007/978-3-030-47318-1_6

and metabolic diseases, oxidative stress leads to pathologically enhancement of vascular leak, contributing to thrombogenesis, and other deleterious effects. Thus, understanding the development and consequences of oxidant-induced pathological vascular leak is of paramount importance to the development of effective therapies. A variety of experimental models and measurement techniques are available for the investigation of oxidant-induced vascular leak in preparations ranging from cells to whole organisms. In this chapter, we discuss the mechanisms underlying oxidant modulation of vascular leak, various experimental models of this phenomenon, and methods to assess structure and function of the vascular endothelium. Particular emphasis is placed on selecting the experimental models and measurement techniques that are most appropriate to the research question under investigation.

Keywords Oxidant-induced vascular leak · Endothelial leak oxidant model · Oxidative stress and blood vessel leak model

The Vasculature

In all vertebrates, a closed circulatory system is a life-critical organ system responsible for continuous delivery of metabolic support, hormonal regulation, and removal of waste for the entire organism. Along with a pump in the form of the heart, the circulatory system is comprised of a variety of blood vessels, each specialized for their particular functions. Arteries carry blood away from the heart and serve as a pressure reservoir, while veins return blood to the heart and serve as a volume reservoir. A vast network of microvessels and capillaries link the arterial and venous circulation to facilitate perfusion of specific tissues.

The Vascular Endothelium

Whereas larger blood vessels are essentially closed conduits for blood, microvessels and capillaries more readily exchange fluid within the interstitium of surrounding tissue (vascular "leak"). Such vascular leak and the consequent mass transfer are obligate for several key functions of circulation, including nutrient distribution, metabolic waste clearance, and hormonal regulation. Mass transfer between blood and tissues takes place across the endothelium, a monolayer of specialized cells that form the inner lining of macro- and microvessels and capillaries. Fluid filters across pores in the endothelium, which are highly permeable to small molecules (<60 kDa), with only the rate of blood flow limiting the transport of such molecules. In contrast, the transport of larger molecules ($\geq$ 60 kDa) is limited by the rate of diffusion across the endothelium. That said, capillary permeability varies significantly between different tissues. Thus, capillaries are classified based on the structure and permeability of their walls into three categories: (1) *continuous capillaries*, which exhibit the lowest permeability, are found in locations such as the blood-brain barrier and skeletal muscle, (2) *fenestrated capillaries* have pores that confer intermediate permeability, and are found in locations such as the intestines and kidneys, and (3) *discontinuous/sinusoidal capillaries* have larger pores, which make them the most permeable, occur in locations such as the liver, lymph nodes, and bone marrow. Control and maintenance of vascular permeability are key physiological functions performed by the endothelium, and dynamic alterations of these processes have important implications in health and in disease.

Structural Determinants of Vascular Endothelial Permeability

Cell–Cell Junctions Endothelial permeability is most directly determined by the composition and integrity of cell–cell junctions between endothelial cells [1] (Fig. 6.1). Foremost in regulating vascular endothelial permeability are tight junctions, which are composed of claudins, occludins, and junctional adhesion molecules (JAMs), and interact with the actin cytoskeleton via zonula occludens-1 (ZO-1), 2 (ZO-2), and 3 (ZO-3) to mechanically couple endothelial cells [2–5]. Equally important are adherens junctions, which are composed of vascular endothelial cadherin (VE-cadherin), and mediate contact inhibition of endothelial growth. Additionally, endothelial cells are also connected via gap junctions composed of connexins 37 (Cx37), 40 (Cx40), and 43 (Cx43), which provide electrical and

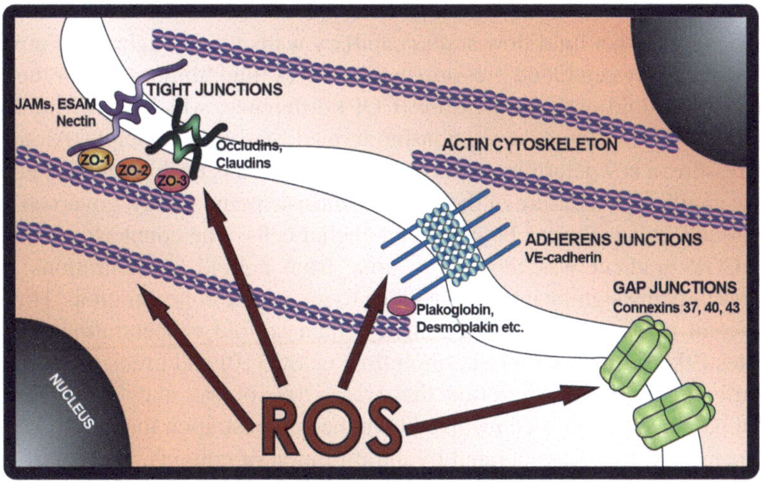

Fig. 6.1 Schematic depicting the major types of cell–cell junctions for regulating endothelial communication and barrier function. Abbreviations are as follows: *ESAM* endothelial cell-selective adhesion molecule, *JAM* junctional adhesion molecule, *VE-cadherin* vascular endothelial cadherin

chemical coupling [3, 5]. Under normal conditions, these structures enable the vascular endothelium to tightly control fluid and solute exchange between blood and tissues. However, under pathophysiological conditions, particularly increased oxidative stress, a variety of signaling pathways are activated, resulting in the disruption of endothelial cell–cell junctions. Specifically, this process involves the reorganization of the actin cytoskeleton into stress fibers, the phosphorylation and internalization of adherens junction components, and finally, the dissolution of tight junctions.

The ultimate consequence of these changes is the breakdown of vascular endothelial barrier function, an effect implicated in a variety of disease states as discussed below. Additionally, cell–cell uncoupling of endothelial cells may increase expression/activity of connexin and pannexin hemichannels, which connect intra- and extracellular spaces, allowing for the release of paracrine signals [6]. Such paracrine signaling has been demonstrated to be involved in the regulation of vascular barrier function in response to inflammatory cues, as during ischemia and hypoxia.

The Endothelial Glycocalyx Another key structure responsible for vascular endothelial barrier function is the endothelial glycocalyx (EG), a carbohydrate-rich, fibrous matrix composed of proteoglycans, glycosaminoglycans, and glycolipids that covers the apical surface of the vascular endothelium [2, 7, 8]. The EG forms the interface between blood within the vascular lumen and the vessel wall, and is thus, the primary barrier to fluid and solute movement out of the vasculature. Additionally, the EG protects the endothelium from shear stress and also serves as a shear stress-sensor for the endothelium. The dynamic balance between the synthesis of EG components and their enzymatic degradation is a key determinant of vascular barrier function.

Classically, the net fluid flow across capillary walls was thought to be governed by the balance between blood pressure, which drives fluid filtration out of the capillary, and the colloid osmotic pressure (COP) difference, which drives fluid reabsorption into the capillary—a relationship termed Starling's principle. Along with blood pressure, a key determinant of the rate of filtration is the hydraulic conductivity of the capillary, a measure of the vessel's intrinsic permeability, governed by the nature of cell–cell junctions between endothelial cells. The counterforce afforded by the COP gradient was thought to arise from colloid concentrations within microvessels being significantly greater than tissue colloid concentrations. However, experimental measurements show roughly equal colloid concentrations between these sites [9]. Yet, the measured lymph flow is low [10] and predicts an effective COP difference opposing filtration that was ~70% of vascular COP [9]. These apparent paradoxes have been resolved with the demonstration that the COP gradient opposing filtration is generated by ultrafiltration of colloids, particularly albumin by the EG [8, 11].

As with cell–cell junctions, damage to the EG also plays an important role in the compromise of vascular barrier function in multiple pathophysiological states. Inflammatory signals and reactive oxygen species (ROS) cause rapid degradation of the EG leading to vascular leak, edema, and increased risk of thrombogenesis. For more detailed discussion of the cellular mechanisms involved in ROS-induced breakdown of the vascular endothelial barrier, readers are referred to previous review articles on the subject [12–14].

Vascular Leak in Disease

Endothelial dysfunction is a key component in the progression of a diverse array of cardiovascular, renal, and metabolic diseases, where it results in pathologically enhanced vascular leak, thrombogenesis, and other deleterious effects [15, 16]. Yet, we are only beginning to grasp its precise role in many of these conditions.

Diabetes Diabetes is part of a continuum of metabolic disorders characterized by hyperglycemia. This can result from inadequate insulin production (type 1) or develop due to cells failing to respond to insulin (type 2). Vascular dysfunction is a well-established sequela of these metabolic disorders [15, 17, 18] and is thought to result from pathologically enhanced enzymatic production of reactive oxygen species (ROS). The resulting oxidative stress, in conjunction with hyperglycemia, prompts vascular inflammation via multiple endothelial cell signaling pathways. Ultimately this leads to degradation of the endothelial glycocalyx and cell–cell junctions, compromising vascular barrier function, and increasing vascular leak [8]. Over time, these effects contribute to compromised wound healing, and increases patients' risk of cardiovascular disease, stroke, etc. Given that over 8% of the world's adult population suffer from diabetes [19], there is an urgent need for effective vascular-management strategies as part of the treatment of diabetes.

Ischemic Heart Disease The leading cause of death worldwide is ischemic heart disease (IHD), which afflicts over 110 million people and accounts for over 8.9 million deaths annually [20–22]. IHD results from compromised blood flow to the heart, which can lead to myocardial infarction (MI), the death of cardiac muscle. Aside from immediate risk to the patients' lives, IHD can lead to complications down the road including heart failure (HF) or sudden arrhythmic death (SAD). Critical in the treatment and management of IHD following an acute MI is minimization of the size of the infarct through successful reperfusion of the heart, in turn the most effective means of improving clinical outcomes [23]. However, reperfusion can itself cause further injury to the heart, an effect termed myocardial reperfusion injury. Thus, there is a critical need to develop effective methods for reperfusion that minimize injury to the heart. A key component of myocardial ischemia/reperfusion (I/R) injury is endothelial dysfunction, which results from damage to the vascular endothelial cells by multiple mechanisms including pH change and oxidative stress due to pathologically elevated ROS production by cardiac myocytes, coronary vascular endothelium, and inflammatory cells. Endothelial dysfunction, in turn, contributes to intravascular microthrombosis, reduced blood flow, and activation of inflammatory cells, particularly neutrophils. Additionally, I/R injury increases ROS production from myocytes, endothelial cells, and activated leukocytes, which acts to compromise vascular endothelial barrier function via multiple mechanisms, namely a) cytoskeletal alterations within endothelial cells prompted [24], b) degradation of endothelial cell–cell junction [5, 25], c) denudation of the EG [8, 26, 27], and d) contractile activation of endothelial cells [28]. These processes promote inflammatory cell migration to the injury cite with important consequences for the severity of I/R injury. Thus, vascular endothelial barrier function may be an important therapeutic target in this widespread pathology as well.

Atrial Fibrillation Atrial fibrillation (AF) is the most common type of arrhythmia, affecting 2–3% of the population [29]. Although not immediately life-threatening, AF is progressive, growing in severity over time, and elevates patients' risk of stroke and further cardiovascular disease. Given its high prevalence and annual cost of over $26 billion to the U.S. healthcare system [30], there is an urgent need to understand the mechanisms of AF and to develop effective therapies against it. Multiple studies have demonstrated AF patients to have elevated levels of inflammatory cytokines known to compromise vascular barrier function [31–36]. Indeed, AF patients show evidence of circulating biomarkers of oxidative stress and endothelial dysfunction and increased levels of these biomarkers associate with adverse outcomes including thrombogenesis [33, 37]. Notably, biomarkers of endothelial dysfunction also predict the onset of AF in patients [38–40], suggesting that endothelial dysfunction may directly contribute to the genesis and progression of AF. Thus, there is much to be gained from elucidating the mechanisms and consequences of vascular endothelial dysfunction in AF.

Heart Failure Heart failure (HF) is a complex, progressive disorder wherein the heart is unable to pump sufficient blood to meet the demands of the body. It afflicts 1–2% of adults worldwide and 6–10% of those aged over the age of 65 [41]. Given the debilitating impact of HF on patients' lives, high levels of mortality, and $39.2 billion contribution to annual US healthcare costs [42], there is a clear need for effective therapies. Notably, HF is often associated with multiple comorbidities including the pathologies discussed above (diabetes, ischemic heart disease, and AF). Thus, the association between HF and endothelial dysfunction should come as no surprise. Indeed, inflammation, oxidative stress, and consequent endothelial dysfunction are associated with HF with preserved (HFpEF) as well as reduced (HFrEF) ejection fraction [43, 44]. In point of fact, the endothelial dysfunction is thought to be a central, causative factor and predictor of mortality in HFpEF [45, 46] and has shown promise as a therapeutic target [43]. Overall, the ongoing healthcare challenge posed by heart failure represents yet another reason to understand the mechanisms of oxidative stress-induced vascular endothelial dysfunction.

Measurement of Vascular Leak

Given the widespread association between cardiac disease and increased vascular leak, there is a great need for reliable and reproducible systems for assessing barrier function from in vitro to in vivo (Fig. 6.2).

Macromolecule Leak This method makes use of monolayers of vascular endothelial cells cultured on permeable supports in order to assess their barrier function [47, 48]. Briefly, endothelial cells are cultured on inserts containing a porous membrane made of nitrocellulose, polyester, or collagen-pretreated polyethylene terephthalate, coated with fibronectin or gelatin, thus generating a two-compartment system. Once the cells become confluent, the culture medium in the upper compartment is replaced with a tracer solution containing macromolecules conjugated with fluorescent labels (e.g. FITC- conjugated albumin, RITC- conjugated dextran) or radiolabeled (^{125}I-conjugated albumin). The culture medium in the lower chamber is then sampled at different time points (1, 6, 24 h) and the concentration of the fluorescent tracer quantified using a spectrofluorometer or microplate reader. By comparing these values to the calculated equilibrium concentration, the tracer flux and the permeability of the endothelial cell monolayer are quantified. Since these methods measure paracellular permeability across the endothelial cell monolayer as a whole, it is critical to ensure uniform, regular monolayers. This approach can be adapted to ex vivo studies, where the extravasation of dye tracers from isolated and perfused organs such as lungs is assessed as a metric of endothelial barrier integrity [49].

Electrical Methods A second class of methods uses electrical measurements to quantify ion permeability of endothelial cell monolayers. Simply, an AC square wave current is applied across the endothelial cell monolayer grown on a permeable support with one pair of electrodes, while the resulting voltage deflection is measured by a second pair. From these, the transepithelial electrical resistance (TER) is

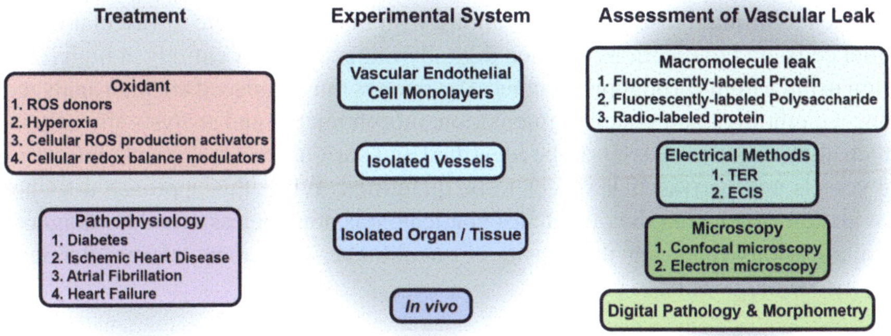

Fig. 6.2 Schematic depicting the common treatments, experimental systems, and experimental techniques used to study vascular leak. Abbreviations are as follows: *ECIS* electric cell-substrate impedance sensing, *ROS* reactive oxygen species, *TER* trans-epithelial electrical resistance

calculated [47]. Although TER is a composite measure of both transcellular and paracellular resistances, the former typically exceeds the latter by a wide margin. Thus, TER serves as a sensitive measure of paracellular resistance, and thereby, the permeability of endothelial cell–cell junctions. Furthermore, by varying the ionic composition of the buffer used, TER can be used to assess ion selectivity.

A further refinement of this approach has been developed in the form of electric cell-substrate impedance sensing (ECIS) [50–52]. Here, cells are grown on gold electrode(s) and subjected to an alternating electric field at multiple frequencies. It has been demonstrated that measurements at lower frequencies (<2 kHz) primarily reflect paracellular resistance, while measurements at higher frequencies (>40 kHz) reflect transcellular resistance. By fitting these measurements to a circuit model, the ECIS technique provides quantitative measures of the barrier function (permeability) of the cell layer, the degree of constricted flow of current under the cells, and the cell membrane capacitance. Additionally, the use of an array of small electrodes in place of a single large electrode can reveal heterogeneities within a monolayer.

Microscopy In addition to functional interrogations of endothelial barriers, structural remodeling, particularly to the actin cytoskeleton, adherens junctions, and tight junctions, can be assessed as structural markers of barrier integrity. Fluorescence microscopy, especially confocal microscopy, is widely used to assess rearrangement of the actin cytoskeleton, as well as breakdown of cell–cell junctions (tight junctions, adherens junctions) and their translocation from cell–cell borders. Nanoscale structural alterations to cell–cell junctions can also be assessed via electron microscopy. These approaches can be applied to in vitro studies in endothelial cell monolayers [53], as well as to tissue samples collected from in vivo studies [49].

Digital Pathology and Morphometry Last but not least, immunohistochemical staining of microvessels is a technique that has long been in use. However, it has been transformed from a qualitative, low-throughput assessment to a quantitative, high-throughput technique with the advent of digital pathology, and advanced morphometry algorithms [54, 55]. Briefly, sections cut from fixed tissue samples are subjected to immunohistochemical staining with markers specific to vascular endothelial cells, and the sections are then digitally imaged in their entirely at high resolution using a high-throughput scanner. The images thus produced are then analyzed using a plethora of automated segmentation, morphometry, and analysis algorithms, which allow for microvessels to be identified and various morphological features of the vessels and perivascular spaces to be quantified. While this approach does not provide information on the rate of vascular leak, it is nonetheless suited for diagnostic applications requiring the sensitive, quantitative assessment of vascular leak from tissue biopsies from patients.

Oxidant-Induced Models of Vascular Leak

As noted above, ROS serve as key mediators of vascular leak in health, and in disease, oxidative stress induced by abnormally high ROS levels leads to pathological enhancement of vascular leak. ROS involved in vascular leak include superoxide (O_2^-) and hydrogen peroxide (H_2O_2) [15, 56]. At high concentrations (>200 μM), as occurs in multiple disease states, ROS recruit a bevy of cell-signaling pathways, which in turn prompt (a) phosphorylation and internalization of VE-cadherin, compromising adherens junctions, (b) reorganization of the actin cytoskeleton into stress fibers, (c) disruption of claudin, occludin tight junctions, and (d) degradation of the EG. Detailed below are various experimental models of oxidant-induced vascular leak.

In vitro Cell culture models including both immortalized endothelial cell lines and primary cultures of endothelial cells are widely used for in vitro studies of vascular biology [57, 58]. Cultured endothelial cells such as human umbilical vein endothelial cells (HUVECs) typically exhibit a life span of ~10 serial passages allowing them to be maintained in culture for up to 5 months. Additionally, a wide selection of immortalized endothelial cell lines have been developed which retain stable phenotypes in culture over even longer periods. Cells can be treated with direct ROS donors such as H_2O_2 [56, 59] or a combination of xanthine oxidase and hypoxanthine (which produce superoxide and other ROS) [60]. Alternatively, cells can be exposed to treatments that prompt ROS generation: These include (1) exposure to hyperoxia [53], (2) ethyl alcohol [61], (3) Diperoxovanadate (DPV; generated by H_2O_2 application following pretreatment with vanadate) [59], and (4) biologically active aldehydes such as 4-Hydroxy-2-nonenal (4-HNE) [62]. In yet another approach, oxidative stress can be induced by targeting the endogenous mechanisms that maintain redox balance within cells: (1) Diamide treatment achieves this by reducing intracellular glutathione [63]. (2) Treatment with NCX-4016, a derivative of aspirin containing a nitro moiety, operates by prompting sustained nitric oxide (NO) release [64]. (3) Glucose-derived oxoaldehydes, such as glyoxal and methylglyoxal, which are released during diabetes, attack the vascular endothelial barrier through the formation of advanced glycation end-products (AGEs) [53]. (4) Cytotoxic agents, such as bleomycin, can upset the thiol-redox balance within vascular endothelial cells, thereby, compromising barrier function [65]. ROS-induced effects on endothelial cells can be further exacerbated by additional treatment with agents that disrupt the cytoskeleton, such as cytochalasin D and latrunculin A [61]. The consequent changes in endothelial cell barrier function can be assessed using methods detailed above while accompanying effects on cell structure and biology can respectively be studied using microscopy and molecular biological techniques.

In vivo/Ex vivo It is possible to cannulate major blood vessels in intact, isolated organs such as the heart, mesentery, stomach, and kidneys, allowing these organs to be utilized for ex vivo investigation of blood vessel function [66]. It is also possible to expose animals to a hyperoxic atmosphere (in vivo treatment) to promote ROS-induced vascular leak, followed by ex vivo assessment of vascular leak via dye extravasation, and fluorescent microscopic examination of cytoskeletal reorganization [61]. Although such preparations pose surgical challenges and have a limited lifetime, they allow blood vessels to be retained intact for study. For studies that demand more direct access to blood vessels, ring, spiral strip, perfused blood vessel preparations are used [66–68]. All of these preparations offer the experimentalist control over tissue perfusion/superfusion, and, thereby, provide a convenient pathway for drug delivery.

Conclusion and Future Perspectives

In summary, vascular leak is a critical physiological process, the dysregulation of which is a common attribute across a multitude of pathologies. There is a pressing need to uncover the mechanisms that modulate vascular leak in health and in disease as well as for novel therapeutics to be developed to target pathological vascular leak. In the pursuit of these goals, it is imperative that the specific attributes of different experimental models of vascular leak and measurement techniques be considered, and the most appropriate models chosen for any study.

References

1. Radeva MY, Waschke J. Mind the gap: mechanisms regulating the endothelial barrier. Acta Physiol (Oxf). 2018;222(1).
2. Schlingmann B, Molina SA, Koval M. Claudins: gatekeepers of lung epithelial function. Semin Cell Dev Biol. 2015;42:47–57.
3. Sukriti S, Tauseef M, Yazbeck P, Mehta D. Mechanisms regulating endothelial permeability. Pulm Circ. 2014;4:535–51.
4. Bravi L, Dejana E, Lampugnani MG. VE-cadherin at a glance. Cell Tissue Res. 2014;355:515–22.
5. Mehta D, Malik AB. Signaling mechanisms regulating endothelial permeability. Physiol Rev. 2006;86:279–367.
6. Begandt D, Good ME, Keller AS, DeLalio LJ, Rowley C, Isakson BE, Figueroa XF. Pannexin channel and connexin hemichannel expression in vascular function and inflammation. BMC Cell Biol. 2017;18:2.
7. Tarbell JM, Cancel LM. The glycocalyx and its significance in human medicine. J Intern Med. 2016;280:97–113.
8. Alphonsus CS, Rodseth RN. The endothelial glycocalyx: a review of the vascular barrier. Anaesthesia. 2014;69:777–84.
9. Adamson RH, Lenz JF, Zhang X, Adamson GN, Weinbaum S, Curry FE. Oncotic pressures opposing filtration across non-fenestrated rat microvessels. J Physiol. 2004;557:889–907.
10. Levick JR. A two-dimensional morphometry-based model of interstitial and transcapillary flow in rabbit synovium. Exp Physiol. 1991;76:905–21.
11. Jacob M, Bruegger D, Rehm M, Stoeckelhuber M, Welsch U, Conzen P, Becker BF. The endothelial glycocalyx affords compatibility of Starling's principle and high cardiac interstitial albumin levels. Cardiovasc Res. 2007;73:575–86.

12. Di A, Mehta D, Malik AB. ROS-activated calcium signaling mechanisms regulating endothelial barrier function. Cell Calcium. 2016;60:163–71.
13. Usatyuk PV, Natarajan V. Hydroxyalkenals and oxidized phospholipids modulation of endothelial cytoskeleton, focal adhesion and adherens junction proteins in regulating endothelial barrier function. Microvasc Res. 2012;83:45–55.
14. Boueiz A, Hassoun PM. Regulation of endothelial barrier function by reactive oxygen and nitrogen species. Microvasc Res. 2009;77:26–34.
15. Zhang X, Sun D, Song JW, Zullo J, Lipphardt M, Coneh-Gould L, Goligorsky MS. Endothelial cell dysfunction and glycocalyx—a vicious circle. Matrix Biol. 2018;71–72:421–31.
16. Cai H, Harrison DG. Endothelial dysfunction in cardiovascular diseases: the role of oxidant stress. Circ Res. 2000;87:840–4.
17. Petrie JR, Guzik TJ, Touyz RM. Diabetes, hypertension, and cardiovascular disease: clinical insights and vascular mechanisms. Can J Cardiol. 2018;34:575–84.
18. Pechlivani N, Ajjan RA. Thrombosis and vascular inflammation in diabetes: mechanisms and potential therapeutic targets. Front Cardiovasc Med. 2018;5:1.
19. Shi Y, Hu FB. The global implications of diabetes and cancer. Lancet. 2014;383:1947–8.
20. Disease GBD, Injury I, Prevalence C. Global, regional, and national incidence, prevalence, and years lived with disability for 310 diseases and injuries, 1990-2015: a systematic analysis for the Global Burden of Disease Study 2015. Lancet. 2016;388:1545–602.
21. Mortality GBD and Causes of Death Collaborators. Global, regional, and national life expectancy, all-cause mortality, and cause-specific mortality for 249 causes of death, 1980–2015: a systematic analysis for the Global Burden of Disease Study 2015. Lancet. 2016;388:1459–544.
22. Lozano R, Naghavi M, Foreman K, Lim S, Shibuya K, Aboyans V, Abraham J, Adair T, Aggarwal R, Ahn SY, Alvarado M, Anderson HR, Anderson LM, Andrews KG, Atkinson C, Baddour LM, Barker-Collo S, Bartels DH, Bell ML, Benjamin EJ, Bennett D, Bhalla K, Bikbov B, Bin Abdulhak A, Birbeck G, Blyth F, Bolliger I, Boufous S, Bucello C, Burch M, Burney P, Carapetis J, Chen H, Chou D, Chugh SS, Coffeng LE, Colan SD, Colquhoun S, Colson KE, Condon J, Connor MD, Cooper LT, Corriere M, Cortinovis M, de Vaccaro KC, Couser W, Cowie BC, Criqui MH, Cross M, Dabhadkar KC, Dahodwala N, De Leo D, Degenhardt L, Delossantos A, Denenberg J, Des Jarlais DC, Dharmaratne SD, Dorsey ER, Driscoll T, Duber H, Ebel B, Erwin PJ, Espindola P, Ezzati M, Feigin V, Flaxman AD, Forouzanfar MH, Fowkes FG, Franklin R, Fransen M, Freeman MK, Gabriel SE, Gakidou E, Gaspari F, Gillum RF, Gonzalez-Medina D, Halasa YA, Haring D, Harrison JE, Havmoeller R, Hay RJ, Hoen B, Hotez PJ, Hoy D, Jacobsen KH, James SL, Jasrasaria R, Jayaraman S, Johns N, Karthikeyan G, Kassebaum N, Keren A, Khoo JP, Knowlton LM, Kobusingye O, Koranteng A, Krishnamurthi R, Lipnick M, Lipshultz SE, Ohno SL, Mabweijano J, MacIntyre MF, Mallinger L, March L, Marks GB, Marks R, Matsumori A, Matzopoulos R, Mayosi BM, McAnulty JH, McDermott MM, McGrath J, Mensah GA, Merriman TR, Michaud C, Miller M, Miller TR, Mock C, Mocumbi AO, Mokdad AA, Moran A, Mulholland K, Nair MN, Naldi L, Narayan KM, Nasseri K, Norman P, O'Donnell M, Omer SB, Ortblad K, Osborne R, Ozgediz D, Pahari B, Pandian JD, Rivero AP, Padilla RP, Perez-Ruiz F, Perico N, Phillips D, Pierce K, Pope CA III, Porrini E, Pourmalek F, Raju M, Ranganathan D, Rehm JT, Rein DB, Remuzzi G, Rivara FP, Roberts T, De Leon FR, Rosenfeld LC, Rushton L, Sacco RL, Salomon JA, Sampson U, Sanman E, Schwebel DC, Segui-Gomez M, Shepard DS, Singh D, Singleton J, Sliwa K, Smith E, Steer A, Taylor JA, Thomas B, Tleyjeh IM, Towbin JA, Truelsen T, Undurraga EA, Venketasubramanian N, Vijayakumar L, Vos T, Wagner GR, Wang M, Wang W, Watt K, Weinstock MA, Weintraub R, Wilkinson JD, Woolf AD, Wulf S, Yeh PH, Yip P, Zabetian A, Zheng ZJ, Lopez AD, Murray CJ, AlMazroa MA, Memish ZA. Global and regional mortality from 235 causes of death for 20 age groups in 1990 and 2010: a systematic analysis for the Global Burden of Disease Study 2010. Lancet. 2012;380:2095–128.
23. Yellon DM, Hausenloy DJ. Myocardial reperfusion injury. N Engl J Med. 2007;357:1121–35.
24. Hastie LE, Patton WF, Hechtman HB, Shepro D. Filamin redistribution in an endothelial cell reoxygenation injury model. Free Radic Biol Med. 1997;22:955–66.

25. Gavard J. Endothelial permeability and VE-cadherin: a wacky comradeship. Cell Adhes Migr. 2014;8:158–64.
26. Kurzelewski M, Czarnowska E, Beresewicz A. Superoxide- and nitric oxide-derived species mediate endothelial dysfunction, endothelial glycocalyx disruption, and enhanced neutrophil adhesion in the post-ischemic guinea-pig heart. J Physiol Pharmacol. 2005;56:163–78.
27. Czarnowska E, Karwatowska-Prokopczuk E. Ultrastructural demonstration of endothelial glycocalyx disruption in the reperfused rat heart. Involvement of oxygen free radicals. Basic Res Cardiol. 1995;90:357–64.
28. Schafer C, Walther S, Schafer M, Dieterich L, Kasseckert S, Abdallah Y, Piper HM. Inhibition of contractile activation reduces reoxygenation-induced endothelial gap formation. Cardiovasc Res. 2003;58:149–55.
29. Zoni-Berisso M, Lercari F, Carazza T, Domenicucci S. Epidemiology of atrial fibrillation: European perspective. Clin Epidemiol. 2014;6:213–20.
30. Kim MH, Johnston SS, Chu BC, Dalal MR, Schulman KL. Estimation of total incremental health care costs in patients with atrial fibrillation in the United States. Circ Cardiovasc Qual Outcomes. 2011;4:313–20.
31. Gramley F, Lorenzen J, Jedamzik B, Gatter K, Koellensperger E, Munzel T, Pezzella F. Atrial fibrillation is associated with cardiac hypoxia. Cardiovasc Pathol. 2010;19:102–11.
32. Kimura T, Takatsuki S, Inagawa K, Katsumata Y, Nishiyama T, Nishiyama N, Fukumoto K, Aizawa Y, Tanimoto Y, Tanimoto K, Fukuda K. Serum inflammation markers predicting successful initial catheter ablation for atrial fibrillation. Heart Lung Circ. 2014;23:636–43.
33. Li J, Solus J, Chen Q, Rho YH, Milne G, Stein CM, Darbar D. Role of inflammation and oxidative stress in atrial fibrillation. Heart Rhythm. 2010;7:438–44.
34. Ogi H, Nakano Y, Niida S, Dote K, Hirai Y, Suenari K, Tonouchi Y, Oda N, Makita Y, Ueda S, Kajihara K, Imai K, Sueda T, Chayama K, Kihara Y. Is structural remodeling of fibrillated atria the consequence of tissue hypoxia? Circ J. 2010;74:1815–21.
35. Scridon A, Morel E, Nonin-Babary E, Girerd N, Fernandez C, Chevalier P. Increased intracardiac vascular endothelial growth factor levels in patients with paroxysmal, but not persistent atrial fibrillation. Europace. 2012;14:948–53.
36. Seko Y, Nishimura H, Takahashi N, Ashida T, Nagai R. Serum levels of vascular endothelial growth factor and transforming growth factor-beta1 in patients with atrial fibrillation undergoing defibrillation therapy. Jpn Heart J. 2000;41:27–32.
37. Polovina MM, Lip GY, Potpara TS. Endothelial (dys)function in lone atrial fibrillation. Curr Pharm Des. 2015;21:622–45.
38. Harling L, Lambert J, Ashrafian H, Darzi A, Gooderham NJ, Athanasiou T. Pre-operative serum VCAM-1 as a biomarker of atrial fibrillation after coronary artery bypass grafting. J Cardiothorac Surg. 2017;12:70.
39. Willeit K, Pechlaner R, Willeit P, Skroblin P, Paulweber B, Schernthaner C, Toell T, Egger G, Weger S, Oberhollenzer M, Kedenko L, Iglseder B, Bonora E, Schett G, Mayr M, Willeit J, Kiechl S. Association between vascular cell adhesion molecule 1 and atrial fibrillation. JAMA Cardiol. 2017;2:516–23.
40. Verdejo H, Roldan J, Garcia L, Del Campo A, Becerra E, Chiong M, Mellado R, Garcia A, Zalaquett R, Braun S, Garayar B, Gonzalez S, Lavandero S, Corbalan R. Systemic vascular cell adhesion molecule-1 predicts the occurrence of post-operative atrial fibrillation. Int J Cardiol. 2011;150:270–6.
41. McMurray JJ, Pfeffer MA. Heart failure. Lancet. 2005;365:1877–89.
42. Voigt J, Sasha John M, Taylor A, Krucoff M, Reynolds MR, Michael Gibson C. A reevaluation of the costs of heart failure and its implications for allocation of health resources in the United States. Clin Cardiol. 2014;37:312–21.
43. Gevaert AB, Lemmens K, Vrints CJ, Van Craenenbroeck EM. Targeting endothelial function to treat heart failure with preserved ejection fraction: the promise of exercise training. Oxidative Med Cell Longev. 2017;2017:4865756.

44. Wray DW, Amann M, Richardson RS. Peripheral vascular function, oxygen delivery and utilization: the impact of oxidative stress in aging and heart failure with reduced ejection fraction. Heart Fail Rev. 2017;22:149–66.
45. Lewis GA, Schelbert EB, Williams SG, Cunnington C, Ahmed F, McDonagh TA, Miller CA. Biological phenotypes of heart failure with preserved ejection fraction. J Am Coll Cardiol. 2017;70:2186–200.
46. Roe AT, Sjaastad I, Louch WE. Heart failure with preserved ejection fraction. Tidsskr Nor Laegeforen. 2017;137.
47. Matter K, Balda MS. Functional analysis of tight junctions. Methods. 2003;30:228–34.
48. Nooteboom A, Hendriks T, Otteholler I, van der Linden CJ. Permeability characteristics of human endothelial monolayers seeded on different extracellular matrix proteins. Mediat Inflamm. 2000;9:235–41.
49. Sliman SM, Patel RB, Cruff JP, Kotha SR, Newland CA, Schrader CA, Sherwani SI, Gurney TO, Magalang UJ, Parinandi NL. Adiponectin protects against hyperoxic lung injury and vascular leak. Cell Biochem Biophys. 2013;67:399–414.
50. McRae M, LaFratta LM, Nguyen BM, Paris JJ, Hauser KF, Conway DE. Characterization of cell-cell junction changes associated with the formation of a strong endothelial barrier. Tissue Barriers. 2018;6:e1405774.
51. Szulcek R, Bogaard HJ, van Nieuw Amerongen GP. Electric cell-substrate impedance sensing for the quantification of endothelial proliferation, barrier function, and motility. J Vis Exp. 2014;(85):51300.
52. Giaever I, Keese CR. Monitoring fibroblast behavior in tissue culture with an applied electric field. Proc Natl Acad Sci U S A. 1984;81:3761–4.
53. Sliman SM, Eubank TD, Kotha SR, Kuppusamy ML, Sherwani SI, Butler ES, Kuppusamy P, Roy S, Marsh CB, Stern DM, Parinandi NL. Hyperglycemic oxoaldehyde, glyoxal, causes barrier dysfunction, cytoskeletal alterations, and inhibition of angiogenesis in vascular endothelial cells: aminoguanidine protection. Mol Cell Biochem. 2010;333:9–26.
54. Duarte D, Hawkins ED, Akinduro O, Ang H, De Filippo K, Kong IY, Haltalli M, Ruivo N, Straszkowski L, Vervoort SJ, McLean C, Weber TS, Khorshed R, Pirillo C, Wei A, Ramasamy SK, Kusumbe AP, Duffy K, Adams RH, Purton LE, Carlin LM, Lo Celso C. Inhibition of endosteal vascular niche remodeling rescues hematopoietic stem cell loss in AML. Cell Stem Cell. 2018;22:64–77.e6.
55. Gomez-Gelvez JC, Salama ME, Perkins SL, Leavitt M, Inamdar KV. Prognostic impact of tumor microenvironment in diffuse large B-cell lymphoma uniformly treated with R-CHOP chemotherapy. Am J Clin Pathol. 2016;145:514–23.
56. Cai H. Hydrogen peroxide regulation of endothelial function: origins, mechanisms, and consequences. Cardiovasc Res. 2005;68:26–36.
57. Rahman NA, Rasil A, Meyding-Lamade U, Craemer EM, Diah S, Tuah AA, Muharram SH. Immortalized endothelial cell lines for in vitro blood-brain barrier models: a systematic review. Brain Res. 1642;2016:532–45.
58. Bouis D, Hospers GA, Meijer C, Molema G, Mulder NH. Endothelium in vitro: a review of human vascular endothelial cell lines for blood vessel-related research. Angiogenesis. 2001;4:91–102.
59. Haorah J, Knipe B, Leibhart J, Ghorpade A, Persidsky Y. Alcohol-induced oxidative stress in brain endothelial cells causes blood-brain barrier dysfunction. J Leukoc Biol. 2005;78:1223–32.
60. Schreibelt G, Kooij G, Reijerkerk A, van Doorn R, Gringhuis SI, van der Pol S, Weksler BB, Romero IA, Couraud PO, Piontek J, Blasig IE, Dijkstra CD, Ronken E, de Vries HE. Reactive oxygen species alter brain endothelial tight junction dynamics via RhoA, PI3 kinase, and PKB signaling. FASEB J. 2007;21:3666–76.
61. Usatyuk PV, Romer LH, He D, Parinandi NL, Kleinberg ME, Zhan S, Jacobson JR, Dudek SM, Pendyala S, Garcia JG, Natarajan V. Regulation of hyperoxia-induced NADPH oxidase activation in human lung endothelial cells by the actin cytoskeleton and cortactin. J Biol Chem. 2007;282:23284–95.

62. Usatyuk PV, Vepa S, Watkins T, He D, Parinandi NL, Natarajan V. Redox regulation of reactive oxygen species-induced p38 MAP kinase activation and barrier dysfunction in lung microvascular endothelial cells. Antioxid Redox Signal. 2003;5:723–30.
63. Shi S, Garcia JG, Roy S, Parinandi NL, Natarajan V. Involvement of c-Src in diperoxovanadate-induced endothelial cell barrier dysfunction. Am J Physiol Lung Cell Mol Physiol. 2000;279:L441–51.
64. Patel RB, Kotha SR, Sauers LA, Malireddy S, Gurney TO, Gupta NN, Elton TS, Magalang UJ, Marsh CB, Haley BE, Parinandi NL. Thiol-redox antioxidants protect against lung vascular endothelial cytoskeletal alterations caused by pulmonary fibrosis inducer, bleomycin: comparison between classical thiol-protectant, N-acetyl-L-cysteine, and novel thiol antioxidant, N,N′-bis-2-mercaptoethyl isophthalamide. Toxicol Mech Methods. 2012;22:383–96.
65. Parinandi NL, Sharma A, Eubank TD, Kaufman BF, Kutala VK, Marsh CB, Ignarro LJ, Kuppusamy P. Nitroaspirin (NCX-4016), an NO donor, is antiangiogenic through induction of loss of redox-dependent viability and cytoskeletal reorganization in endothelial cells. Antioxid Redox Signal. 2007;9:1837–49.
66. Kenakin T. Isolated blood vessel assays. Curr Protoc Pharmacol. 2001;Chapter 4:Unit4 4.
67. Mochizuki S, Vink H, Hiramatsu O, Kajita T, Shigeto F, Spaan JA, Kajiya F. Role of hyaluronic acid glycosaminoglycans in shear-induced endothelium-derived nitric oxide release. Am J Physiol Heart Circ Physiol. 2003;285:H722–6.
68. Shreim SG, Steward E, Botvinick EL. Extending vaterite microviscometry to ex vivo blood vessels by serial calibration. Biomed Opt Express. 2012;3:37–47.

Chapter 7
'Ozone-Specific' Oxysterols and Neuronal Cell Signaling

Achuthan C. Raghavamenon, Xueli Gao, Deidra S. Atkins-Ball, Sanjay Varikuti, Narasimham L. Parinandi, and Rao M. Uppu

Abstract Cholesterol is an important plasma membrane component, precursor for hormones and vitamins, and is a regulator of metabolism. However, the oxidized forms of cholesterol (oxysterols) can cause toxicity and induce pro-inflammatory responses and are implicated in chronic degenerative diseases. In general, oxysterols with a modified sidechain serve in various physiological and/or pathophysiological functions. The source of these oxysterols may be exogenous, from the food we ingest, or endogenous, as the by-product of normal cholesterol metabolism, free radical-mediated oxidation, or autoxidation of cholesterol. This chapter discusses the nature of oxysterols as oxidized cholesterol species, oxysterol signaling and pathophysiology, oxysterols and neurodegenerative diseases, ozone-oxidized cholesterol as a new class of oxysterols, detection of 3β-hydroxy-5-oxo-5,6-secocholestan-6-al (cholesterol secoaldehyde, ChSeco or atheronal-A) at sites of inflammation and evi-

---

A. C. Raghavamenon
Department of Environmental Toxicology and the Health Research Center, Southern University and A&M College, Baton Rouge, LA, USA

Present address: Department of Biochemistry, Amala Cancer Research Center, Thrissur, Kerala, India

X. Gao
Department of Environmental Toxicology and the Health Research Center, Southern University and A&M College, Baton Rouge, LA, USA

Present address: American Clinical Solutions, Sun City Center, FL, USA

D. S. Atkins-Ball
Department of Biological Sciences and Chemistry, Southern University and A&M College, Baton Rouge, LA, USA

S. Varikuti · N. L. Parinandi
Division of Pulmonary, Critical Care and Sleep Medicine, Department of Internal Medicine, The Ohio State University, Columbus, OH, USA

R. M. Uppu (✉)
Department of Environmental Toxicology and the Health Research Center, Southern University and A&M College, Baton Rouge, LA, USA
e-mail: rao_uppu@subr.edu

L. J. Berliner, N. L. Parinandi (eds.), *Measuring Oxidants and Oxidative Stress in Biological Systems*, Biological Magnetic Resonance 34,
https://doi.org/10.1007/978-3-030-47318-1_7

dence for in vivo existence, cytotoxicity of ChSeco and pro-inflammatory actions in the cells of mammalian systems, and ChSeco signaling in neuronal cells and implications in Alzheimer's pathology.

Keywords Apoptotic cell signaling · Atheronals A and B · Cholesterol secoaldehyde · Neurodegenerative diseases · Neuronal cell signaling · Neuronal damage · Oxidized cholesterol · Oxysterols · Ozone · Ozone oxidation of cholesterol and neurotoxicity

Introduction

Oxysterols Are Oxidized Cholesterol Species

Cholesterol is a nontoxic lipid, essential to life as a plasma membrane component, precursor for hormones and vitamins, and as a regulator of metabolism. However, the oxidized forms of cholesterol generally induce toxic and pro-inflammatory responses and have been increasingly implicated in chronic degenerative diseases [1–4]. Although structurally similar to cholesterol with the intact cyclopentanophenanthrene sterol ring system, most oxysterols differ among themselves by the presence of oxidized functional groups, such as hydroxyl and epoxy moieties attached to side chains (Fig. 7.1). There are other sets of oxysterols in which the sterol ring structure is modified (Fig. 7.2) with or without additional changes to the sidechain (Fig. 7.3) [5–12]. In general, oxysterols with modified sidechain serve in various physiological and/or pathophysiological functions. The source of these oxysterols may be exogenous, from the food we ingest, or endogenous, as the byproduct of normal cholesterol metabolism, free radical-mediated oxidation, or autoxidation of cholesterol [6]. Many studies have established the structure and formation of oxysterols [1, 6, 8, 13–19].

Oxysterol Signaling and Pathophysiology

The cytotoxic potential of oxysterol is well studied. Oxysterols have shown to trigger cell death, induce oxidative stress and inflammation [4, 20], and modulate lipid homeostasis [21]. It is suggested that secondary to oxidative stress and/or inflammatory responses, oxysterols do induce phospholipidosis, which is considered important in pathological changes [19]. These pathophysiological changes are thought to be a result of oxysterol-induced activation of nuclear receptors and subsequent gene expression. For over a decade, the oxysterol-mediated gene expression has been extensively studied and a suggestion has been put forth that oxysterol forms the natural ligand activator for a class of nuclear receptors called Liver X receptors

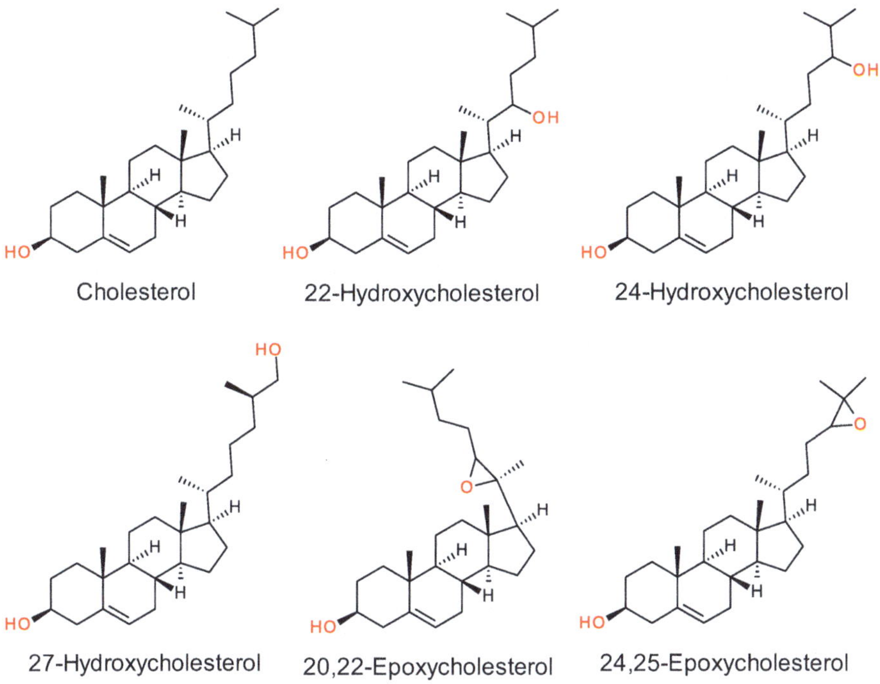

Fig. 7.1 Structures of cholesterol and some representative oxysterols in which the side chain is modified

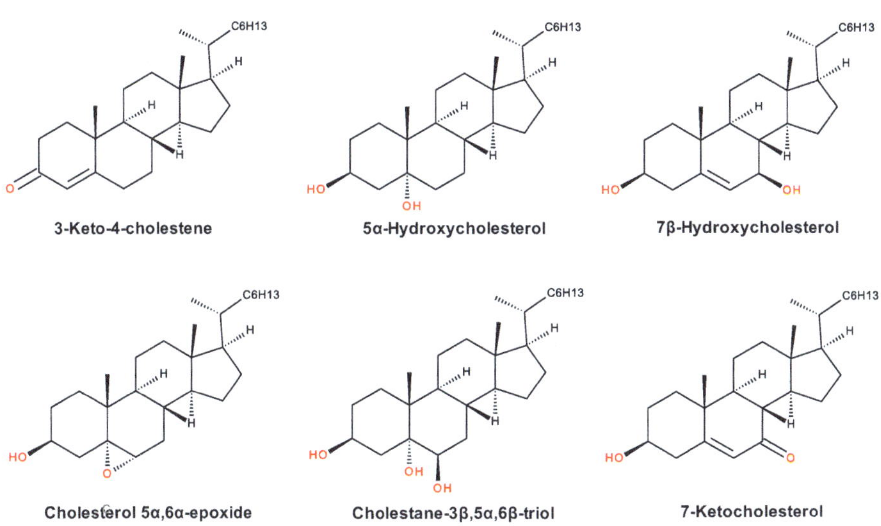

Fig. 7.2 Structures of some representative oxysterols in which the sterol ring is modified

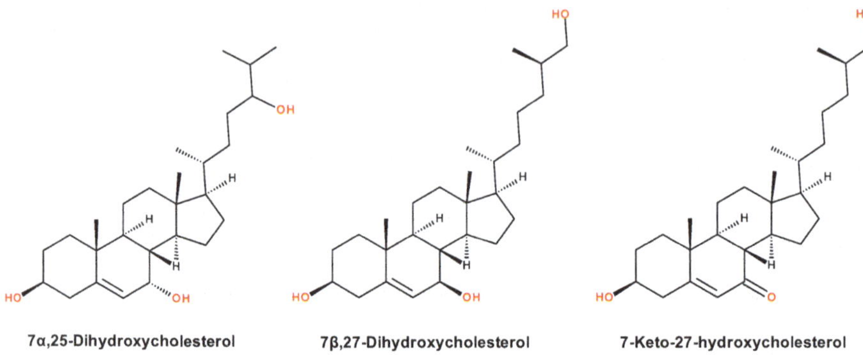

Fig. 7.3 Structures of some representative oxysterols in which both the sterol ring and the side chain are modified

(LXRs) [22]. Normally LXRs act as sterol sensors in the body and likely protect the tissues from cholesterol overload by enhancing cholesterol efflux and transport to the liver with subsequent conversion to bile acids [23, 24].

Upon oxysterol binding, LXRs dimerize with retinoic acid X receptors (RXR) and bind to LXR responsive elements on the promoter sequence of target genes to regulate their expression [23, 24]. In liver macrophages, oxysterols play a major role in lipid and carbohydrate metabolism [21] presumably by inducing expression of ABC A1 and apolipoprotein E and further enhancing cholesterol efflux [25]. Activators of LXRs are also known to induce the proteolytic maturation of sterol regulatory element-binding protein 1c (SREBP1c) that is attached to membranes of the endoplasmic reticulum. This allows SREBP1c to bind to sterol regulatory elements on the promoter region of genes responsible for fatty acid and sterol biosynthesis. Thus, oxysterols can induce deregulation in both the lipid and carbohydrate metabolism. Oxysterols impart stress-associated signaling, especially in cases where there is enhanced activity of the NADPH oxidase system and the associated pro-oxidant effect. MEK/ERK system of MAP kinases is known to be common in these situations [3].

Oxysterols and Neurodegenerative Diseases

Elevated levels of oxysterols, from exogenous and/or endogenous sources, are often seen in circulation when associated with degenerative diseases. Although differing in chemical structure and biological responses, most of these oxysterol species are known to induce a toxic response, oxidative and inflammatory changes, and regulate lipid metabolism [19]. Therefore, an enormous interest has been devoted to understanding the role of oxysterols in lipid disorders, especially degenerative diseases such as atherosclerosis, Alzheimer's disease, and Parkinsonism, where various lipids classes have a succinct role [13, 26–28].

The human brain possesses 25% of the total body cholesterol and has a very high turnover rate. As cholesterol cannot cross the blood-brain barrier, de novo synthesis is the major route for its formation in brain tissue. This process depends on CYP46A1 to generate 24-hydroxy cholesterol (24-OHC) for its effective efflux to circulation in order to maintain cholesterol homeostasis [29]. The amount of 24-OHC that can cross the blood-brain barrier (BBB) is considered the only source of 24-OHC for the normal level in circulation and a reduced level is considered an indication of neuronal loss or atrophy [30]. Recently, it has been shown that 27-hydroxycholesterol (27-OHC) can also cross the BBB [26, 27, 29] and is suggested to be involved in various degenerative changes in the brain.

These age-related degenerative changes are manifested by progressive loss in neurons and it is suggested that oxidative stress plays a major role. Under circumstances of pro-oxidant stress, it is likely that free radical-mediated oxidation of cholesterol may disrupt cholesterol homeostasis leading to the formation of other forms of oxysterols in brain tissues. Identification of different forms of oxysterols other than 24-OH cholesterol and 27-OHC in aging brain tissues and their involvement in enhanced neurodegenerative changes are important aspects to study.

Ozone-Oxidized Cholesterol: A New Class of Oxysterols

Ozone (trioxygen; O_3) is a strong oxidizing agent. Exposure to ozone in the air or from pollutants that produce it affects lung function and irritates respiratory systems in humans and animals [31, 32]. In humans, ozone exposure accounts for premature death, asthma, bronchitis, heart attack, and other cardiopulmonary problems [31–38]. During the years 1985–2000, William Pryor and his coworkers were the pioneers of studies of ozone-mediated oxidation of biological target molecules, in particular, those classes of molecules (unsaturated fatty acids, cholesterol and antioxidants) that reside in the lung lining fluids and the epithelial cell membranes present at the air–lung interface [39–55]. The basic assumption in these studies is that the critical targets for ozone reaction in the breathing air reside in the biomolecules present in the lung lining fluid and the pulmonary epithelium and that the reactions of unsaturated fatty acids and cholesterol produce ozone-specific oxidation products that act as signal-transduction products [42, 43, 49]. In ozone-exposed rats and rabbits, Pryor and his coworkers [48, 50, 55–57] and Pulfer et al. [58, 59] were able to detect several ozone-specific aldehydic products, in which 3β-hydroxy-5-oxo-5,6-secocholestan-6-al (cholesterol secoaldehyde, ChSeco or atheronal-A; Fig. 7.4) was identified as one of the products. An earlier study by Wang et al. [55] and several other investigators demonstrated that ChSeco was one of the major products formed in ozone-mediated oxidation of cholesterol in aqueous environments, and that it was not formed in any of the other known oxidations of cholesterol [60]. These observations lead Pryor and his coworkers to propose that ChSeco is a marker for ozone exposure in biological systems [48, 61].

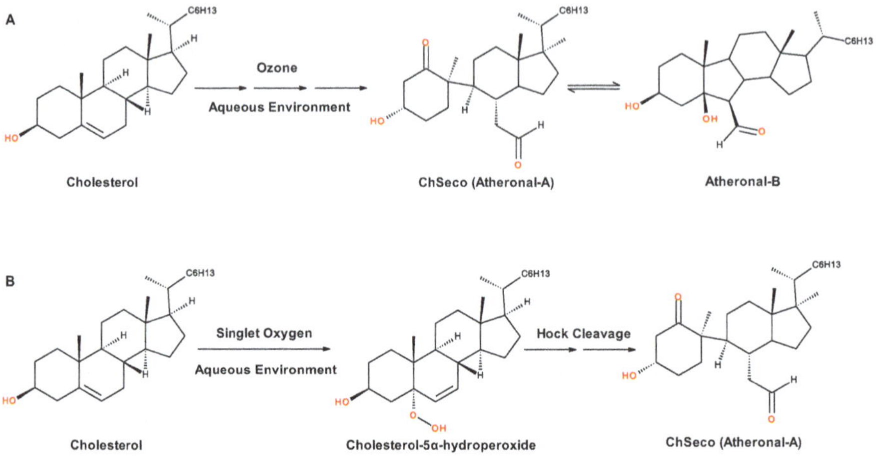

Fig. 7.4 Formation of ChSeco and atheronal-B (aldolized ChSeco) in (**a**) ozone- and (**b**) singlet oxygen-mediated oxidations of cholesterol in aqueous environments

Detection of ChSeco at Sites of Inflammation and Evidence for In Vivo Existence

Parallel to the studies of ChSeco formation in ozone-mediated oxidations of cholesterol [48, 55, 58, 59], ChSeco received a great deal of attention in the early 2000s in view of its formation in tissues and organs where ozone itself cannot reach if it were to be absorbed from the breathing air in the lungs [42, 43, 49]. For example, it has been shown that ChSeco (also referred to as atheronal-A by investigators), by inference, and its aldolized product atheronal-B (Fig. 7.4) are present in the arterial plaque at concentrations as high as 60 μM [62]. Similarly, ChSeco has been shown to be present in the brain samples of patients with Alzheimer's disease [63] and Lewy body dementia [64]. Dantas and his colleagues [65], who examined the presence of secosterol aldehydes in the blood plasma, spinal cord, and the motor cortex of ALS rats (ALS SOD1-G93A), showed that atheronal-B was significantly increased in the plasma of symptomatic ALS rats compared to pre-symptomatic animals, suggesting an association between the levels of atheronals-A and -B and the progression of ALS disease. Tomono et al. [66], who developed a highly sensitive LC-ESI-MS/MS method with detection limits as low as 0.01–0.05 fmol for quantitation of ChSeco and atheronal-B as derivatives with 2-hydrazino-1-methylpyridine, showed the presence of ChSeco and atheronal-B in the blood plasma and several tissues (such as the brain, liver, and the lung) of normal C57BL/6 mice. According to these investigators, the typical concentrations of atheronal-A and -B (range: 1–4 nM) in the blood plasma are about 1 to 2 orders magnitude smaller than those in the brain, liver, and lung tissues (10–160 nM; calculated based on the assumption that a tissue mass of 1 g equals 1 mL by volume).

In the beginning, it was thought that ozone is produced at these and, possibly, other inflammatory sites as a result of neutrophil-derived singlet oxygen through an antibody-catalyzed water-oxidation pathway [62, 67–69]. An editorial that appeared in *Science* [70] speculated that ozone may be produced as part of the neutrophil armamentarium and that ChSeco produced as a result of the reaction of ozone with cholesterol plays a role in the atherosclerotic plaque formation. The biological formation of ozone was, however, questioned by several investigators based on theoretical as well as practical considerations [12, 61, 71–75]. Some of the practical considerations include the lack of specificity of probes originally employed for the purpose of demonstration of ozone. Kettle et al. [72] have demonstrated that the oxidation of indigo carmine to isatin sulfonic acid can be brought about by superoxide anion under physiologically relevant conditions. Similarly, Rangan et al. [74] have shown that indigo carmine can be oxidized by the free radicals of nitrogen dioxide and carbonate radical derived from the reactions of peroxynitrite anion with carbon dioxide.

Brinkhorst et al. [76] have shown that Hock cleavage of cholesterol-5α-hydroperoxide can result in the formation of ChSeco and atheronal-B (Fig. 7.4). This observation was subsequently confirmed by Uemi et al. [77] and Tomono et al. [78]. While Uemi and his colleagues used a chemical system, Tomono and his colleagues used the model system of singlet oxygen produced in reactions of myeloperoxidase-derived hypochlorous acid and hydrogen peroxide. Thus, given due consideration to various lines of evidence, the most probable mechanism for the formation of ChSeco at inflammatory sites other than the air–lung interface (including the deeper lung stratus and all extrapulmonary tissues) could involve the oxidation of cholesterol and its 3-acyl esters by singlet oxygen followed by the Hock cleavage of one of the hydroperoxide products, cholesterol-5α-hydroperoxide.

In summary, ChSeco formation at the air–lung interface is most probably mediated through the direct reactions of ozone in the breathing air with cholesterol and its 3-acyl esters present in the lung lining fluids and epithelial cell membranes. In the deeper tissue strata of the lung and in most extrapulmonary tissues and organs where ozone has limited or no penetration due to its extreme reactivity with components of the lung lining fluids and the epithelial cell membranes, ChSeco is most likely to be formed via the reactions of singlet oxygen with cholesterol and its 3-acyl esters. Among various sources of singlet oxygen, the myeloperoxidase/hydrogen peroxide/halide ion the system may be the most contributory to the formation of ChSeco and its aldolized form, atheronal-B.

ChSeco Is Cytotoxic and Proinflammatory to Cells of Mammalian Systems

Similar to other oxysterols, ChSeco is found to be highly toxic to many mammalian cell lines including humans [79–86]. Studies from our own laboratory have shown that ChSeco induces apoptosis in H9c2 cardiomyocytes [80–82], GT1-7 hypothalamic neurons [83–85], THP monocytes and J774 macrophages [86, 87], and primary cortical neurons [88]. This process involves the use of reactive oxygen species (ROS) as mediators and therefore the cytotoxicity could be reversed either partially or completely by antioxidants such a Trolox (a water-soluble analog of vitamin E) and *N*-acetyl-L-cysteine (NAC). The reversal of cytotoxicity by Trolox, in general, is partial, whereas NAC completely attenuates the ChSeco-induced cytotoxicity. Detailed mechanistic studies indicated that at low-concentration ChSeco, the cell death confirms to apoptotic type [80, 81] with the involvement of both the death receptor and mitochondrial pathways. It is also interesting to note that ChSeco leads to enhanced H_2O_2 production through the plasma membrane NADPH oxidase system [89] and that H_2O_2 and the secondary oxidants resulting thereafter are responsible for the activation of stress-related signaling pathways along with P38 MAPK [82].

However, in immune cells such as THP-1 monocytes and Mouse J774 macrophages, the ChSeco failed to produce ROS yet revealed cytotoxicity [86, 87]. The IC_{50} concentrations observed with immune cells such as monocytes and macrophages (20–30 μM) were higher than the IC_{50} concentrations of cardiomyocytes. It is thus obvious that in actively dividing cells, ChSeco was needed in a higher amount to elicit a toxic response and that it is independent of ROS generated from the plasma membrane NADPH oxidase system. In cells that are noncancerous, ChSeco activates or upregulates the NADPH oxidase system to generate increased H_2O_2 and other oxidants [82, 89]. It is not clear if LPS-induced oxidative burst and ChSeco-induced ROS generation use a similar signaling pathway; however, it seems that normal cells may be more susceptible to oxidative stress by these unique sterols. A variety of inflammatory genes and proteins were found to be upregulated, while anti-inflammatory proteins were found to be downregulated when macrophages and monocytes were exposed to this unique oxysterol. It has been hypothesized that ChSeco has the capability of enhancing the NADPH oxidative burst and subsequent inflammation in normal cells [89].

ChSeco Signaling in Neuronal Cells Show Implications in Alzheimer's Pathology

Previous studies from our laboratory have shown that ChSeco induces cytotoxicity in murine immortalized GT1-7 hypothalamic neuronal cells [83–85] and rat primary cortical neurons [88]. In both cell types, exposure to ChSeco results in a marked increase in the formation of intracellular H_2O_2 or peroxide-like substances along with depletion of cellular GSH and a significant loss of mitochondrial transmembrane potential, all indicative of excessive oxidant stress status. In the case of GT1-7 hypothalamic neurons, there was the formation of amyloid foci, possibly aggregated amyloid-beta (Aβ), which enlarged in size and numbers with an increase in the concentration of ChSeco. With primary cortical neurons, no such amyloid foci were observed but there was evidence for a decrease in the concentration of intracellular Aβ as revealed by immunohistochemical staining for Aβ42. For instance, the Aβ42 staining (presumably indicative of the monomeric form) in ChSeco-exposed primary neurons appeared to be more diffuse, and it decreased with increasing concentration of ChSeco used in the original cultures. Based on associated changes in the axonal outgrowth and dendritic branches in the ChSeco-exposed primary neurons, it was reasoned that the decrease in the intracellular Aβ42 was mainly a result of increased permeability that results from the loss of integrity of dendrites, the plasma membrane, or both [88].

The amyloid formation observed in ChSeco-exposed GT1-7 hypothalamic neurons [83] was thought to be analogous to the metabolite-induced protein misfolding enumerated using Aβ peptides, α-synuclein, and apolipoprotein-B_{100} with atheronals A- and -B and certain other lipid-derived aldehydes [63, 64, 90–94]. Interestingly, the formation of intracellular amyloids is least contributory to the cytotoxicity observed in the ChSeco-exposed GT1-7 neurons. For example, co-incubation with Aβ40 and Aβ42 fibrillogenesis inhibitors which inhibited the development of amyloid foci was with any effect on the ChSeco-induced cytotoxicity in GT1-7 neurons [83]. In both primary neurons and GT-17 neuronal cells, pretreatment with Trolox offered partial but significant protection against the ChSeco-induced cytotoxicity, whereas *N*-acetyl-L-cysteine completely attenuated the cytotoxic effects of ChSeco, meaning that increased oxidative stress status is the primary contributor to the observed cytotoxicity. However, since Alzheimer's and other neurodegenerative diseases are slow and progressive, the formation of Aβ aggregates in vivo may have long-term pathological consequences. Strenuous efforts are necessary to study the involvement of oxidized cholesterol species like ChSeco to unravel the pathophysiology of these and other degenerative diseases. Recent studies of the adduct formation between atheronals and the basic myelin protein resulting in exposure of immunodominant epitope [93], experimental demonstration of the presence of excessive amounts of atheronals in the blood plasma of symptomatic ALS rats (ALS SOD1-G93A) compared to the pre-symptomatic animals [65], and loss of function of the wild-type tumor protein p53 and associated impairment in chromatin binding and transcription in ChSeco and atheronal-B exposed lung carcinoma cells [94] appear to be some of the promising approaches towards unraveling the role of atheronals in neurodegenerative diseases.

Acknowledgments The authors acknowledge the support from the National Institutes of Health (NIH) through the National Institute of General Medical Science (NIGMS) Grant 5 P2O GM103424-17 and the US Department of Education (US DoE; Title III, HBGI Part B grant number P031B040030). Its contents are solely the responsibility of authors and do not represent the official views of NIH, NIGMS, or US DoE.

References

1. Leoni V. Oxysterols as markers of neurological disease—a review. Scand J Clin Lab Invest. 2009;69:22–5.
2. Poli G, Sottero B, Gargiulo S, Leonarduzzi G. Cholesterol oxidation products in the vascular remodeling due to atherosclerosis. Mol Aspects Med. 2009;30:180–9.
3. Sottero B, Gamba P, Gargiulo S, Leonarduzzi G, Poli G. Cholesterol oxidation products and disease: an emerging topic of interest in medicinal chemistry. Curr Med Chem. 2009;16:685–705.

4. Vejux A, Lizard G. Cytotoxic effects of oxysterols associated with human diseases: Induction of cell death (apoptosis and/or oncosis), oxidative and inflammatory activities, and phospholipidosis. Mol Aspects Med. 2009;30:153–70.
5. Beck KR, Kanagaratnam S, Kratschmar DV, Birk J, Yamaguchi H, Sailer AW, Seuwen K, Odermatt A. Enzymatic interconversion of the oxysterols 7β,25-dihydroxycholesterol and 7-keto,25-hydroxycholesterol by 11β-hydroxysteroid dehydrogenase type 1 and 2. J Steroid Biochem Mol Biol. 2019;190:19–28.
6. Brown AJ, Jessup W. Oxysterols: sources, cellular storage and metabolism, and new insights into their roles in cholesterol homeostasis. Mol Aspects Med. 2009;30:111–22.
7. Diczfalusy U. Analysis of cholesterol oxidation products in biological samples. J AOAC Int. 2004;87:467–73.
8. Gill S, Chow R, Brown AJ. Sterol regulators of cholesterol homeostasis and beyond: the oxysterol hypothesis revisited and revised. Prog Lipid Res. 2008;47:391–404.
9. Haigh WG, Lee SP. Identification of oxysterols in human bile and pigment gallstones. Gastroenterology. 2001;121:118–23.
10. Kurschus FC, Wanke F. EBI2—sensor for dihydroxycholesterol gradients in neuroinflammation. Biochimie. 2018;153:52–5.
11. Smith LL. Review of progress in sterol oxidations: 1987–1995. Lipids. 1996;31:453–87.
12. Smith LL. Oxygen, oxysterols, ouabain, and ozone: a cautionary tale. Free Radic Biol Med. 2004;37:318–24.
13. Bjorkhem I, Diczfalusy U. Oxysterols: friends, foes, or just fellow passengers? Arterioscler Thromb Vasc Biol. 2002;22:734–42.
14. Bjorkhem I, Starck L, Andersson U, Lutjohann D, von Bahr S, Pikuleva I, Babiker A, Diczfalusy U. Oxysterols in the circulation of patients with the Smith-Lemli-Opitz syndrome: abnormal levels of 24S- and 27-hydroxycholesterol. J Lipid Res. 2001;42:366–71.
15. Garenc C, Julien P, Levy E. Oxysterols in biological systems: the gastrointestinal tract, liver, vascular wall and central nervous system. Free Radic Res. 2010;44:47–73.
16. Luu B, Moog C. Oxysterols: biological activities and physicochemical studies. Biochimie. 1991;73:1317–20.
17. Olkkonen VM, Lehto M. Oxysterols and oxysterol binding proteins: role in lipid metabolism and atherosclerosis. Ann Med. 2004;36:562–72.
18. Schroepfer GJ Jr. Oxysterols: modulators of cholesterol metabolism and other processes. Physiol Rev. 2000;80:361–554.
19. Vejux A, Malvitte L, Lizard G. Side effects of oxysterols: cytotoxicity, oxidation, inflammation, and phospholipidosis. Braz J Med Biol Res. 2008;41:545–56.
20. Töröcsik D, Szanto A, Nagy L. Oxysterol signaling links cholesterol metabolism and inflammation via the liver X receptor in macrophages. Mol Aspects Med. 2009;30:134–52.
21. Edwards PA, Kennedy MA, Mak PA. LXRs: oxysterol—activated nuclear receptors that regulate genes controlling lipid homeostasis. Vascul Pharmacol. 2002;38:249–56.
22. Makishima M. Nuclear receptors as targets for drug development: regulation of cholesterol and bile acid metabolism by nuclear receptors. J Pharmacol Sci. 2005;97:177–83.
23. Baranowski M. Biological role of liver X receptors. J Physiol Pharmacol. 2008;59(Suppl. 7):31–55.
24. Liu Y, Chang YS, Fang FD. [Liver X receptor: crucial mediator in lipid and carbohydrate metabolism]. Zhongguo Yi Xue Ke Xue Yuan Xue Bao. 2007;29:430–5.
25. Beltowski J. Liver X receptors (LXR) as therapeutic targets in dyslipidemia. Cardiovasc Ther. 2008;26:297–316.
26. Bjorkhem I, Cedazo-Minguez A, Leoni V, Meaney S. Oxysterols and neurodegenerative diseases. Mol Aspects Med. 2009;30:171–9.
27. Bjorkhem I. Crossing the barrier: oxysterols as cholesterol transporters and metabolic modulators in the brain. J Intern Med. 2006;260:493–508.
28. Brown AJ, Jessup W. Oxysterols and atherosclerosis. Atherosclerosis. 1999;142:1–28.

29. Bjorkhem I, Meaney S. Brain cholesterol: long secret life behind a barrier. Arterioscler Thromb Vasc Biol. 2004;24:806–15.
30. Kolsch H, Lutjohann D, von Bergmann K, Heun R. The role of 24S-hydroxycholesterol in Alzheimer's disease. J Nutr Health Aging. 2003;7:37–41.
31. Taylor-Clark TE, Undem BJ. Ozone activates airway nerves via the selective stimulation of TRPA1 ion channels. J Physiol. 2010;588:423–33.
32. Wolkoff P, Clausen PA, Larsen K, Hammer M, Larsen ST, Nielsen GD. Acute airway effects of ozone-initiated d-limonene chemistry: importance of gaseous products. Toxicol Lett. 2008;181:171–6.
33. Fauroux B, Sampil M, Quenel P, Lemoullec Y. Ozone: a trigger for hospital pediatric asthma emergency room visits. Pediatr Pulmonol. 2000;30:41–6.
34. Lin S, Bell EM, Liu W, Walker RJ, Kim NK, Hwang SA. Ambient ozone concentration and hospital admissions due to childhood respiratory diseases in New York State, 1991–2001. Environ Res. 2008;108:42–7.
35. Loomis DP, Borja-Aburto VH, Bangdiwala SI, Shy CM. Ozone exposure and daily mortality in Mexico City: a time-series analysis. Res Rep Health Eff Inst. 1996;(75):1–37; discussion 39–45.
36. Peng KJ, Huang YS, An LN, Han XQ, Zhang JG, Wang QL, Sun J, Wang SR. Effect of ozone produced from antibody-catalyzed water oxidation on pathogenesis of atherosclerosis. Acta Biochim Biophys Sin (Shanghai). 2006;38:417–22.
37. Srebot V, Gianicolo EA, Rainaldi G, Trivella MG, Sicari R. Ozone and cardiovascular injury. Cardiovasc Ultrasound. 2009;7:30.
38. Zhang Y, Huang W, London SJ, Song G, Chen G, Jiang L, Zhao N, Chen B, Kan H. Ozone and daily mortality in Shanghai, China. Environ Health Perspect. 2006;114:1227–32.
39. Giamalva D, Church DF, Pryor WA. A comparison of the rates of ozonation of biological antioxidants and oleate and linoleate esters. Biochem Biophys Res Commun. 1985;133:773–9.
40. Giamalva DH, Church DF, Pryor WA. Effect of bilayer structure on the rates of reaction of ozone with polyunsaturated fatty acids in phosphatidylcholine liposomes. Chem Res Toxicol. 1988;1:144–5.
41. Postlethwait EM, Cueto R, Velsor LW, Pryor WA. O3-induced formation of bioactive lipids: estimated surface concentrations and lining layer effects. Am J Physiol. 1998;274:L1006–16.
42. Pryor WA. How far does ozone penetrate into the pulmonary air/tissue boundary before it reacts? Free Radic Biol Med. 1992;12:83–8.
43. Pryor WA. Ozone in all its reactive splendor. J Lab Clin Med. 1993;122:483–6.
44. Pryor WA. Mechanisms of radical formation from reactions of ozone with target molecules in the lung. Free Radic Biol Med. 1994;17:451–65.
45. Pryor WA, Church DF. Aldehydes, hydrogen peroxide, and organic radicals as mediators of ozone toxicity. Free Radic Biol Med. 1991;11:41–6. Review. Erratum in: Free Radic Biol Med 12:451 (1992).
46. Pryor WA, Uppu RM. A kinetic model for the competitive reactions of ozone with amino acid residues in proteins in reverse micelles. J Biol Chem. 1993;268:3120–6.
47. Pryor WA, Das B, Church DF. The ozonation of unsaturated fatty acids: aldehydes and hydrogen peroxide as products and possible mediators of ozone toxicity. Chem Res Toxicol. 1991;4:341–8.
48. Pryor WA, Wang K, Bermudez E. Cholesterol ozonation products as biomarkers for ozone exposure in rats. Biochem Biophys Res Commun. 1992;188:618–23.
49. Pryor WA, Squadrito GL, Friedman M. The cascade mechanism to explain ozone toxicity: the role of lipid ozonation products. Free Radic Biol Med. 1995;19:935–41.
50. Pryor WA, Bermudez E, Cueto R, Squadrito GL. Detection of aldehydes in bronchoalveolar lavage of rats exposed to ozone. Fundam Appl Toxicol. 1996;34:148–56.
51. Squadrito GL, Uppu RM, Cueto R, Pryor WA. Production of the Criegee ozonide during the ozonation of 1-palmitoyl-2-oleoyl-sn-glycero-3-phosphotidylcholine liposomes. Lipids. 1992;27:955–8.

52. Uppu RM, Pryor WA, W.A. Ozonation of lysosome in the presence of oleate in reverse micelles of sodium di-2-ethylhexylsulfosuccinate. Biochem Biophys Res Commun. 1992;187:473–9.
53. Uppu RM, Pryor WA. The reactions of ozone with proteins and unsaturated fatty acids in reverse micelles. Chem Res Toxicol. 1994;7:47–55.
54. Uppu RM, Cueto R, Squadrito GL, Pryor WA. What does ozone react with at the air lung interface-model studies using human red-blood-cell membranes? Arch Biochem Biophys. 1995;319:257–66.
55. Wang K, Bermudez E, Pryor WA. The ozonation of cholesterol: separation and identification of 2,4-dinitrophenylhydrazine derivatization products of 3 beta-hydroxy-5-oxo-5,6-secocholestan-6-al. Steroids. 1993;58:225–9.
56. Frampton MW, Pryor WA, Cueto R, Cox C, Morrow PE, Utell MJ. Ozone exposure increases aldehydes in epithelial lining fluid in human lung. Am J Respir Crit Care Med. 1999;159:1134–7.
57. Frampton MW, Pryor WA, Cueto R, Cox C, Morrow PE, Utell MJ. Aldehydes (nonanal and hexanal) in rat and human bronchoalveolar lavage fluid after ozone exposure. Res Rep Health Eff Inst. 1999;1–15; discussion 17–8.
58. Pulfer MK, Murphy RC. Formation of biologically active oxysterols during ozonolysis of cholesterol present in the lung surfactant. J Biol Chem. 2004;279:26331–8.
59. Pulfer MK, Taube C, Gelfand E, Murphy RC. Ozone exposure in vivo and formation of biologically active oxysterols in the lung. J Pharmacol Exp Ther. 2005;312:256–65.
60. Smith LL. Cholesterol autoxidation. New York: Plenum Press; 1981. See also the references therein.
61. Pryor WA, Houk KN, Foote CS, Fukuto JM, Ignarro LJ, Squadrito GL, Davis KAJ. It's a gas, man! Free Radic Biol Med. 2006;291:491–511.
62. Wentworth P Jr, Nieva J, Takeuchi C, Galve R, Wentworth AD, Dilley RB, DeLaria GA, Saven A, Babior BM, Janda KD, Eschenmoser A, Lerner RA. Evidence for ozone formation in human atherosclerotic arteries. Science. 2003;302:1053–6.
63. Zhang Q, Powers ET, Nieva J, Huff ME, Dendle MA, Bieschke J, Glabe CG, Eschenmoser A, Wentworth P Jr, Lerner RA, Kelley JW. Metabolite-initiated protein misfolding may trigger Alzheimer's disease. Proc Natl Acad Sci U S A. 2004;101:4752–7.
64. Bosco DA, Fowler DM, Zhang Q, Nieva J, Powers ET, Wentworth P Jr, Lerner RA, Kelly JW. Elevated levels of oxidized cholesterol metabolites in Lewy body disease brains accelerate alpha-synuclein fibrilization. Nat Chem Biol. 2006;2:249–53.
65. Dantas LS, Chaves-Filho AB, Coelho FR, Genaro-Mattos TC, Tallman KA, Porter NA, Augusto O, Miyamoto S. Cholesterol secosterol aldehyde adduction and aggregation of Cu, Zn-superoxide dismutase: potential implications in ALS. Redox Biol. 2018;19:105–15.
66. Tomono S, Miyoshi N, Ito M, Higashi T, Ohshima H. A highly sensitive LC-ESI-MS/MS method for the quantification of cholesterol ozonolysis products secosterol-A and secosterol-B after derivatization with 2-hydrazino-1-methylpyridine. J Chromatogr B Analyt Technol Biomed Life Sci. 2011;879:2802–8.
67. Wentworth P Jr, McDunn JE, Wentworth AD, Tekeuchi C, Nieva J, Jones T, Bautista C, Ruedi JM, Gutierrez A, Janda KD, Babior BM, Eschenmoser A, Lerner RA. Evidence for antibody-catalyzed ozone formation in bacterial killing and inflammation. Science. 2002;298:2195–9.
68. Wentworth P Jr, Wentworth AD, Zhu X, Wilson IA, Janda KD, Eschenmoser A, Lerner RA. Evidence for the production of trioxygen species during antibody-catalyzed chemical modification of antigens. Proc Natl Acad Sci U S A. 2003;100:1490–3.
69. Babior BM, Tekeuchi C, Ruedi J, Gutierrez A, Wentworth P Jr. Investigating antibody-catalyzed ozone generation by human neutrophils. Proc Natl Acad Sci U S A. 2003;100:3031–43.
70. Marx J. Ozone may be secret ingredient in plaques' inflammatory stew. Science. 2004;302:965.
71. Drahl C. Probing for in-body ozone. Chem Eng News. 2009;87:40–2.
72. Kettle AJ, Clark BM, Winterbourn CC. Superoxide converts indigo carmine to isatin sulfonic acid: implications for the hypothesis that neutrophils produce ozone. J Biol Chem. 2004;279:18521–5.

73. Kettle AJ, Winterbourn CC. Do neutrophils produce ozone? An appraisal of current evidence. Biofactors. 2005;24:41–5.
74. Rangan V, Perumal TE, Sathishkumar K, Uppu RM. Oxidation of indigo carmine by peroxynitrite (±CO_2): implications for the hypothesis on ozone production by neutrophils. In: 45th annual meeting of the Society of Toxicology, San Diego, CA, March 5–9, 2006.
75. Yamashita K, Miyoshi T, Arai T, Endo N, Itoh H, Makino K, Mizugishik K, Uchiyama T, Sasada M. Ozone production by amino acids contributes to killing of bacteria. Proc Natl Acad Sci U S A. 2008;105:16912–7.
76. Brinkhorst J, Nara SJ, Pratt DA. Hock cleavage of cholesterol 5alpha-hydroperoxide: an ozone-free pathway to the cholesterol ozonolysis products identified in arterial plaque and brain tissue. J Am Chem Soc. 2008;130:12224–5.
77. Uemi M, Ronsein GE, Miyamoto S, Medeiros MH, Di Mascio P. Generation of cholesterol carboxaldehyde by the reaction of singlet molecular oxygen [O2 (1Δg)] as well as ozone with cholesterol. Chem Res Toxicol. 2009;22:875–84.
78. Tomono S, Miyoshi N, Sato K, Ohba Y, Ohshima H. Formation of cholesterol ozonolysis products through an ozone-free mechanism mediated by the myeloperoxidase-H2O2-chloride system. Biochem Biophys Res Commun. 2009;383:222–7.
79. Takeuchi C, Galve R, Nieva J, Witter DP, Wentworth AD, Troseth RP, Lerner RA, Wentworth P Jr. Proatherogenic effects of the cholesterol ozonolysis products, atheronal-A and atheronal-B. Biochemistry. 2006;45:7162–70.
80. Sathishkumar K, Haque M, Perumal TE, Francis J, Uppu RM. A major ozonation product of cholesterol, 3beta-hydroxy-5-oxo-5,6-secocholestan-6-al, induces apoptosis in H9c2 cardiomyoblast. FEBS Lett. 2005;579:6444–50.
81. Sathishkumar K, Gao X, Raghavamenon AC, Parinandi N, Pryor WA, Uppu RM. Cholesterol secoaldehyde induces apoptosis in H9c2 cardiomyoblasts through reactive oxygen species involving mitochondrial and death receptor pathways. Free Radic Biol Med. 2009;47:548–58.
82. Laynes L, Raghavamenon AC, D'Auvergne O, Achuthan V, Uppu RM. MAPK signaling in H9c2 cardiomyoblasts exposed to cholesterol secoaldehyde—role of hydrogen peroxide. Biochem Biophys Res Commun. 2011;404:90–5.
83. Sathishkumar K, Xi X, Martin R, Uppu RM. Cholesterol secoaldehyde, an ozonation product of cholesterol, induces amyloid aggregation and apoptosis in murine GT1-7 hypothalamic neurons. J Alzheimers Dis. 2007;11:261–74.
84. Sathishkumar K, Murthy SN, Uppu RM. Cytotoxic effects of oxysterols produced during ozonolysis of cholesterol in murine GT1-7 hypothalamic neurons. Free Radic Res. 2007;41:82–8.
85. Sathishkumar K, Raghavamenon AC, Ganeshkumar K, Telaprolu R, Parinandi NL, Uppu RM. Simultaneous analysis of expression of multiple redox-sensitive and apoptotic genes in hypothalamic neurons exposed to cholesterol secoaldehyde. Methods Mol Biol. 2010;610:263–84.
86. Gao X, Raghavamenon AC, D'Auvergne O, Uppu RM. Cholesterol secoaldehyde promotes adhesion of THP-1 monocytes to human vascular smooth muscle cells and induces release of PDGF by cultured monocytes. In: 16th annual meeting of the Society for Free Radical Biology and Medicine, San Francisco, CA, November 18–22, 2009.
87. Gao X, Raghavamenon AC, D'Auvergne O, Uppu RM. Cholesterol secoaldehyde induces apoptosis in J774 macrophages via mitochondrial pathway but not involving reactive oxygen species as mediators. Biochem Biophys Res Commun. 2009;389:382–7.
88. Raghavamenon AC, Gernapudi R, Babu S, D'Auvergne O, Murthy SN, Kadowitz PJ, Uppu RM. Intracellular oxidative stress and cytotoxicity in rat primary cortical neurons exposed to cholesterol secoaldehyde. Biochem Biophys Res Commun. 2009;386:170–4.
89. Laynes L, Raghavamenon AC, Atkins-Ball D, Uppu RM. NADPH oxidase system contributes to cholesterol secoaldehyde-induced oxidative stress in H9C2 cardiomyoblasts. 2020; (In preparation).

90. Bieschke J, Zhang Q, Powers ET, Lerner RA, Kelly JW. Oxidative metabolites accelerate Alzheimer's amyloidogenesis by a two-step mechanism, eliminating the requirement for nucleation. Biochemistry. 2005;44:4977–83.
91. Nieva J, Shafton A, Altobell LJ III, Tripuraneni S, Rogel JK, Wentworth AD, Lerner RA, Wentworth P Jr. Lipid-derived aldehydes accelerate light chain amyloid and amorphous aggregation. Biochemistry. 2008;47:7695–705.
92. Scheinost JC, Wang H, Boldt GE, Offer J, Wentworth P Jr. Cholesterol secosterol-induced aggregation of methylated amyloid-beta peptides—insights into aldehyde-initiated fibrillization of amyloid-beta. Angew Chem Int Ed Engl. 2008;47:3919–22.
93. Cygan NK, Scheinost JC, Butters TD, Wentworth P Jr. Adduction of cholesterol 5,6-secosterol aldehyde to membrane-bound myelin basic protein exposes an immunodominant epitope. Biochemistry. 2011;50:2092–100.
94. Nieva J, Song BD, Rogel JK, Kujawara D, Altobel L III, Izharrudin A, Boldt GE, Grover RK, Wentworth AD, Wentworth P Jr. Cholesterol secosterol aldehydes induce amyloidogenesis and dysfunction of wild-type tumor protein p53. Chem Biol. 2011;18:920–7.

Chapter 8
Measurement of Oxidative Stress Status in Human Populations: A Critical Need for a Metabolomic Profiling

Rao M. Uppu, Danial Woods, and Narasimham L. Parinandi

Abstract Oxidative stress (OS) is characteristic of a diverse set of physiological and pathophysiological states. For example, human health problems associated with oxidative stress include Parkinson's disease, Alzheimer's disease, myocardial infarction, cancer, diabetes, various inflammations, renal failure, and atherosclerosis as well as aging. It has become routine and convenient to screen body fluids, including blood (serum or plasma), saliva, and urine, as well as exhaled breath for small molecules that are biomarkers of oxidative stress to ascertain the oxidative stress status (OSS) of a particular targeted organ or the whole body. Unfortunately, circulating levels of oxidation products and/or antioxidants often do not truly represent the tissue/organ/whole body state of oxidative stress or antioxidant status due to the diverse nature of oxidative reactions, metabolic status, and tissue retention. Hence, the analyst has to bear several important points in mind while ascertaining the state of oxidative stress or antioxidant status by measuring one or two chosen biomarkers in one or two selected sampling sites at any given time.

Keywords Biomarkers · Biomarkers of oxidative stress · Epidemiological studies · Metabolomic approaches · Metabolomics of oxidative stress in human · Oxidative stress determination in human · Oxidative stress status

R. M. Uppu (✉)
Department of Environmental Toxicology and the Health Research Center, Colleges of Sciences and Engineering, Southern University and A&M College, Baton Rouge, LA, USA
e-mail: rao_uppu@subr.edu

D. Woods
Daniel Woods, Inscent, Inc., Irvine, CA, USA

N. L. Parinandi
Division of Pulmonary, Critical Care and Sleep Medicine, Department of Internal Medicine, The Ohio State University, Columbus, OH, USA

L. J. Berliner, N. L. Parinandi (eds.), *Measuring Oxidants and Oxidative Stress in Biological Systems*, Biological Magnetic Resonance 34,
https://doi.org/10.1007/978-3-030-47318-1_8

Oxidative stress (OS) is a characteristic of a diverse set of physiological and pathophysiological states. For example, human health problems associated with oxidative stress include Parkinson's disease, Alzheimer's disease, myocardial infarction, cancer, diabetes, various inflammations, renal failure, and atherosclerosis [1–10] as well as aging [11–13].

The oxidants that lead to OS are mainly reactive oxygen and nitrogen species (RONs) [14–16]. These oxidants have been identified as the key initiators of the cellular oxidative reactions wherein crucial molecules such as proteins, lipids, carbohydrates, nucleic acids, and small molecules (amino acids, sugars, biogenic amines, nucleic acid bases, etc.) are attacked, resulting in the formation of oxidation products with diverse metabolic fates. Oxidation products can cause a series of abnormal metabolic cascades, especially those leading to pathological manifestations [8–10]. The formation of the oxidation products depends on the type of RON, the nature of the target molecule(s), and the cellular antioxidant(s) capable of counteracting the oxidant reactions. The half-life of the oxidation products and antioxidants is regulated at the cellular level by their reactivity and metabolism by cellular enzymatic and nonenzymatic machinery [9, 10, 17]. The antioxidant status of the cell also appears to control the extent of oxidant reactions and the formation of oxidation products. Not all antioxidants act effectively on all of the molecular targets that are vulnerable to oxidative attack. This is due to antioxidants having different chemical reactivity than the cellular oxidants and their metabolic fates that eventually dictate their biological half-lives [18, 19]. These interactions greatly complicate attempts to corroborate the OS level of a particular tissue at a given time through sampling with the levels of one or two chosen species of either cellular small molecule oxidation products or antioxidant markers. These are mainly based on the analytical convenience and availability of suitable technology [20–22].

In complex biological systems, including humans, oxidative events in tissues are dynamic and strictly controlled by several factors such as the oxidant flux, antioxidant status, detoxification mechanisms, and the organ-to-organ relationships [8, 17]. It has become routine and convenient to screen body fluids including blood (serum or plasma), saliva, and urine as well as exhaled breath for small molecules that are biomarkers of oxidative stress to ascertain the oxidative stress status (OSS) of a particular targeted organ or the whole body. Unfortunately, circulating levels of oxidation products and/or antioxidants often do not truly represent the tissue/organ/whole body state of oxidative stress or antioxidant status due to the diverse nature of oxidative reactions, metabolic status, and tissue retention. Hence, the analyst has to bear several important points in mind while ascertaining the state of oxidative stress or antioxidant status by measuring one or two chosen biomarkers in one or two selected sampling sites at any given time. These points include:

1. The need to include the determination of as many small-molecule oxidation products in the analyte as possible;
2. The need to determine the small molecule oxidation products at different times of sampling to include the metabolic state (diurnal variations) and nutritional state of the individual;
3. The need to screen as many sample sources as possible, such as blood (plasma and serum), urine, saliva, and exhaled breath for the small molecule oxidation products. This will offer a true picture of the whole body/tissue state of oxidative stress and/or antioxidant status, because each sample source is tapped for a different oxidative stress biomarker. However, measuring only one specific biomarker utilizing a convenient technology will not provide a true reflection of the state of the oxidative stress and/or antioxidant status of the whole body or tissue;
4. Unlike screening the plasma or serum to determine the oxidative stress biomarker or antioxidant levels, analyzing white blood cells, red blood cells, and platelets for small molecule oxidation products and antioxidants will provide a closer reflection of the whole-body status of the oxidative events;
5. High-throughput analysis of several small-molecule oxidation products and antioxidants from several sample sources (blood, blood cells, saliva, and urine) will provide a more comprehensive and a truer picture of the state of the oxidative stress and antioxidant status of the whole body.

Thus, an efficient and effective means of rapidly identifying numerous oxidative stress biomarkers simultaneously from diverse samples is needed.

Screening for OSS has become a key challenge in the early diagnosis of numerous diseases [1–10]. Almost three decades ago, Pryor and Godber compiled the literature on the techniques for measurement of OSS listing numerous biomarkers [10]. It has been subsequently demonstrated that many of these, including the then newly discovered isoprostanes, are of questionable significance to OSS. This, together with the fact that clinical trials of antioxidants including vitamin E and β-carotene have repeatedly failed to show beneficial effects [23–25], indicates that a new technique for examining biomarkers of OSS could provide valuable new tools in the identification of several disease processes. A novel implementation of protein arrays or metabolomic profiling not hindered by the limitations of existing approaches has the potential to greatly benefit public health. The list of metabolites currently generated (previously identified in different oxidative processes) can be included in studies of metabolomics for a better understanding of the oxidative stress status in human populations (Table 8.1 and Fig. 8.1). This effort will greatly help in identifying (oxidative stress-related) new and hitherto unknown biomarkers of different diseases and aid in not only the diagnosis and prognosis but also in the management of the disease processes.

Table 8.1 Analytes that could be included in metabolic profiling for assessment of oxidative stress status in human populations

Group	Biomarker		Ref.
1	Acrolein	Hexanal	[26–28]
2	Allantoin	Triuret	[29]
3	Biopyrrin-a	Biopyrrin-b	[30, 31]
4	Creatol	Methylguanidine	[32, 33]
5	Dimethyl disulfide	Hydrogen (or methyl) sulfide	[34–38]
6	Glyoxal	Methylglyoxal	[39]
7	Guanidinohydantoin	5-Nitro-4-guanidinohydantoin	[40–42]
8	Homocysteine	Cysteinylglycine	[43, 44]
9	7α-Hydroxycholesterol,	27-Hydroxycholesterol	[45]
10	3β-Hydroxy-5-oxo-5,6-secocholestan-6-al	5α-Hydroperoxycholesterol	[46–52]
11	15-F2t-Isoprostane	2,3-Dinor-5,6-dihydro-15-F2t-isoprostane	[1, 15, 53, 54]
12	Melatonin	Pinoline	[55]
13	3-Nitro-L-Tyrosine	3-chloro-3-nitro-L-Tyrosine	[56–58]

Notes

Group 1: Aldehydic products formed when polyunsaturated fatty acids, either free or bound to phospholipids, undergo autoxidation

Group 2: Products of uric acid (UA) oxidation. UA is present blood plasma up to 0.5 mM and constitutes a major scavenging system for peroxyl radicals.

Group 3: These are some of the further oxidation products of heme

Group 4: Creatol is the hydroxylated product of creatinine. Methylguanidine is formed when creatol undergoes further oxidation

Group 5: Dimethyl sulfide and hydrogen sulfide are formed as minor products when methionine (Met) and cysteine (Cys), either free or bound to proteins, undergo oxidation by 1-electron oxidants

Group 6: These are two advanced glycation end products (AGE) often measured as indicators of OSS in diabetes

Group 7: Guanidinohydantoin and 5-guanidino-4-nitroimidazole are two major products identified when 8-oxoguanine undergoes further oxidation by free radicals in the peroxynitrite/CO_2 system

Group 8: Homocysteine and cysteinylglycine are two well-studied markers of OSS

Group 9: 7α- and 27-Hydroxycholesterols are the hydroxylated products of cholesterol

Group 10: 3β-Hydroxy-5-oxo-secocholestan-6-al is a product of Hock cleavage of cholesterol-5α-hydroperoxide, the latter being a product in singlet oxygen-mediated oxidation of cholesterol

Group 11: 15-F2t-Isoprostane (15-F2t-IsoP) is one of the most extensively studied isoprostanes formed during free radical-catalyzed peroxidation of essential fatty acids (primarily arachidonic acid) without the involvement of cyclooxygenase. 2,3-Dinor-5,6-dihydro-15-F2t-IsoP (15-F2t-IsoPM) is further oxidation product of 15-F2t-IsoP

Group 12: Melatonin is a hormone secreted by the pineal gland. Pinoline is the methoxylated form of tryptoline formed in the pineal gland during the metabolism of melatonin. Both melatonin and pinoline are potent free radical scavengers

Group 13: 3-Nitro-L-tyrosine is a marker of peroxynitrite formation in vivo. 5-chloro-3-nitro-L-tyrosine is an oxidation product of free or protein-bound L-tyrosine and 5-chloro-3-nitro-L-tyrosine at inflammatory sites where there is co-production of peroxynitrite and hypochlorous acid

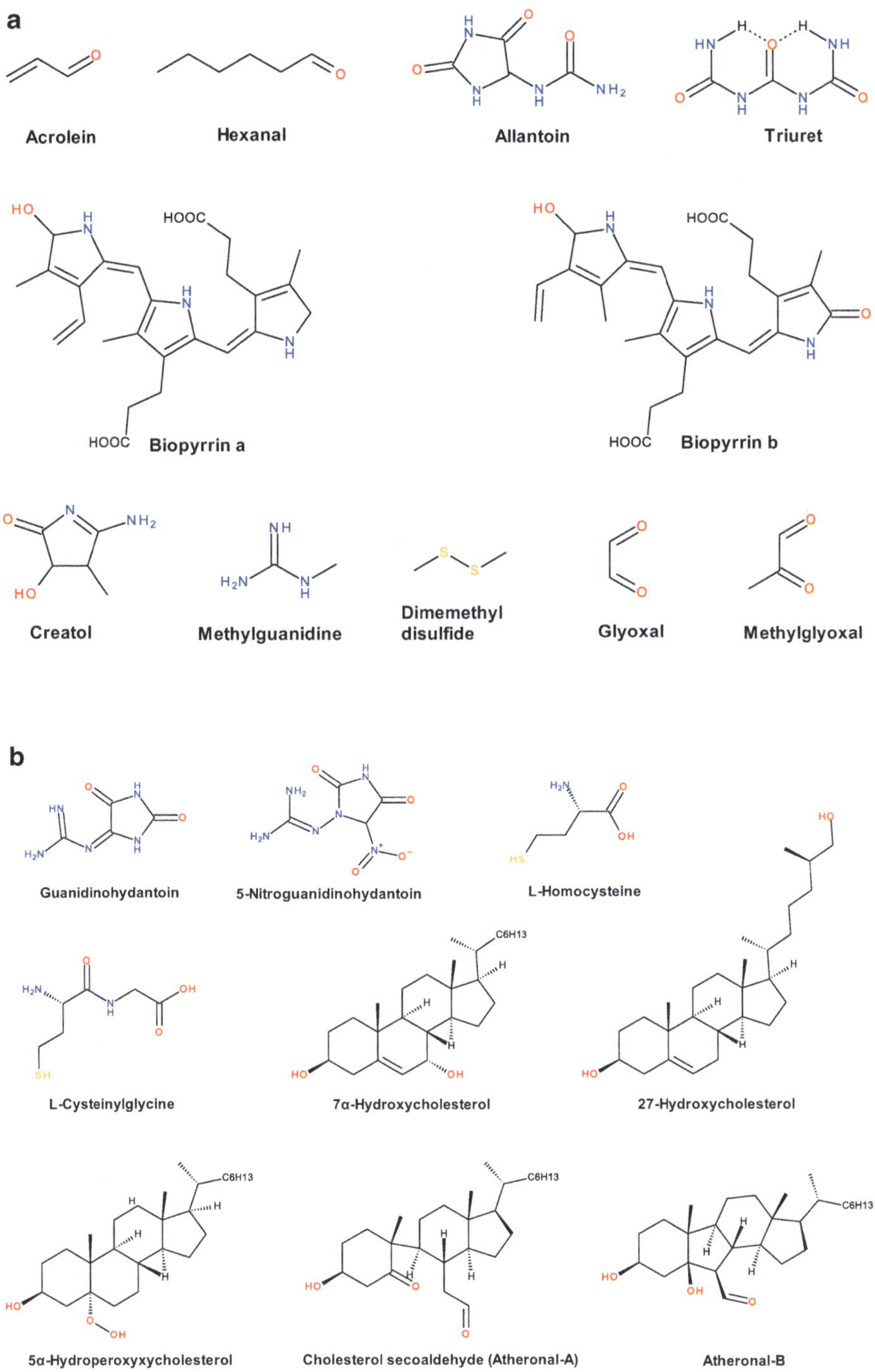

Fig. 8.1 (**a–c**) Structures of some representative oxidation products of that can be included in studies of metabolomic profiling for assessment of oxidative stress status in human studies

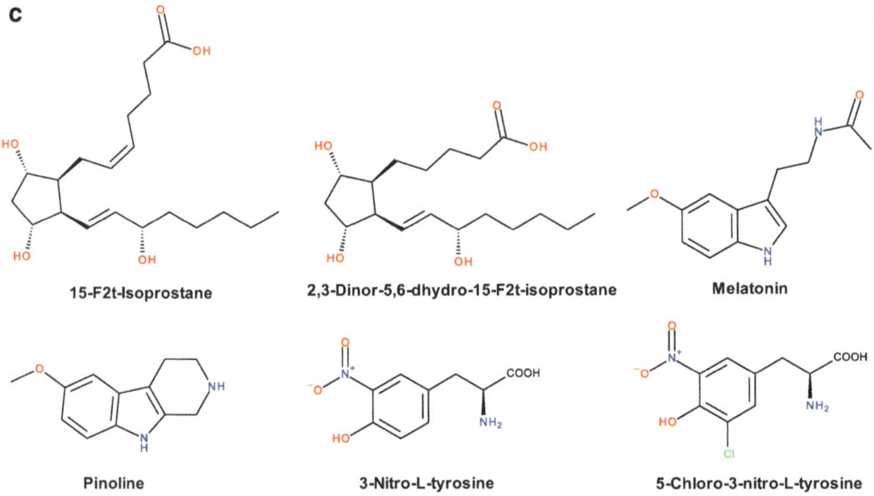

Fig. 8.1 (continued)

Acknowledgments The authors acknowledge the support from the National Institutes of Health (NIH) through the National Institute of General Medical Science (NIGMS) Grant 5 P2O GM103424-17 and the US Department of Education (US DoE; Title III, HBGI Part B grant number P031B040030). Its contents are solely the responsibility of authors and do not represent the official views of NIH, NIGMS, or US DoE.

References

1. Dai Q, Gao YT, Shu XO, Yang G, Milne G, Cai Q, Wen W, Rothman N, Cai H, Li H, et al. Oxidative stress, obesity, and breast cancer risk: results from the Shanghai Women's Health Study. J Clin Oncol. 2009;27(15):2482–8.
2. Kharitonov SA, Barnes PJ. Biomarkers of some pulmonary diseases in exhaled breath. Biomarkers. 2002;7:1–32.
3. Kharitonov SA, Barnes PJ. Exhaled markers of inflammation. Curr Opin Allergy Clin Immunol. 2001;1:217–24.
4. Phillips M, Cataneo RN, Cheema T, Greenberg J. Increased breath biomarkers of oxidative stress in diabetes mellitus. Clin Chim Acta. 2004;344:189–94.
5. Phillips M, Cataneo RN, Ditkoff BA, Fisher P, Greenberg J, Gunawardena R, Kwon CS, Rahbari-Oskoui F, Wong C. Volatile markers of breast cancer in the breath. Breast J. 2003;9:184–91.
6. Phillips M, Cataneo RN, Greenberg J, Grodman R, Gunawardena R, Naidu A. Effect of oxygen on breath markers of oxidative stress. Eur Respir J. 2003;21:48–51.
7. Phillips M, Cataneo RN, Greenberg J, Grodman R, Salazar M. Breath markers of oxidative stress in patients with unstable angina. Heart Dis. 2003;5:95–9.
8. Pryor WA. Measurement of oxidative stress status in humans. Cancer Epidemiol Biomark Prev. 1993;2:289–92.
9. Pryor WA, Godber SS. Oxidative stress status: an introduction. Free Radic Biol Med. 1991;10:173.

10. Pryor WA, Godber SS. Noninvasive measures of oxidative stress status in humans. Free Radic Biol Med. 1991;10(3–4):177–84.
11. Phillips M, Cataneo RN, Greenberg J, Gunawardena R, Rahbari-Oskoui F. Increased oxidative stress in younger as well as in older humans. Clin Chim Acta. 2003;328:83–6.
12. Phillips M, Greenberg J, Cataneo RN. Effect of age on the profile of alkanes in normal human breath. Free Radic Res. 2000;33:57–63.
13. Wei YH, Lee HC. Oxidative stress, mitochondrial DNA mutation, and impairment of antioxidant enzymes in aging. Exp Biol Med (Maywood). 2002;227:671–82.
14. Knez WL, Jenkins DG, Coombes JS. Oxidative stress in half and full Ironman triathletes. Med Sci Sports Exerc. 2007;39:283–8.
15. Morrow JD, Harris TM, Roberts LJ II. Noncyclooxygenase oxidative formation of a series of novel prostaglandins: analytical ramifications for measurement of eicosanoids. Anal Biochem. 1990;184:1–10.
16. Lucantoni G, Pietraforte D, Matarrese P, Gambardella L, Metere A, Paone G, Bianchi EL, Straface E. The red blood cell as a biosensor for monitoring oxidative imbalance in chronic obstructive pulmonary disease: an ex vivo and in vitro study. Antioxid Redox Signal. 2006;8:1171–82.
17. Cutler RG, Plummer J, Chowdhury K, Heward C. Oxidative stress profiling: part II. Theory, technology, and practice. Ann N Y Acad Sci. 2005;1055:136–58.
18. Noguchi N, Niki E. Phenolic antioxidants: a rationale for design and evaluation of novel antioxidant drug for atherosclerosis. Free Radic Biol Med. 2000;28:1538–46.
19. Yin H, Porter NA. New insights regarding the autoxidation of polyunsaturated fatty acids. Antioxid Redox Signal. 2005;7:170–84.
20. Floyd RA. Development of a sensitive analysis for 8-hyrdoxy-2′deoxyguanosine. Free Radic Res Comm. 1990;8:139–41.
21. Moller L, Hofer T, Zeisig M. Methodological considerations and factors affecting 8-hydroxy2-deoxyguanosine analysis. Free Radic Res. 1998;28:511–24.
22. Uppu RM, Cuerto R, Squadrito GL, Salgo MG, Pryor WA. Competitive reactions of peroxynitrite with 2′-deoxyguanosine and 7,8-dihydro-8-oxo-2′-deoxyguanosine (8- oxodG): relevance to the formation of 8-oxodG in DNA exposed to peroxynitrite. Free Radic Biol Med. 1996;21:407–11.
23. Robinson I, de Serna DG, Gutierrez A, Schade DS. Vitamin E in humans: an explanation of clinical trial failure. Endocr Pract. 2006;12:576–82.
24. Thomson MJ, Puntmann V, Kaski JC. Atherosclerosis and oxidant stress: the end of the road for antioxidant vitamin treatment? Cardiovasc Drugs Ther. 2007;21:95–210.
25. Steinhubl SR. Why have antioxidants failed in clinical trials? Am J Cardiol. 2008;101:14D–9D.
26. Sousa BC, Pitt AR, Spickett CM. Chemistry and analysis of HNE and other prominent carbonyl-containing lipid oxidation compounds. Free Radic Biol Med. 2017;11:294–308.
27. Pan J, Keffer J, Emami A, Ma X, Lan R, Goldman R, Chung FL. Acrolein-derived DNA adduct formation in human colon cancer cells: its role in apoptosis induction by docosahexaenoic acid. Chem Res Toxicol. 2009;22:798–806.
28. Tamura S, Tsukahara H, Ueno M, Maeda M, Kawakami H, Sekine K, Mayumi M. Evaluation of a urinary multi-parameter biomarker set for oxidative stress in children, adolescents and young adults. Free Radic Res. 2006;40:1198–205.
29. Gruber J, Tang SY, Jenner AM, Mudway I, Blomberg A, Behndig A, Kasiman K, Lee CY, Seet RC, Zhang W, Chen C, Kelly FJ, Halliwell B. Allantoin in human plasma, serum and nasal lining fluids as a biomarker of oxidative stress; avoiding artifacts and establishing real in vivo concentrations. Antioxid Redox Signal. 2009;11:1767–76.
30. Ihara H, Matsumoto T, Morita Y, Hirano A, Okada M, Hashizume N, Shioji I, Yoshimura H. Diurnal variation of biopyrrin excretion in random urine specimens is not corrected by creatinine. J Clin Lab Anal. 2007;21:1–6.

31. Kunii H, Ishikawa K, Yamaguchi T, Komatsu N, Ichihara T, Maruyama Y. Bilirubin and its oxidative metabolite biopyrrins in patients with acute myocardial infarction. Fukushima J Med Sci. 2009;55:39–51.
32. Ienaga K, Nakamura K, Fujisawa T, Fukunaga Y, Nihei H, Narita M, Tomino Y, Sanaka T, Aoyagi K, Nakano K, et al. Urinary excretion of creatol, an in vivo biomarker of hydroxyl radical, in patients with chronic renal failure. Ren Fail. 2007;29:279–83.
33. Nakamura K, Ienaga K, Nakano K, Nakai M, Nakamura Y, Hasegawa G, Sawada M, Kondo M, Mori H, Kanatsuna T. Diabetic renal failure and serum accumulation of the creatinine oxidative metabolites creatol and methylguanidine. Nephron. 1996;73:520–5.
34. Al Mardini H, Leonard J, Bartlett K, Lloyd S, Record CO. Effect of methionine loading and endogenous hypermethioninaemia on blood mercaptans in man. Clin Chim Acta. 1988;176:83–9.
35. Castro GD, Díaz Gómez MI, Castro JA. Dimethyldisulfide formation during trichloromethyl radical attack on methionine. Biochem Pharmacol. 1989;38:4145–7.
36. Couch RD, Dailey A, Zaidi F, Navarro K, Forsyth CB, Mutlu E, Engen PA, Keshavarzian A. Alcohol induced alterations to the human fecal VOC metabolome. PLoS One. 2015;10:e0119362.
37. Wagner CA. Hydrogen sulfide: a new gaseous signal molecule and blood pressure regulator. J Nephrol. 2009;22:173–6.
38. Yang G, An SS, Ji Y, Zhang W, Pei Y. Hydrogen sulfide signaling in oxidative stress and aging development. Oxidative Med Cell Longev. 2015;2015:357824.
39. Lapolla A, Reitano R, Seraglia R, Sartore G, Ragazzi E, Traldi P. Evaluation of advanced glycation end products and carbonyl compounds in patients with different conditions of oxidative stress. Mol Nutr Food Res. 2005;49:685–90.
40. Niles JC, Wishnok JS, Tannenbaum SR. Spiroiminodihydantoin and guanidinohydantoin are the dominant products of 8-oxoguanosine oxidation at low fluxes of peroxynitrite: mechanistic studies with 18O. Chem Res Toxicol. 2004;17:1510–9.
41. Niles JC, Wishnok JS, Tannenbaum SR. Peroxynitrite-induced oxidation and nitration products of guanine and 8-oxoguanine: structures and mechanisms of product formation. Nitric Oxide. 2006;14:109–21.
42. Jena NR, Mishra PC. Formation of ring-opened and rearranged products of guanine: mechanisms and biological significance. Free Radic Biol Med. 2012;53:81–94.
43. Blasco H, Garcon G, Patin F, Veyrat-Durebex C, Boyer J, Devos D, Vourc'h P, Andres CR, Corcia P. Panel of oxidative stress and inflammatory biomarkers in ALS: a pilot study. Can J Neurol Sci. 2017;44:90–5.
44. Hobbs CA, Cleves MA, Zhao W, Melnyk S, James SJ. Congenital heart defects and maternal biomarkers of oxidative stress. Am J Clin Nutr. 2005;82:598–604.
45. Schroepfer GJ Jr. Oxysterols: modulators of cholesterol metabolism and other processes. Physiol Rev. 2000;80:361–554.
46. Pryor WA, Wang K, Bermudez E. Cholesterol ozonation products as biomarkers for ozone exposure in rats. Biochem Biophys Res Commun. 1992;188:618–23.
47. Sathishkumar K, Xi X, Martin R, Uppu RM. Cholesterol secoaldehyde, an ozonation product of cholesterol, induces amyloid aggregation and apoptosis in murine GT1-7 hypothalamic neurons. J Alzheimers Dis. 2007;11:261–74.
48. Raghavamenon AC, Gernapudi R, Babu S, D'Auvergne O, Murthy SN, Kadowitz PJ, Uppu RM. Intracellular oxidative stress and cytotoxicity in rat primary cortical neurons exposed to cholesterol secoaldehyde. Biochem Biophys Res Commun. 2009;386:170–4.
49. Sathishkumar K, Gao X, Raghavamenon AC, Parinandi N, Pryor WA, Uppu RM. Cholesterol secoaldehyde induces apoptosis in H9c2 cardiomyoblasts through reactive oxygen species involving mitochondrial and death receptor pathways. Free Radic Biol Med. 2009;47:548–58.
50. Brinkhorst J, Nara SJ, Pratt DA. Hock cleavage of cholesterol 5alpha-hydroperoxide: an ozone-free pathway to the cholesterol ozonolysis products identified in arterial plaque and brain tissue. J Am Chem Soc. 2008;130:12224–5.

51. Tomono S, Miyoshi N, Sato K, Ohba Y, Ohshima H. Formation of cholesterol ozonolysis products through an ozone-free mechanism mediated by the myeloperoxidase-H_2O_2-chloride system. Biochem Biophys Res Commun. 2009;383:222–7.
52. Prapainop K, Wentworth P Jr. A shotgun proteomic study of the protein corona associated with cholesterol and atheronal-B surface-modified quantum dots. Eur J Pharm Biopharm. 2011;77:353–9.
53. Morrow JD, Hill KE, Burk RF, Nammour TM, Badr KF, Roberts LJ II. A series of prostaglandin F2-like compounds are produced in vivo in humans by a noncyclooxygenase, free radical-catalyzed mechanism. Proc Natl Acad Sci U S A. 1990;87:9383–97.
54. Il'yasova D, Morrow JD, Ivanova A, Wagenknecht LE. Epidemiological marker for oxidant status: comparison of the ELISA and the gas chromatography/mass spectrometry assay for urine 2,3-dinor-5,6-dihydro-15-F2t-isoprostane. Ann Epidemiol. 2004;14:793–7.
55. Rimón R, Airaksinen MM, Kari I, Gynther J, Venäläinen E, Heikkilä L, Ryyppö J, Palo J. Pinoline, a beta-carboline derivative in the serum and cerebrospinal fluid of patients with schizophrenia. Ann Clin Res. 1984;16:171–5.
56. Uppu RM, Winston GW, Pryor WA. Reactions of peroxynitrite with aldehydes as probes for the reactive intermediates responsible for biological nitration. Chem Res Toxicol. 1997;10:1331–7.
57. Saimanen I, Rahkola D, Kuosmanen V, Kärkkäinen J, Selander T, Holopainen A, Aspinen S, Eskelinen M. Nitrotyrosine (NT), a nitrosative stress biomarker, plasma concentrations in gallstone disease and cancer patients. Anticancer Res. 2019;39:809–14.
58. Mutahi TT, Edagwa BJ, Fronczek FR, Uppu RM. N-Acetyl-5-chloro-3-nitro-L-tyrosine ethyl ester. Acta Crystallogr Sect E Struct Rep Online. 2012;68(Pt 9):o2810–1.

Part IV
Instrumental Methods

Chapter 9
Sense and Sensibility of Oxygen in Pathophysiology Using EPR Oximetry

Periannan Kuppusamy

Abstract Physiological homeostasis in aerobic organisms is strictly maintained by optimal cellular and tissue oxygen levels through intricate oxygen-sensing mechanisms, signaling cascades, and transport processes. Molecular oxygen is at the center of oxygenation, oxidative phosphorylation, and oxidative stress. An increase (hyperoxia) or decrease (hypoxia) in cellular oxygen level may result in altered cell-signaling cascades and redox imbalance leading to pathophysiological processes including cell death and tissue damage. Hypoxia has been implicated as a critical factor influencing the outcomes for several diseases, including cardiovascular diseases (myocardial infarction, ischemic stroke, and peripheral arterial disease), cancer, wound healing, and diabetic foot ulcer. The capability to measure tissue oxygenation in a reliable and repeated manner will be immensely useful for correct prognosis and treatment. This chapter focuses on the methods, particularly electron paramagnetic resonance (EPR) oximetry for quantitative measurement of tissue oxygenation using implantable oxygen sensors. Representative examples for cardiovascular and cancer applications are presented.

Keywords Oxygen · Oximetry · Electron paramagnetic resonance · EPR · Oxygen sensor · OxyChip · Tumor · Myocardial infarction · Oxygen cycling · Reactive oxygen species

Abbreviations

AMI	Acute myocardial infarction
ATA	Atmospheric absolute
BOLD	Blood oxygen level dependent
CAD	Coronary artery disease

P. Kuppusamy (✉)
Departments of Radiology and Medicine, Geisel School of Medicine, Dartmouth College, Lebanon, NH, USA
e-mail: kuppu@dartmouth.edu

L. J. Berliner, N. L. Parinandi (eds.), *Measuring Oxidants and Oxidative Stress in Biological Systems*, Biological Magnetic Resonance 34,
https://doi.org/10.1007/978-3-030-47318-1_9

CuZnSOD	Copper zinc superoxide dismutase
CYP	Cytochrome p450 reductase
eNOS	Endothelial nitric oxide synthase
EPR	Electron paramagnetic resonance
EPRI	EPR imaging
H&E	Hematoxylin and eosin
H_2O_2	Hydrogen peroxide
HBO	Hyperbaric oxygen
HBOT	Hyperbaric oxygen therapy
$HO^{\cdot}$	Hydroxyl radical
I-R	Ischemia-reperfusion
iNOS	Inducible nitric oxide synthase
L-NAME	N^G-nitro-L-arginine methyl ester
LAD	Left anterior-descending artery
LCA	Left coronary artery
LiNc-BuO	Lithium octa-*n*-butoxy-naphthalocyanine
LiPc	Lithium phthalocyanine
LV	Left ventricular
MI	Myocardial infarction
MnSOD	Manganese superoxide dismutase
MRI	Magnetic resonance imaging
MSC	Mesenchymal stem cell
NIR	Near infrared
NMR	Nuclear magnetic resonance
NO	Nitric oxide
NOS	Nitric oxide synthase
NOS3	Nitric oxide synthase 3
$O_2^{\cdot -}$	Superoxide anion radical
OMRI	Overhauser-enhanced magnetic resonance imaging
$ONOO^-$	Peroxynitrite anion
OxCy	Oxygen cycling
OxyChip	Probe (sensor) for measuring oxygen
OxyChip-EL	OxyChip with extended loop
PDMS	Polydimethylsiloxane
pO_2	Partial pressure of oxygen
RIF-1	Radiation-induced fibrosarcoma-1
ROS	Reactive oxygen species
SM	Skeletal myoblast
SOD	Superoxide dismutase
SPOT Chip	Probe (sensor) for measuring superficial perfusion oxygen tension
SPZ	Sulfaphenazole
TcOM	Transcutaneous oxygen measurement
$TcpO_2$	Transcutaneous pO_2
VEGF	Vascular endothelial growth factor

Oxygen in Pathophysiology

Oxygen is indispensable for the life of aerobic organisms. It is hailed as the "*Elixir of Life*," a wonder tonic, a cure for aging, a beauty treatment, and a potent therapeutic [1]. Respiratory process in mammals, including humans, allows the inspired oxygen to be transported to the target tissue by blood, especially hemoglobin. At the target tissue, oxygen is released from the oxygen-bound hemoglobin (oxyhemoglobin) and used for oxidative phosphorylation in metabolically active cells. Under conditions of metabolic homeostasis, a dynamic equilibrium is established between oxygen supply and demand resulting in a "normoxic" level of tissue oxygen. The level of oxygen in "normoxia" varies considerably among tissues within the same organism due to varying levels of blood flow (oxygen supply) and metabolic demand (oxygen utilization). This leads to the establishment of a unique normoxic level (physiological oxygen or "physioxia") for each tissue. The physiological homeostasis of the tissue is strictly maintained by optimal cellular and tissue oxygen status through complex oxygen-sensing mechanisms, signaling cascades, and transport processes. Any imbalance in oxygen levels may affect metabolic homeostasis and lead to pathophysiological conditions [2]. In the event of fluctuating oxygen levels leading to either an increase (*hyperoxia*) or decrease (*hypoxia*) in cellular oxygen, the organism faces a crisis involving depletion of energy reserves, altered cell-signaling cascades, oxidative events, and cell death. The role of hypoxia in cardiovascular diseases, particularly ischemic stroke or myocardial infarction [3, 4], cancer treatment [5–8], and wound healing [9–13], has been well documented.

As the oxygen level is an important determinant of the disease pathophysiology and treatment prognosis, it is critical to accurately quantify the level of oxygen. A precise knowledge of the level of oxygen in the tissue of interest will be of paramount importance in our ability to understand the mechanism of pathogenesis and to develop strategies to correct the imbalance. This would require methods capable of quantifying the level of tissue oxygen with good spatial and temporal resolution. The information gained will enable better understanding of various metabolic and disease states and will assist in making effective clinical decisions regarding treatment and therapy options. This chapter focuses on the methods for determination of oxygen concentration in tissues with an emphasis on methods based on electron paramagnetic resonance (EPR) oximetry using implantable oxygen sensors for preclinical and clinical applications, carried out in the author's laboratory.

Measurement of Tissue Oxygenation

Although the discovery of oxygen was made in the eighteenth century, measurements of oxygen concentration (*oximetry*) in living systems (in vivo) are only a recent phenomenon. Early attempts were made in the 1960s, but it was in the late 1980s that the computerized polarographic needle electrode system was used extensively to assess the oxygenation in tumors clinically. The use of this technique helped to establish the role of hypoxia in the efficacy of radiation therapy [14]. Now, there are several potentially clinically useful methods [15] that are based on other principles, including fluorescence-quenching, phosphorescence-quenching, near-infrared (NIR) absorption, immunohistochemical, and magnetic resonance techniques. Although several methods have been utilized to measure oxygen concentration, a suitable technique for direct and repeated measurements of oxygen in the clinical setting is not currently available. While electrode-based methods have evolved as the standard for measuring oxygen, they generate analytical artifacts during assay procedures at the freshly probed sites and are not suitable for repeated measurements [16]. Near-infrared and magnetic resonance techniques—such as nuclear magnetic resonance (NMR), blood-oxygen-level-dependent (BOLD) magnetic resonance imaging (MRI), and Overhauser-enhanced magnetic resonance imaging (OMRI)—are noninvasive methods; however, they do not report the absolute values of oxygen concentration in the tissues [17–22]. Electron paramagnetic resonance (EPR) oximetry, closely related to the aforementioned magnetic resonance techniques, enables direct, accurate, reliable, and repeated measurements of oxygen concentration in tissues [23].

Some of the important criteria for improvements in the ability to make successful clinical measurements of oxygenation include minimal or no invasiveness, capability to make repeated measurements, accessibility to the region of interest, appropriate spatial resolution, adequate depth of measurement, accuracy and robustness of measurements, usefulness of the parameter reported for clinical purposes, measurement time consistent with use in patients, ease of use in the clinical setting, and potential for the instrumentation to be commercially available. It is especially important that the method should enable repeated measurements from the region of interest in order to follow the changes in oxygenation over a period of time, preferably for up to several months, or longer [24, 25]. The technique should also provide appropriate spatial and temporal resolution. The depth of measurement (penetration) and accessibility to the region of interest are some of the important factors for establishing the scope of applicability of the technique.

EPR Oximetry

EPR oximetry refers to the measurement of oxygen concentration (e.g., partial pressure of oxygen, pO_2) using EPR spectroscopy [23]. The principle of EPR oximetry is based on the paramagnetic characteristics of molecular oxygen, which, in its ground state, has two unpaired electrons and undergoes spin-exchange interaction with an exogenous paramagnetic EPR spin-probe placed at the region of interest (Fig. 9.1). This process is sensitive to the concentration of oxygen present in the local environment, with the relaxation rate of the spin-probe increasing as a function of oxygen content (concentration or pressure) [23, 26]. The increased spin-spin relaxation rate results in increased line-broadening. The fact that the linewidths of

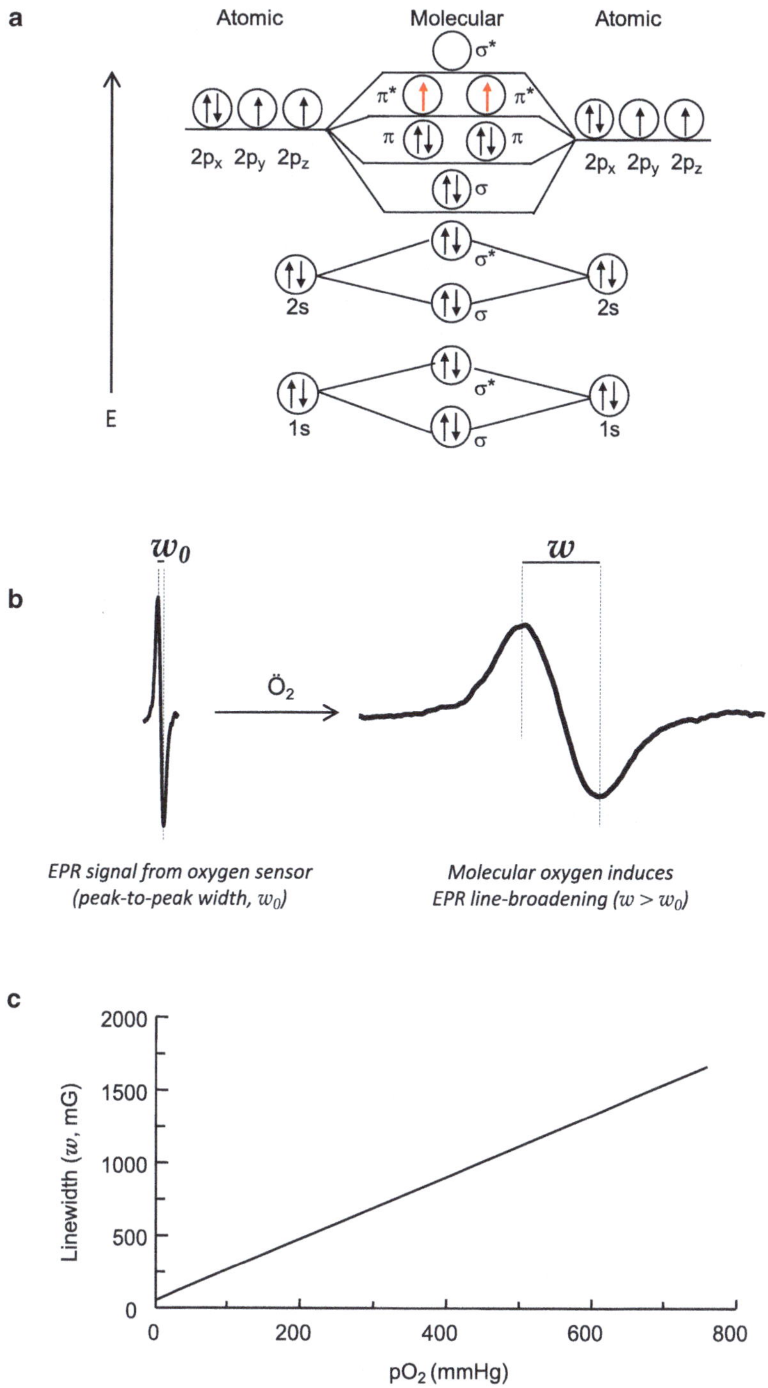

Fig. 9.1 EPR oximetry. (**a**) Molecular orbital diagram showing the distribution of electrons in (diatomic) oxygen molecule. The doubly degenerate π^* orbital makes the two electrons occupy separate levels resulting in two unpaired electrons in the molecule. As a result, diatomic oxygen is paramagnetic. (**b**) Principle of EPR oximetry. Paramagnetic oxygen undergoes spin exchange with probe (oxygen sensor) resulting in broadening of the EPR spectrum of the probe. The broadening, measured as peak-to-peak linewidth, w, is proportionate to oxygen content. (**c**) Typical calibration curve exhibiting a linear relationship of EPR linewidth with partial pressure of oxygen (pO_2)

the EPR spectrum linearly correlate with oxygen concentration has been exploited in a variety of biological oximetry studies [27–40].

Typical in vivo EPR scanners, either commercially available ones or those custom-built for laboratory research, operate at about 1.2-GHz (L-band) frequency, which, due to nonresonant microwave absorption in tissues, limits the measurement depth to about 1 cm. The depth can be substantially increased, e.g., to 7 cm, by lowering the frequency to 300 MHz [41, 42]. The tradeoff, however, is that the signal-to-noise ratio (SNR) decreases at lower frequencies. Hence, we make a prudent decision on the choice of operating frequency as a compromise between measurement depth versus SNR. In clinical EPR oximetry, we must select operating frequencies that allow adequate depth of RF penetration to probe regions deep in the tissue.

Oxygen-Sensing Probes for EPR Oximetry

Measurement of oxygen-induced line-broadening requires the presence of a suitable EPR probe in the tissue region of interest. This approach uses implants of particulate oximetry probes such as lithium phthalocyanine (LiPc) or lithium octa-*n*-butoxy-naphthalocyanine (LiNc-BuO) whose EPR linewidths are highly sensitive to the local oxygen concentration [43, 44]. The oxygen-induced line-broadening is usually linear with respect to the partial pressure of oxygen Fig. 9.1c. The measured linewidth can be converted to the oxygen concentration using appropriate standard curves. The probes are implanted at the desired site [45] or internalized in cells enabling highly accurate measurements of intracellular pO_2 [46]. These probes are stable in tissues, nontoxic, and biocompatible [43]. The measurements can be performed noninvasively, and repeatedly, over periods of months or longer at the same site [45].

The most notable drawback and potential barrier to the use of these materials in clinical EPR applications is the need to leave these particulate probe materials permanently in the tissue, which may present practical barriers for obtaining approval for use in human subjects. New approaches are therefore sought to eliminate this obstacle. One approach is to use the oxygen-sensing probe embedded in a biocompatible polymer substrate that has high oxygen permeability [47–50]. The probes would effectively be shielded from interaction with the biological *milieu* that could result in biochemical degradation and breakdown, as well as local and/or systemic toxicity effects that may occur due to these reactions. The implants could be left in the tissue or removed when no longer needed. An optimal biomaterial membrane intended for encapsulation in oximetry applications should confer rapid and stable responsiveness to oxygen to the sensing probes. The membrane matrix must be nontoxic and biocompatible with potential host cells and tissues, and as well have sufficient mechanical strength to withstand physiological pressures in vivo. To this end, we have identified PDMS as a potential material for probe encapsulation. PDMS has been well characterized, and was one of two primary reference materials made available by the National Heart Lung and Blood Institute for standardized biocompatibility testing [51, 52]. One characteristic of PDMS that makes it especially suited for oximetry applications is that it is known to have very high gas permeability [53–55]. In addition, PDMS has been the elastomer of choice in specialized medical and healthcare applications since the 1960s. PDMS elastomers are inexpensive and have been used in a wide range of biomedical applications due to their biocompatibility, low toxicity, and excellent optical transparency, and good thermal and oxidative stability [56]. Medical devices based on PDMS include blood pumps, cardiac pace leads, mammary prostheses, drainage implants for use in glaucoma treatment, artificial skin, contact lenses, oxygenators, catheters, drug delivery systems, etc. [56, 57]. PDMS is amenable to rapid prototyping and with the use of several polymeric fabrication techniques they can be formed in a variety of desired device sizes and shapes [58–60].

What is considered a limitation of EPR oximetry for clinical application—EPR oximetry requires an external/implanted spin probe to report oxygen—is also a potential advantage, in that the probe can be customized for specific applications in terms of depth in tissue, oxygen sensitivity (desired resolution of pO_2), oxygen level (desired pO_2 range), and repeated measurements over months or longer on the same selected sites. The probes use PDMS-encased lithium octa-*n*-butoxy-naphthalocyanine (LiNc-BuO) crystals, called "OxyChip," (Fig. 9.2) which have been thoroughly characterized, in terms of biocompatibility [61, 62] and long-term stability [63, 64], and established as robust sensor of tissue oxygen using animal models [49, 61, 64–69] as well as for the measurement of oxygen in human tumors [70, 71]. The following sections provide a brief description of the type of OxyChip-based probes for EPR oximetry, especially for measurement in human subjects.

Topical Oximetry

To establish EPR oximetry for measuring skin oxygen (transcutaneous pO_2), we have developed an oxygen sensor, named SPOT (skin perfusion oxygen tension) chip [72, 73]. Microcrystals of LiNc-BuO are embedded in PDMS and fabricated into circular films of 3-mm diameter and ~0.06-mm thickness (Fig. 9.2a). One side of the chip is covered with an oxygen-impermeable film, while the other side is permeable to oxygen. The isolation is necessary to ensure that only the oxygen that diffuses through the skin surface is measured, and not the oxygen from the ambient environment. The chip is secured to the skin by a medical transfer adhesive. We have demonstrated the functioning of the chip in healthy human subjects. The results established that the SPOT chip can be used for pO_2 measurements when applied topically to the region of interest or multiple anatomical sites. No surgical procedure is required for the placement of the chip. The method is noninvasive, highly sensitive to low oxygen levels (hypoxia), and repeatable for long-term tissue oxygen monitoring.

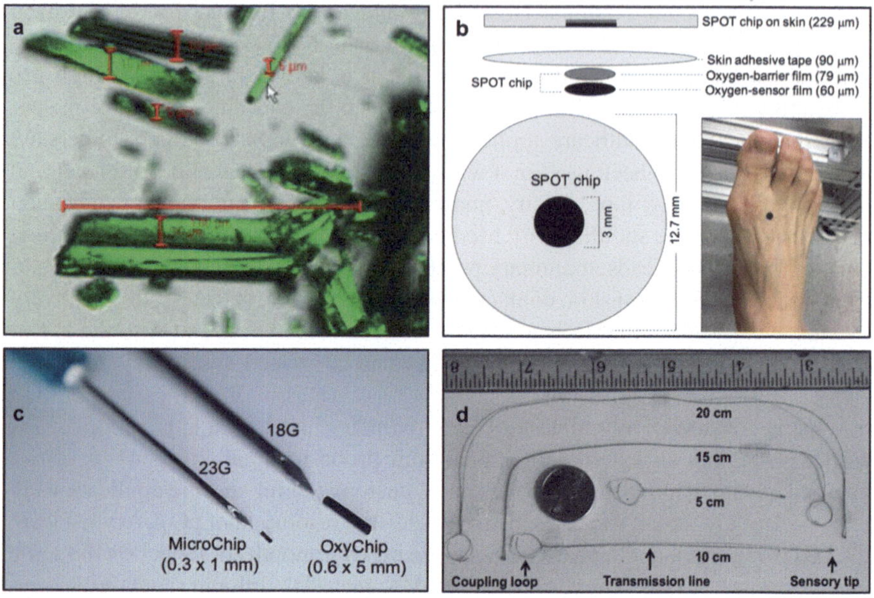

Fig. 9.2 Oxygen-sensing paramagnetic probes for EPR oximetry. (**a**) Raw (un-encased) microcrystals of LiNc-BuO. (**b**) SPOTChip for skin oximetry. Assembly of the SPOTChip, which when applied to the skin will have two layers of oxygen-barrier films to insulate the chip from ambient oxygen. Placement of a SPOTChip on the foot of a healthy volunteer is shown. (**c**) MicroChip and OxyChip are shown with the needles used for implantation in tissues. (**d**) OxyChip-EL with different lengths of transmission line for deep-tissue oximetry

Superficial (1–5 mm-Depth) Tissue Oximetry

We have fabricated and extensively tested a new form of implantable oxygen sensor called MicroChip (Fig. 9.2c). The MicroChip has been thoroughly characterized for sensitivity, safety, integrity, and biocompatibility for measurements in live tissues [74]. The following were established for MicroChips: Linearity of the calibration (linewidth vs. with pO_2); stability of EPR and oxygen sensitivities against autoclaving; exposure to ionizing radiation, and sonication procedures; mechanical stability and integrity of the chip for holding the embedded crystals under adverse conditions; up to 5-mm depth for measurement in superficial tissue; capability for multiple implants (4–10) in a single tissue/tumor and obtaining mean and median pO_2 values using a single scan; repeated measurements of tissue pO_2 over a long term (8 weeks) studied in animal models; and overall, a robust and reliable probe for pO_2 measurements [74].

Shallow-Depth (Up to 10 mm) Tissue Oximetry

For pO_2 measurements for up to 10-mm depth, we use a larger version of a chip, called OxyChip, in the form of a cylindrical pellet with 0.6-mm diameter and 5-mm length (Fig. 9.2c), ideal for implanting using a 18G syringe needle, which is routinely used in the clinic for brachytherapy [49, 61]. The OxyChips are made by cast-molding medical-grade PDMS polymer embedded with LiNc-BuO microcrystals. In vitro evaluation of the OxyChip showed that it is robust and highly oxygen-sensitive (up to 14 mG/mmHg). The dependence of its EPR linewidth to oxygen is linear in the range 0–100% oxygen, stable over time and after autoclave or irradiation. In vivo efficacy of the OxyChips was evaluated by implanting them in rat femoris muscle and following their response to tissue oxygenation for up to 1 year [63]. The results revealed the preservation of the integrity (size and shape) and calibration (oxygen sensitivity) of the OxyChip throughout the implantation period. Further, no inflammatory or unanticipated adverse reactions around the implantation area were observed, thereby establishing its biocompatibility and safety for long-term use.

Deep-Tissue (>10 mm) Oximetry

OxyChips, with an extended sensing-loop, are made with one or more sensory points (LiNc-BuO in PDMS) incorporated into one end of a twisted-pair transmission line, made of MP35N wire (diameter 0.080–0.125 mm; length >10 mm) attached to a coupling loop (10-mm dia.) on the other end of the transmission line (Fig. 9.2d). The coupling loop is designed to be placed subcutaneously to allow inductive coupling to an external surface loop resonator for signal detection. To avoid any toxicity or mechanical damage, the resonators are further coated with PDMS. Standard in vitro and in vivo biocompatibility tests were carried out to establish the safety and efficacy of the sensor for deep-tissue applications [32, 65, 75].

Oxygen Measurements in Animal Models of Diseases

EPR oximetry, based on un-embedded (raw) LiNc-BuO crystals [43, 76], has been extensively used in animal models of cardiovascular diseases and cancer. Typically, the microcrystals are implanted in the tissue of interest using a minimally invasive syringe needle. The oxygen measurements would be carried out 2–3 days after implantation to avoid artifacts due to surgical trauma, inflammation, and microvascular rupture at the implant site. The following sections provide some of the selected studies carried out using raw LiNc-BuO crystals.

Stem-Cell Therapy for Myocardial Infarction

Acute myocardial infarction (AMI) leads to impaired cardiac function, which is a major cause of morbidity and mortality [77]. Although a variety of surgical interventions are available to rescue the failing heart, cellular cardiomyoplasty, wherein the lost cells are replaced by transplantation of stem cells in the affected region, is being perceived as a potential alternative. The grafting of stem cells onto myocardial scar tissue has resulted in marginal improvements in cardiac function and in the limitation of abnormal cardiac remodeling [78–85]. The limited survival of the transplanted cells in the infarcted myocardium is thought to be the reason for such modest improvements [86]. The hypovascular nature of the infarcted tissue may severely compromise the availability of oxygen, nutrients, and growth factors essential for the survival and differentiation of the transplanted cells. The local hypoxic environment in the infarct myocardium may be the main impediment to the survival of the transplanted cells. However, it is not clear whether the oxygen concentration in the ischemic myocardium (infarcted area) is affected by strategies that stimulate angiogenesis and/or by cell transplantation. It is also unknown whether there is a relationship between oxygen concentration and transplanted cell survival. Oxygen tension plays an important role in the growth of stem cells in culture and significantly influences their expansion and differentiation [87–89]. In response to acute hypoxia, cardiomyocytes have been shown to exhibit adaptations that may facilitate cell-survival and develop tolerance to subsequent acute severe hypoxia [90]. Hence, the determination of in situ oxygenation at the transplant site within the ischemic heart tissue is vital for the understanding of the effects of cell therapy.

EPR oximetry was used to study the role of oxygen on the grafting of stem cells and cardiac function [91]. Skeletal myoblast (SM) cells isolated from thigh muscle biopsies of mice were labeled with LiNc-BuO by coculturing the cells with submicron-size (270 ± 120 nm) particulates of the probe. Myocardial infarction was created by ligation of left coronary artery (LCA) in mice. Immediately after ligation, labeled SM cells were transplanted in the ischemic region of the heart. The engraftment of the transplanted cells and in situ pO_2 in the heart were monitored weekly for 4 weeks. EPR measurements revealed the retention of cells in the infarcted tissue. The myocardial pO_2 at the site of SM cell therapy was significantly higher compared with the untreated group throughout the 4-week period (Fig. 9.3). Histological studies revealed differentiation and engraftment of SM cells into myotubes and increased incidence of neovascularization in the infarct region. The infarct size in the treated group was significantly decreased while echocardiography showed an overall improvement in cardiac function when compared to untreated hearts. The increased myocardial pO_2 positively correlated with neoangiogenesis and cardiac function. The EPR oximetry clearly established the feasibility of measuring in situ pO_2 from the engraftment site, in vivo.

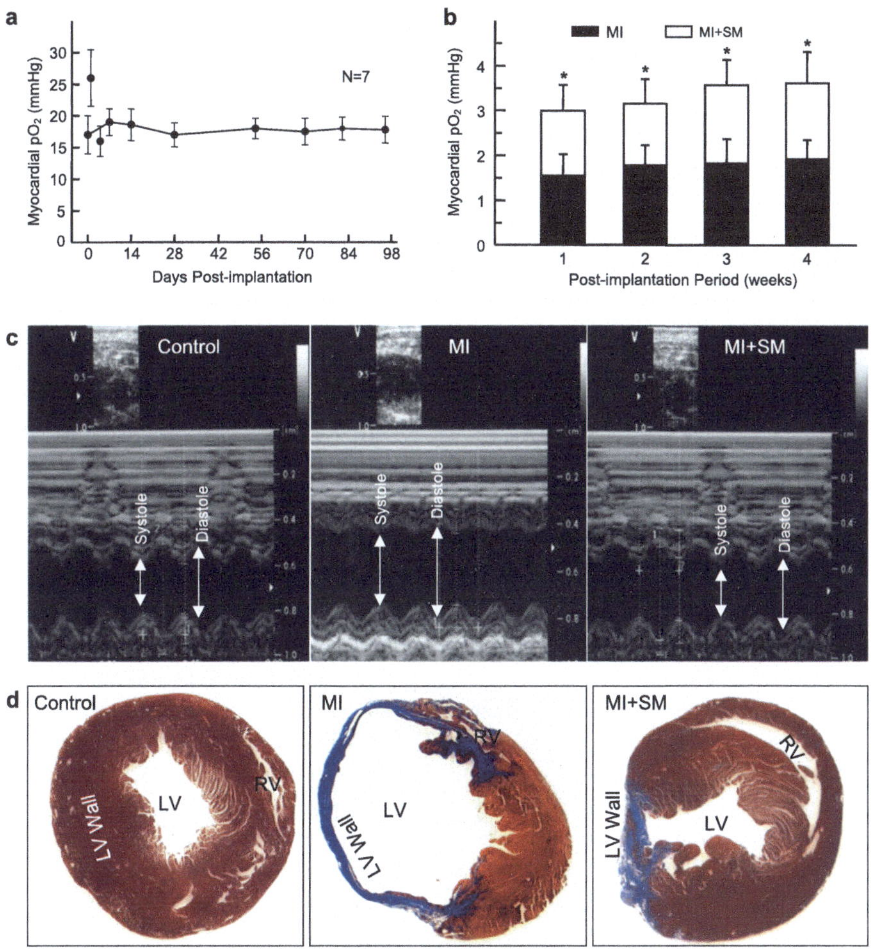

Fig. 9.3 Myocardial pO_2, cardiac function and infarct size in the infarcted hearts treated with skeletal myoblasts. (**a**) Repeated measurements of myocardial pO_2 in the left-ventricular wall of non-infarct (healthy) hearts of male C57BL mice measured using implanted submicron-size crystals of LiNc-BuO for over 3 months. The data show a stable readout of pO_2 values of 19 ± 2 mmHg in the healthy myocardium. (**b**) Repeated measurements of myocardial pO_2 in the infarct hearts of male C57BL mice transplanted with LiNc-BuO-labeled autologous skeletal myoblast (SM) cells for 4 weeks. Values are expressed as mean ± SD ($n = 7$ per group). $^*p < 0.05$ vs. MI group. Tissue pO_2 values after treatment with SM cells (MI+SM) were significantly higher when compared to untreated (MI) hearts. (**c**) Functional assessment (echocardiography) of hearts at 4 weeks after SM cell transplantation. Echocardiography was performed in non-infarcted hearts (Control), infarcted hearts (MI), and infarcted hearts treated with SM cells (MI+SM). Representative recordings of M-mode echocardiography are shown. The images show a substantial improvement of cardiac function in the treated heart. (**d**) Assessment of infarct size and LV remodeling in hearts at 4 weeks after SM cell transplantation. Transverse sections of hearts with triphenyltetrazolium chloride (TTC) staining are shown for non-infarcted hearts (Control), infarcted hearts (MI), and infarcted hearts treated with SM cells (MI+SM). The images show a substantial reduction of infarct in the treated heart. (Recomposed from Khan et al. [91])

Protection of Heart Against Ischemia-Reperfusion Injury by Sulfaphenazole

The pathogenesis of myocardial ischemia-reperfusion (I-R) injury is known to involve interplay of multiple mechanisms. Several studies have implicated reactive oxygen species (ROS), including superoxide radical ($O_2^{·-}$), hydrogen peroxide (H_2O_2), hydroxyl radical ($HO^·$), and peroxynitrite ($ONOO^-$) that are generated upon reperfusion, in the I-R-mediated oxidative damage to the myocardium [92]. The involvement of ROS in mediating I-R injury has been established based on the efficacy of antioxidants and free radical scavengers such as superoxide dismutase (SOD), catalase, melatonin, and vitamin E in minimizing I-R injury [93]. Overexpression of manganese SOD (MnSOD), copper-zinc SOD (CuZnSOD), or glutathione peroxidase has been reported to protect the heart from I-R injury, further supporting the involvement of ROS in the reperfusion injury [94–96]. Unlike ROS, the involvement of nitric oxide (NO) in I-R injury has been controversial. NO, which is produced by a variety of mammalian cells, is an important mediator of both physiological and pathological vascular functions [97, 98]. NO production is catalyzed by nitric oxide synthase (NOS). Enhanced NO generation was observed in the heart during I-R [99, 100]. Decrease of NO production using NOS-inhibitors showed a decrease in the I-R-mediated functional impairment of the heart [101, 102]. However, NO has also been shown to play a cardioprotective role in myocardial I-R injury [103–108]. Supplementation with L-arginine (substrate for NO production by NOS) and NO donors during reperfusion has been shown to be cardioprotective in regional, as well as in global ischemic models [105, 106, 108].

The cytochrome P450 (CYP) family of enzymes play a significant role in normal cardiovascular homeostasis as well as in cardiovascular pathogenesis [109]. Particularly, CYP 2C9 has been implicated in myocardial I-R injury [110–112]. CYP 2C9 has been identified as a potent source of superoxide radicals in the reperfused heart [110–113]. A role for CYP 2C9 in myocardial I-R injury was first demonstrated by Granville et al. [110] in an isolated rat heart model using CYP 2C6/9 inhibitors such as chloramphenicol, sulfaphenazole (SPZ), and cimetidine. The CYP 2C6/9 inhibitors were found to markedly attenuate infarct size and creatine kinase release. Superoxide was also found to be significantly reduced, while post-ischemic coronary flow was increased in the CYP inhibitor-treated hearts, indicating that NO scavenging and oxidative damage are likely to play a role in the protection against I-R-induced injury [110]. These results were also reproduced in a rabbit coronary artery ligation model of focal I-R injury [110]. A recent clinical study showed that SPZ could improve endothelium-dependent, NO-mediated vasodilation in patients with coronary artery disease (CAD) as assessed by increased acetylcholine-induced forearm blood-flow compared with control patients [114]. The beneficial effect of SPZ administration was attributed to an increase in the bioavailability of NO in tissue and circulation during reperfusion. Similarly, a more recent study using SPZ administration in diabetic mice exhibited restoration of endothelium-dependent vasodilation, possibly by decreasing superoxide levels

[115]. However, the exact mechanism by which SPZ attenuates myocardial injury is not well understood.

A study using isolated rat hearts showed that SPZ protected the hearts from I-R injury by scavenging ROS and increasing NO levels [116]. However, it was not clear whether the increased NO levels in the SPZ-treated I-R hearts was due to SPZ-mediated superoxide depletion or activation of endogenous pathways of NO production. This led us to hypothesize that SPZ may, in addition to inhibition and/or scavenging of superoxide generation, induce NO generation by activating endogenous inducible NOS (iNOS) in the reperfused heart. The elevated levels of NO bioavailability in the reperfused heart could have profound effects on the oxidative/nitrative stress, tissue oxygenation, infarction, and functional recovery. Henceforth, we investigated the cardioprotective effect of SPZ and to delineate the involvement of NO, superoxide, oxygenation and also to establish the signaling mechanism involved in cardioprotection in an in vivo rat model of acute myocardial infarction [36]. MI was induced by 30-min ligation of left anterior descending coronary artery followed by 24-h reperfusion (I-R). The study used six groups: I-R (control); SPZ; L-NAME; L-NAME+SPZ; 1400W (an inhibitor of iNOS); 1400W+SPZ. The agents were administered orally through drinking water for 3 days prior to the induction of I-R. Myocardial oxygenation (pO_2) at the I-R site was measured using EPR oximetry (Fig. 9.4). The preischemic pO_2 value was 18 ± 2 mmHg in all groups. At 1-h of reperfusion, the SPZ group showed a significantly higher hyperoxygenation when compared to control (45 ± 1 versus 34 ± 2 mmHg). The SPZ group showed a significant improvement in the contractile functions and reduction in infarct-size [36]. Histochemical staining of SPZ-treated hearts exhibited significantly lower levels of superoxide and peroxynitrite, and markedly increased levels of iNOS activity and nitric oxide (Fig. 9.5). Western-blot analysis indicated upregulation of Akt and attenuation of p38MAPK activities in the reperfused myocardium [36]. The study established that SPZ attenuated myocardial I-R injury through overexpression of iNOS leading to enhancement of nitric-oxide bioavailability and tissue oxygenation.

In Vivo Imaging of Changes in Tumor Oxygenation During Growth and After Treatment with Radiation

Tumor hypoxia is strongly linked to the poor treatment outcome of chemo- or radiotherapy in several human malignancies [117, 118]. Hypoxic tumors are biologically more aggressive—it has been reported that hypoxic sarcoma [119] or cervical cancers [120] tend to metastasize. The fact that poorly oxygenated tumors are more aggressive and less susceptible to treatment suggests that the tumor oxygenation status is an important parameter for cancer treatment [121, 122]. The observation of substantial inter- and intra-tumor heterogeneities among tumors of similar histology and sites further emphasizes the importance of the measurement of hypoxia in

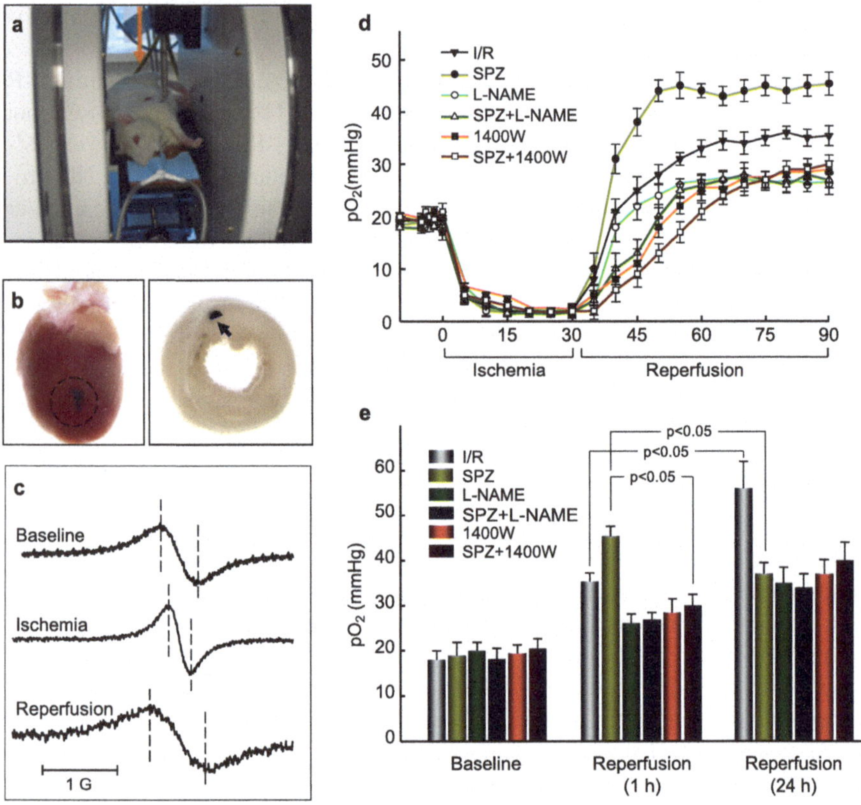

Fig. 9.4 In vivo measurement of pO_2 in the rat heart using EPR oximetry. (**a**) Placement of a rat in the EPR spectrometer for monitoring of myocardial oxygenation. The animal, under isoflurane inhalation anesthesia, is placed in a right-lateral position with the chest open to the loop of a surface-coil resonator. (**b**) Implantation of oxygen-sensing microcrystals of LiNc-BuO in the left ventricular mid-myocardium. The probe particulates are seen as a black implant in the images of the whole heart and a formalin-fixed transverse slice through the left ventricle. The probe, which is nontoxic to tissue can respond to the partial pressure of oxygen (pO_2) at the site of placement. (**c**) Representative EPR signals obtained from a heart during pre-ischemia (baseline), ischemia, and reperfusion. The peak-to-peak (indicated by dashed line) width of the signal is used to calculate pO_2 using a standard curve. (**d**) Changes in pO_2 during a 30-min ischemia followed by 60-min reperfusion in rats pretreated with vehicle (I-R), SPZ, L-NAME, SPZ+L-NAME, 1400 W, or SPZ+1400 W. Data represent mean ± SD, obtained from 6 animals/group. (**e**) Bar-graphical representation pO_2 data (mean ± SD, $N = 6$ animals/group) at the end of 1 h (from panel **d**) and at 24 h of reperfusion period. The "Baseline" data were obtained from preischemic hearts. The results show a significant oxygen overshoot (hyperoxygenation) during reperfusion. (Reproduced from Khan et al. [36])

individual tumor/patients. The ability to monitor changes in pO_2 following treatment could have profound implications in the planning of effective therapeutic strategies [121, 122]. In particular, radiotherapy could benefit from modulated treatment based on regional variations in pO_2.

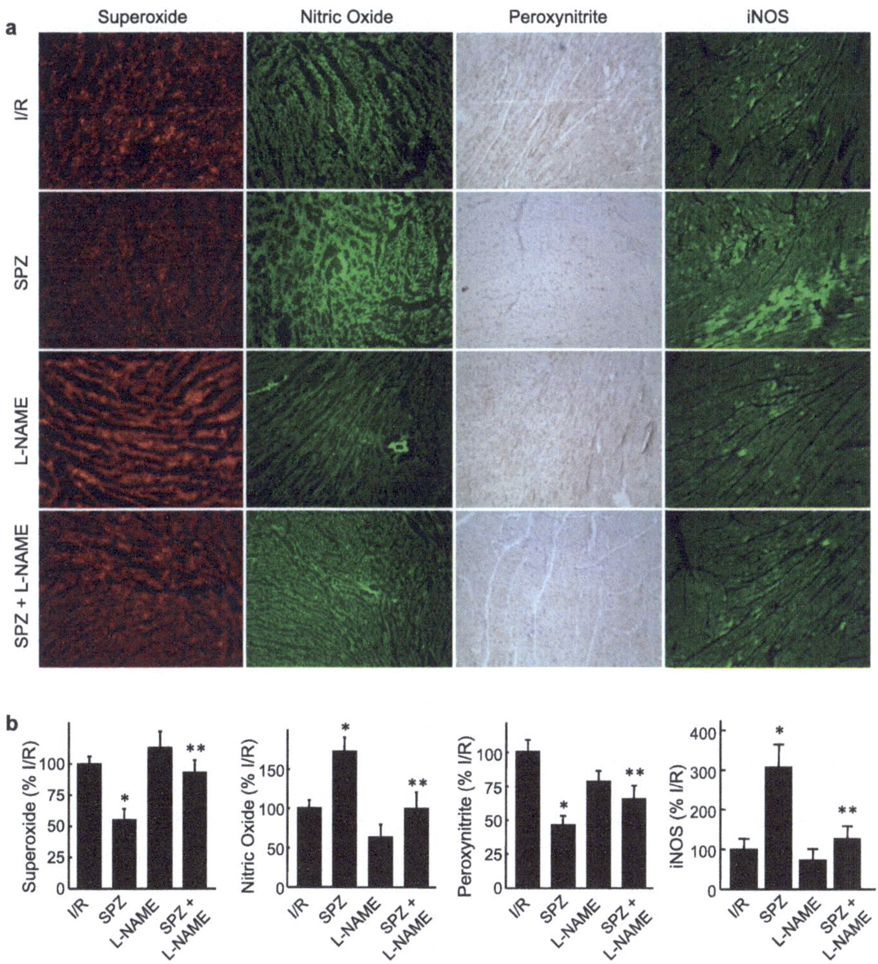

Fig. 9.5 Effect of SPZ on the tissue levels of superoxide, nitric oxide, peroxynitrite, and iNOS in the infarct heart. Superoxide and nitric oxide levels in the excised heart tissue were determined by histochemical staining and fluorescence microscopy at 10 min of reperfusion. Peroxynitrite and iNOS levels in the excised heart tissue were determined by immunohistochemical staining and fluorescence microscopy at 24 h of reperfusion. (**a**) Representative images (at 200x magnification) of superoxide, nitric oxide, peroxynitrite, and iNOS from tissue sections obtained from hearts treated with SPZ±L-NAME. (**b**) Mean fluorescence intensity from triplicate hearts. Data represent mean±SD. $^{*}p < 0.05$ versus I/R group; $^{**}p < 0.05$ versus SPZ group. (Reproduced from Khan et al. [36])

The influence of tumor oxygenation on the treatment outcome has stimulated the development of a variety of methods for measuring tumor oxygenation [123–125]. These methods include: paired-survival curve assays of hypoxic fractions [126], radiation-induced DNA damage measured by "Comet" assay [127], cryospectrophotometric measurements of hemoglobin oxygen saturation [128], immunohistochemical detection of nitroimidazole binding [129, 130], polarographic oxygen electrodes [131], fluorescent and phosphorescent probes based on oxygen-quenching [132, 133], and magnetic resonance methods [20, 24, 134–137]. The polarographic electrode, although invasive, provides a direct measurement of the concentration of oxygen (pO_2). It is the only device (the "pO_2 histograph") approved for use in patients. The magnetic resonance-based methods, such as nuclear magnetic resonance (NMR) and electron paramagnetic resonance (EPR), have the advantage of noninvasive measurement and imaging of the oxygen concentration in tissues [20, 39, 138, 139].

We used EPR oximetry to determine the oxygen concentration in experimental tumors [46]. Unlike the existing methods of oxygen measurement, wherein the probes (needle electrodes, optical probes, or EPR implants) are physically inserted during measurement, we used nanoprobes that were permanently embedded in the tumor [43]. This was done by labeling the cancer cells with the oxygen-sensing nanoprobes at the time of implantation in the animal model. A particular advantage of this procedure is that it is noninvasive, both in terms of implantation of the probe, as well as in obtaining readouts of oxygen concentration. Further, EPR imaging could be performed to map the spatial distribution of oxygen concentration in the tumor. The cellular labeling procedure, although not applicable to preexisting tumors, is useful for studying experimental tumors. The goal of the present study was to demonstrate the feasibility of noninvasive mapping of changes in oxygen concentration in growing experimental tumors subjected to radiation treatment. The study was performed on implanted tumors in C3H mice using radiation-induced fibrosarcoma (RIF-1) cells, labeled with oxygen-sensing nanoprobes [46]. The procedure used electron paramagnetic resonance imaging (EPRI) of oxygen-sensing nanoprobes embedded in the tumor cells. Typical images obtained from a tumor labeled with the spin probe at two different time points of tumor development, namely the 5th and 9th day, after inoculation are shown in Fig. 9.6a, b. The image represents the distribution of the spin probe whose voxel intensity was above the detection threshold. It was observed that the area of distribution of the probe was smaller than the size of tumor. For example, on day 5 (Fig. 9.6a), the probes were detected in about 62% of the tumor. Oxygen imaging, obtained from the same tumor at the same orientation, showed a very heterogeneous distribution of oxygen levels. The mean pO_2 value of the tumor on day 5 (volume: 86 mm^3) was 9.1 mmHg (median: 7.5 mmHg) and on day 9 (volume: 113 mm^3) was 7.4 mmHg (median: 4.9 mmHg). The data suggested a decrease in pO_2 during the tumor growth period. This was also evident from the left-shift of the values on day 9 (skew: 1.25) as compared

Fig. 9.6 (continued) (95% CO_2 + 5% CO_2). The images (10 mm × 10 mm) were obtained on day 12 after inoculation (tumor volume = 125 mm^3) of a mouse with RIF-1 cells internalized with the nanoparticulate oxygen probes. The left panels show the distribution of the probe in the core of the tumor. Oxygen information is obtained only from regions where the particulates are present. The pO_2 image of the air-breathing mouse shows the presence of hypoxic regions in the tumor, which become more oxygenated during carbogen breathing. The histograms reveal a significant increase in tumor oxygenation upon carbogen treatment. (Reproduced from Bratasz et al. [255])

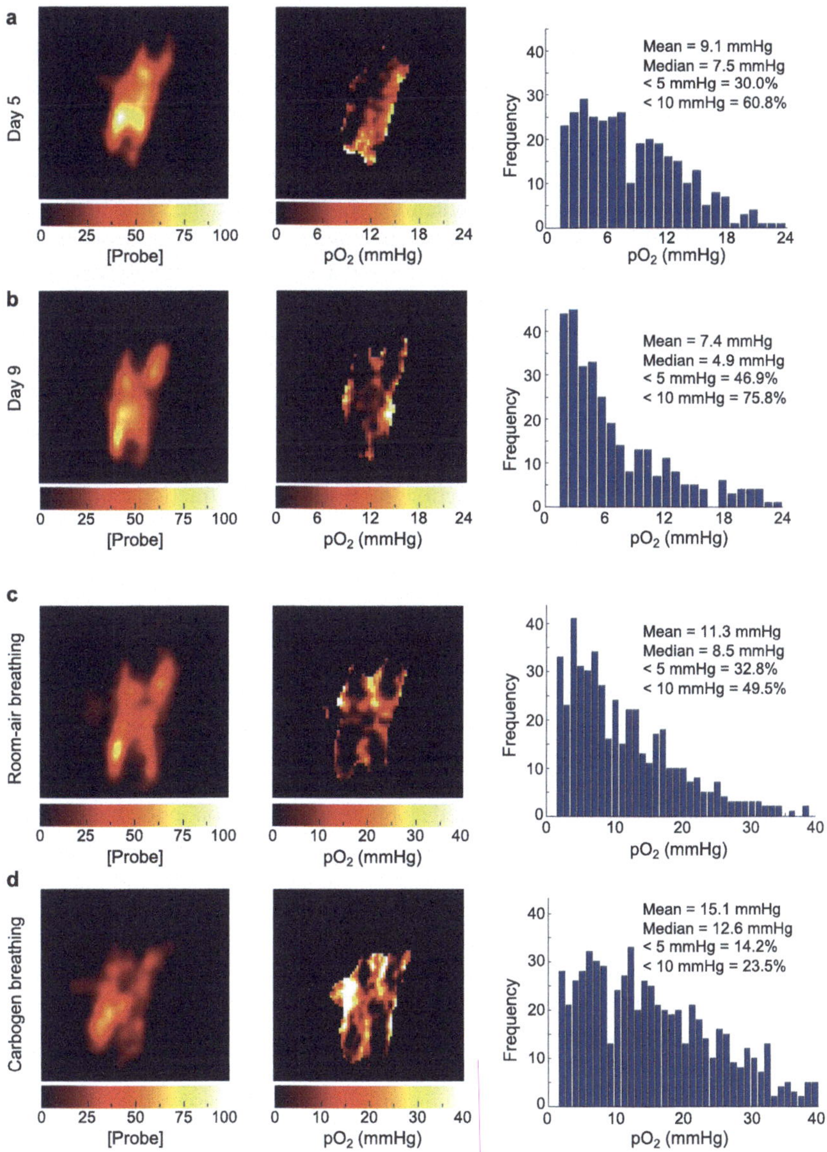

Fig. 9.6 Representative images of oxygen probe and oxygen distribution in a growing RIF-1 tumor. Mice were inoculated with RIF-1 cells internalized with the nanoparticulate oxygen probes. (**a**) and (**b**) show representative example of 2D images (10 mm × 10 mm) of probe and pO_2 distribution, and pO_2 histogram, obtained on day 5 (tumor volume: 86 mm^3) and day 9 (tumor volume: 113 mm^3), respectively, after cancer-cell inoculation. The probe is seen to be distributed in the core of the tumor which is about 62 ± 9% of the tumor volume on day 5, and 42 ± 6% on day 9. Note that the oxygen information is obtained only from the regions where the particulates are present. The pO_2 images show the presence of significantly hypoxic regions in the tumor under examination. (**c**) and (**d**) show representative 2D images (10 mm × 10 mm) of probe distribution and oxygenation (pO_2) in a RIF-1 tumor taken when a tumor-bearing mouse breathed room air or carbogen

to the distribution on day 5 (skew: 0.63). The images showed a considerable increase in the hypoxic fractions (<5 mmHg: 30.0–46.9%; <10 mmHg: 60.8–75.8%) during this period. It should be noted that the distribution of the probe within the tumor did not change significantly between the two time periods. Carbogen (95% O_2 + 5% CO_2) breathing is known to enhance blood flow and hemoglobin-O_2 saturation in the tumor and hence increase tissue pO_2. In order to map the spatially resolved changes in the tumor pO_2, the imaging measurements were repeated on the same tumor when the mice were breathing room air or carbogen (Fig. 9.6c, d). While the probe distribution was intact, except for some minor degradation of image quality due to line-broadening, the oxygen map showed that carbogen breathing by the mice increased the overall tumor pO_2 and that the oxygen distribution further showed a significant increase in the mean pO_2 value. The mean pO_2 values of the tumor before and after the mice breathed carbogen were 11.3 mmHg (median 8.5 mmHg) and 15.1 mmHg (median 12.6 mmHg), respectively. The increase in pO_2 after carbogen-breathing by the mice was also evident from the right-shift of the pO_2 histogram (Fig. 9.6d). The data also showed a substantial reduction in the hypoxic fractions (<5 mmHg: 32.8–14.2%; <10 mmHg: 49.5–23.5%) on carbogen-breathing.

In order to determine the response of pO_2 to tumor treatment, tumors were irradiated on the 9th or 11th day following inoculation of tumor cells internalized with the particulates. Figure 9.7a, b show the oxygen maps obtained from a small tumor (volume: 113 mm^3) with internalized particulates before and 1 h after 30-Gy irradiation [46]. The mean and median pO_2 value of the tumor before treatment was 10.6 and 7.0 mmHg (respectively), which increased to 12.5 and 8.0 mmHg after irradiation. The increase was also evident from the right-shift of the pO_2 values in the frequency plot before treatment (skew: 1.17) as compared to the distribution after treatment (skew: 0.82). The data showed a significant reduction in the hypoxic fractions (<5 mmHg: 32.5–28.1%; <10 mmHg: 59.1–53.8%) following irradiation. Figure 9.7c–e show the distribution of oxygen obtained from a large tumor (volume: 327 mm^3) at 1.5 h and 7.2 h after irradiation. The mean pO_2 value of the tumor before treatment was 6.8 mmHg (median: 6.7 mmHg), which showed a decrease (mean: 4.2 mmHg, median: 2.5 mmHg) at 1.5 h followed by an increase (mean: 5.9 mmHg, median: 6.2 mmHg) at 7.2 h after irradiation. The images as well as the frequency plots clearly showed that there were substantial alterations in the oxygen concentration in the core region of the large tumor during the first several hours following irradiation. The data showed a substantially large increase in the "<5 mmHg" hypoxic fraction (26.8–70.8%) and a significant increase in the "<10 mmHg" fraction (78.0–91.5%) at 1.5 h post-irradiation period. However, the values decreased (<5 mmHg: 36.2%; <10 mmHg: 84.1%) at 7.2 h following irradiation.

Fig. 9.7 (continued) are shown for a small (volume: 113 mm^3) RIF-1 tumor (**a**) before (pre-) and (**b**) 1 h after (post-) X-ray irradiation. Data show a redistribution of pO_2 with modest increase in the pO_2 values. Tumor oxygen levels are shown for a large (volume: 327 mm^3) RIF-1 tumor (**c**) before (pre-), (**d**) 1.5 h and (**e**) 7.2 h after X-ray irradiation. The data show a redistribution of pO_2 in the central core of the tumor. There is a significant decrease in the first 1.0–1.5 h followed by an increase at 7.2 h after irradiation. (Reproduced from Bratasz et al. [255])

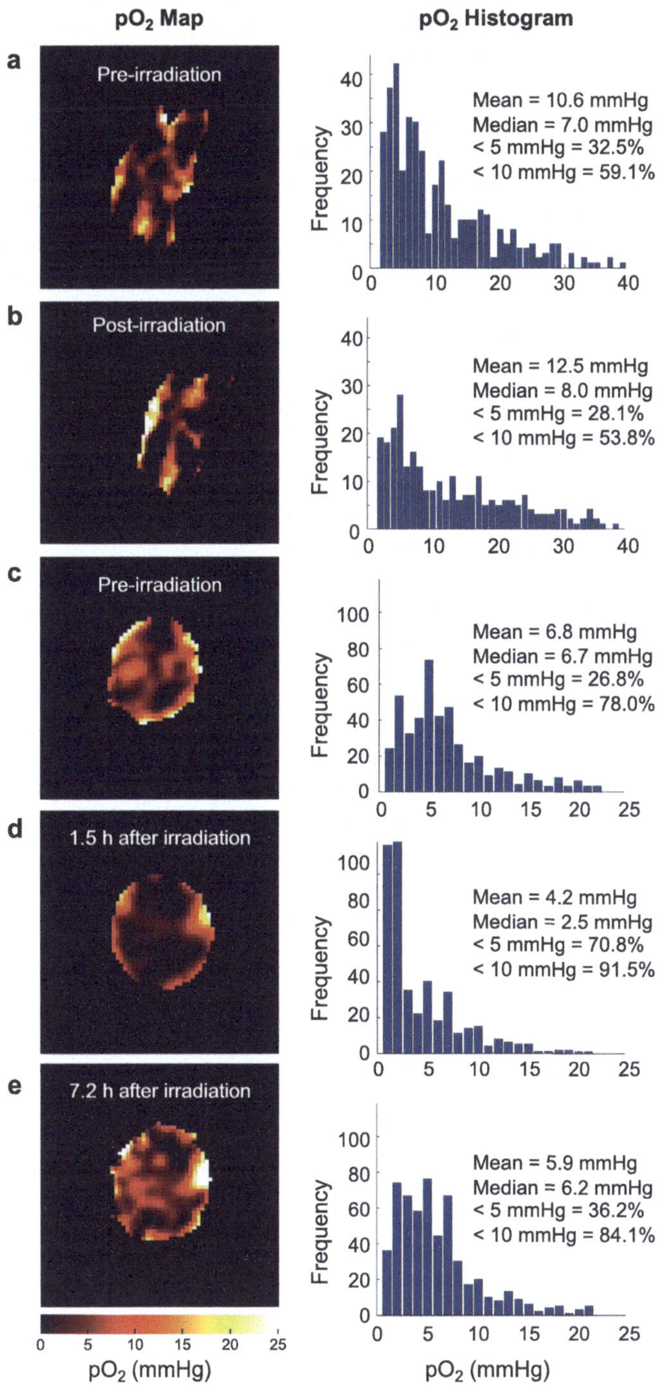

Fig. 9.7 Effect of X-ray irradiation on tumor oxygenation. The tumor was implanted with RIF-1 cells internalized with nanoparticulate LiNc-BuO probe. A dose of 30 Gy was delivered with 6 MeV electrons at a dose rate of 3 Gy/min. Change and redistribution of tumor oxygen levels

The RIF-1 tumor oxygenation data appear to be different from that described by Goda et al. [140]. After a 20-Gy radiation dose the initial pO_2 value of 8.7 ± 1.1 mmHg in RIF-1 tumor decreased in 24 h to 5.4 ± 1.2 mmHg and in 72 h increased to 8.2 ± 1.0 mmHg. The time course for these changes in pO_2 was found to be independent of the dose. In another work, it was shown that an increase in the pO_2 of the tumor can be associated with increased tumor perfusion [141], which can be related to the increase in nitric oxide production [142]. Olive et al. [143] have reported that 6 h after irradiation of a squamous cell carcinoma tumor (SCCVII) with a 10-Gy dose, the hypoxic fraction decreased from 18 to 6%. The authors concluded that the rapid reoxygenation of the tumor following exposure to 10-Gy radiation could be attributed to a decrease in oxygen consumption and an increase in tumor perfusion rather than to cells lost or redistribution. Knowledge of the changes in tumor pO_2 could be used as an indicator of the time to schedule multiple cycles of radiation to improve efficacy. In summary, the studies demonstrated that accurate measurement and imaging of cellular pO_2 can be performed in tumors using permanently embedded oxygen-sensing nanoprobes. The method enables repeated measurements of tissue oxygenation from the same site.

Oximetry in Animal Models Using OxyChips

EPR oximetry, based on PDMS-embedded LiNc-BuO crystals (OxyChip), has been used in animal models. Typically, the OxyChips (microChips) are implanted using a 23G syringe needle and the chip pushed into the tissue using a stylus [74, 144]. The oxygen measurements would be carried out 3–4 days after implantation. The following sections provide a few selected studies carried out using microChips.

Mice

We carried out several studies in mouse models—healthy, cancer, and cardiovascular diseases—to establish the probes for providing reliable oximetry. The studies included oxygen measurements in human triple-negative-breast-cancer xenograft tumors in mice for studying tumor oxygen and hyperoxygenation as a supplement for radiation treatment [144]; evaluation of multiple-implant MicroChips for the measurement of mean pO_2 in tissues [74] ; and pO_2 measurements in human pancreatic xenograft tumors using the MicroChips [145].

Rats

Evaluation and/or application of the oximetry probes in rat models included: Real-time, in vivo determination of dynamic changes in lung and heart oxygenation using EPR oximetry [146]; rat skeletal muscle oxygenation using OxyChip [64];

evaluation of OxyChip-EL (implantable resonator) for measurements of deep-tissue oxygen [75]; preclinical evaluation of OxyChip for long-term EPR oximetry [63]; protection of hearts against acute myocardial infarction by supplemental oxygenation [147]; and measurement of flap oxygen using EPR oximetry [69].

Rabbits

We have performed the following studies using rabbits as a preclinical model: Deep-tissue oxygen monitoring in the brain of rabbits with stroke [148]; dynamic changes in intracerebral oxygen tension during thromboembolism [65]; and measurement of pO_2 in rabbit tumors using OxyChip [68].

Oxygenation as a Therapeutic Adjuvant

Several studies have been performed to determine the beneficial role of hyperoxygen (supplemental oxygen) in cardiovascular and cancer applications [147, 149–151]. The following examples illustrate how administration of supplemental oxygen, that is gas mixtures containing more than 21% oxygen under normobaric or hyperbaric conditions.

Supplemental Oxygen Protects Heart Against Acute Myocardial Infarction

Myocardial infarction (MI) that occurs as a result of coronary artery disease (CAD) often results in progressive loss of viable cardiomyocytes and tissue damage leading to cardiac remodeling, dysfunction and eventual heart failure. Therapeutic approaches that could limit this long-term debilitating effect on the heart would be beneficial toward improving clinical outcomes post-MI. Administration of pure oxygen during and after a cardiac event to increase blood oxygen delivery to the affected heart tissue would reduce infarct size and salvage at-risk myocardial tissue. This, in turn, is expected to lead to improvement in functional recovery. Under most circumstances, provision of supplemental oxygen to suspected MI patients by emergency responders is routine [152]. Until recently, some guidelines recommended regular administration of oxygen for the treatment of MI [153, 154]. This position has changed, however, so that continued treatment with oxygen is only recommended under certain circumstances and not for cases of uncomplicated MI [152, 155]. It is likely that this change in treatment protocol is due to known hemodynamic side effects associated with hyperoxygenation. Hyperoxygenation of patients with acute MI results in a rise in arterial blood pressure and a reduction in cardiac output [156, 157]. These changes have been attributed to decreases in heart rate and stroke volume [158], and an increase in vascular resistance [159–161]. Hyperoxygenation is also a powerful stimulus of coronary circulation and vasoconstriction [162].

In addition to normobaric oxygen, hyperbaric oxygen (HBO) has also been investigated as a potential therapeutic measure for MI. A study by Cameron et al, reported that the hemodynamic effects of oxygen therapy in MI patients at normobaric pressure (1 ATA, atmospheres absolute) were enhanced upon an increase to 2 ATA [159]. In 1998, the "HOT MI" study attributed greater left-ventricular ejection fraction in the HBO-treated group to increased myocardial salvage when compared to the nontreated subjects [163]. In 1969, Ashfield and Gavey enrolled 40 volunteers who were treated with HBO cycling continuously for 4 days in periods of 2-h exposures to 100% O_2 at 2 ATA, followed by 1 h in room air at normobaric pressure [164]. They concluded that the use of frequent, intermittent sessions of hyperbaric oxygen at 2 ATA, during the acute phase of the circulatory crisis shows promise of being a significant advance in treatment [164].

Our group has reported that brief periods of hyperbaric oxygen cycling (OxCy; 100% O_2; 2 ATA pressure; 90 min/day; 5 days/week for 4 weeks) enhanced the retention of transplanted mesenchymal stem cells (cardiomyoplasty) and improved cardiac function in a rat model of MI induced by ischemia-reperfusion [149]. Comparisons were made to MI hearts receiving stem cell transplantation alone, and an additional MI group receiving hyperoxygenation alone. The data from these two comparative groups was intriguing, as both of these "control" procedures appeared to have near-equivalent benefit, albeit not as significant as combined therapy. Nevertheless, this prompted a number of questions, including whether 2 ATA pressure (hyperbaric) is required for the beneficial effect. We determined an optimum oxygenation protocol that can be clinically applicable for treating acute MI [147]. Using EPR oximetry, we studied the effect of exposure to supplemental oxygen cycling (OxCy) administered by inhalation of 21–100% oxygen for brief periods (15–90 min), daily for 5 days, using a rat model of acute MI. Myocardial oxygen tension (pO_2), cardiac function and pro-survival/apoptotic signaling molecules were used as markers of treatment outcome. Myocardial pO_2 was measured by in vivo EPR oximetry in healthy (non-infarcted) and infarcted hearts of rats subjected to HBO. The mean baseline pO_2 in healthy hearts before HBO was 21.1±2.2 mmHg (Fig. 9.8). Immediately following HBO, there was a decrease in myocardial pO_2 at 10 min; however, the pO_2 recovered and showed a significant increase, to 30.0 ± 3.2 mmHg or 42.9% beyond the baseline level, at 60 min after termination of HBO. The pO_2 level returned to baseline value in about 2 h after HBO. Similarly, in HBO-treated infarct rat hearts the myocardial pO_2 was substantially elevated; for example, at 90-min post-HBO the pO_2 level was significantly higher (15.7 ± 3.1 mmHg) compared to the baseline (2.7 ± 0.8 mmHg). The myocardial pO_2 remained significantly elevated up to 2 h in the infarcted region after HBO treatment. The pO_2 results indicated that exposure of rats to HBO temporarily enhanced myocardial oxygenation in the normal and infarcted myocardium.

The effect of increasing concentrations of inspired oxygen on myocardial pO_2 in healthy (non-MI) rats was determined [147]. The myocardial pO_2 increased from a

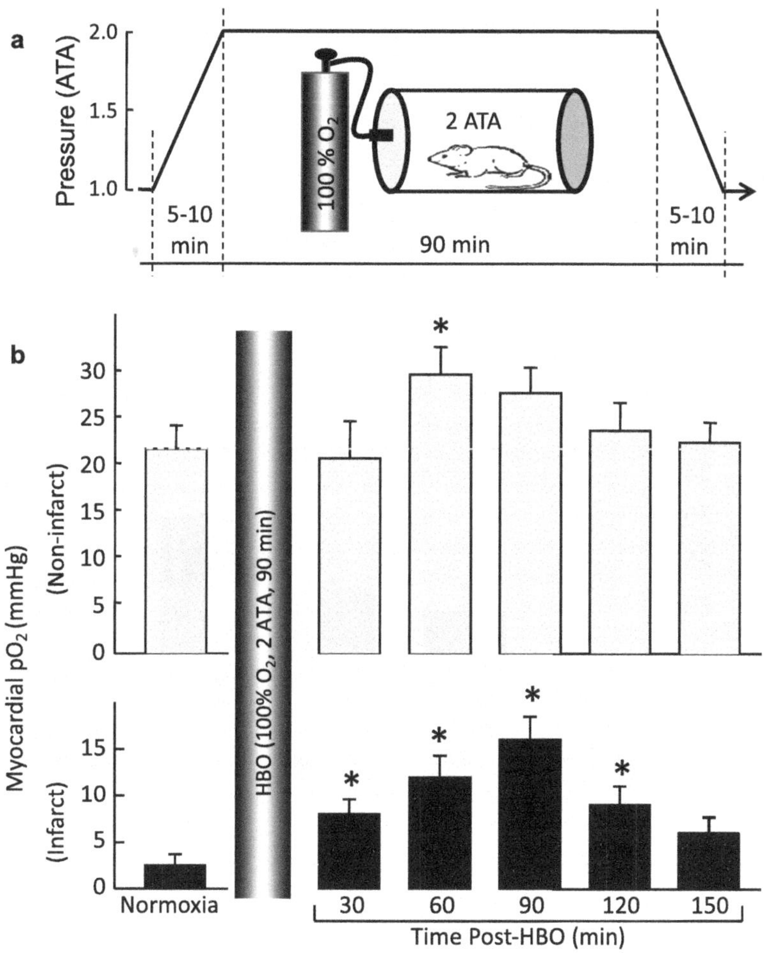

Fig. 9.8 Hyperbaric oxygenation (HBO) and myocardial pO_2, in vivo. (**a**) The HBO protocol used 5-min compression, 90-min maintenance of hyperbaric oxygenation (100% O_2 at 2 ATA), followed by 5-min decompression using a custom-built small-animal chamber. (**b**) Myocardial pO_2 in (non-infarct and infarct) rat hearts measured before (normoxia) and after HBO. Data represent mean ± SD obtained from four rats. The results show an increase in oxygenation levels reaching significance at 60 min followed by returning to baseline in about 2.5 h in the non-infarct hearts. In the infarct hearts the oxygenation levels are significantly higher up to 2 h post-treatment. It is interesting to note that a 5.8-fold increase in oxygenation is achieved in the infarct hearts after 90 min of HBO. (Reproduced from Khan et al. [150])

baseline value of about 14 mmHg to levels peaking around 28 mmHg during exposure to carbogen (95% O_2/5% CO_2) (Fig. 9.9a). Myocardial pO_2 reached its peak in about 12 min after start of carbogen administration and remained elevated above the baseline (room air) level after termination of carbogen and reexposure to room air. We next determined the peak levels of myocardial pO_2 observed during exposure of

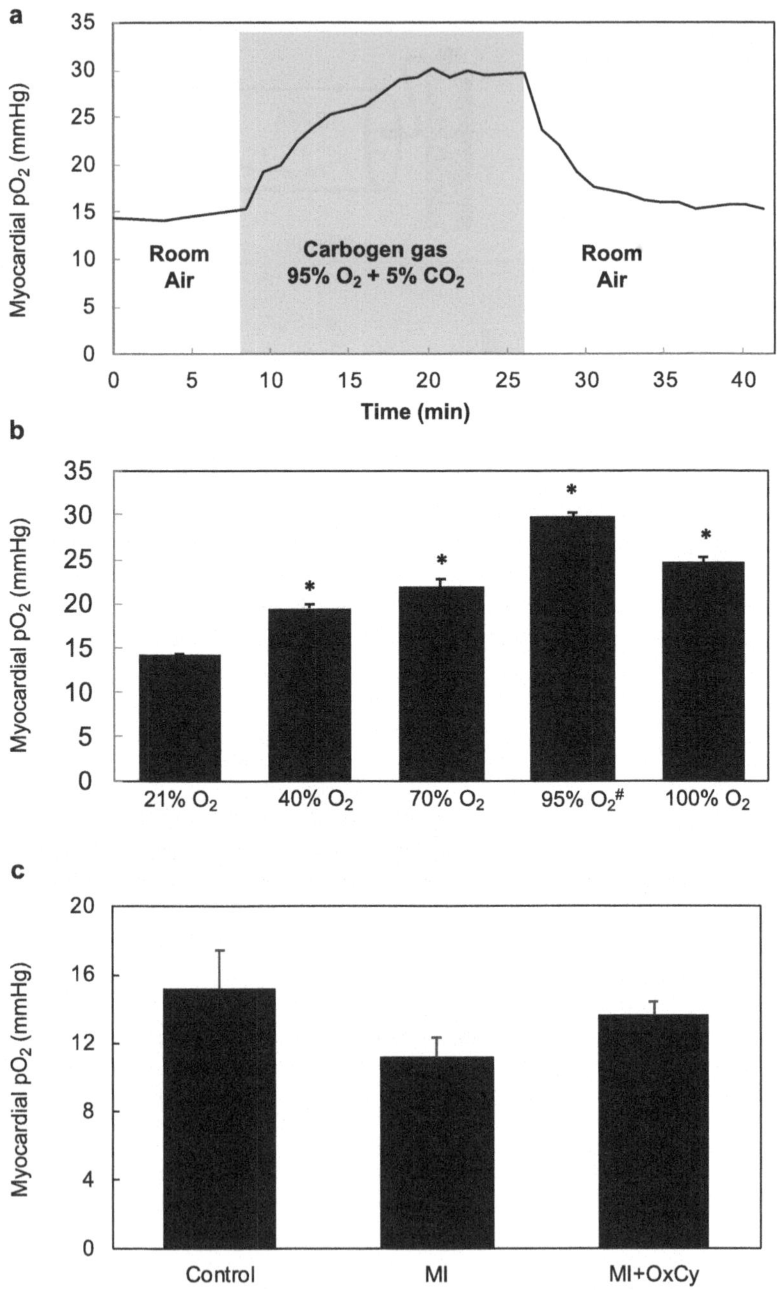

Fig. 9.9 Myocardial pO_2 in rat hearts subjected to hyperoxygenation. LiNc-BuO oxygen-sensing crystals were implanted in the left ventricular myocardium. Myocardial pO_2 was measured using EPR oximetry. Rats were intubated and allowed to breathe gas mixture containing the respective O_2 levels along with 1.5% isoflurane anesthesia during EPR measurements. (**a**) Effect of carbogen

healthy (non-MI) rats to hyperoxygen ranging from 40 to 100% O_2. The results showed a concentration-dependent increase in peak myocardial oxygenation, while carbogen exposure showed a significantly higher increase when compared to all other groups. Interestingly, exposure to 100% O_2 showed a significant decrease in peak oxygenation when compared to carbogen (Fig. 9.9b). Overall, the results indicated that exposure of rats to hyperoxygen increased myocardial pO_2 during the period of administration. Oxygen-cycling in rats at 1-week post-MI showed no significant differences in the myocardial pO_2 values among the control, MI, and MI hearts treated with oxygen (MI+OxCy) (Fig. 9.9c). To determine the effect of supplemental oxygen (oxygen cycling) on cardiac function, M-mode echocardiography was used on day 9 after induction of MI. Significant decreases in ejection fraction and fractional shortening were observed in the infarct (MI) hearts that did not receive oxygen-cycling treatment (Fig. 9.10). Rats treated with all levels of hyperoxygenation, both at normobaric and hyperbaric pressures, showed variable levels of recovery of cardiac function depending on the inspired oxygen level. Rats subjected to hyperoxygenation (100% O_2) at 1.5 or 2 ATA pressure, 1 h/day for 5 days, demonstrated notable recovery of cardiac function; however, improvement at these hyperbaric levels was not significantly better than that at ambient pressure. In fact, their respective ejection fraction and fractional shortening means were less than those at 1 ATA. Interestingly, rats subjected to either normobaric 100% O_2 or 95% O_2/5% CO_2 (carbogen) for the same time duration showed comparable levels of cardiac functional recovery after MI. Recovery of MI hearts treated with 1 h of carbogen were not significantly different when compared to hearts treated with 1 h of 100% oxygen. In addition, the results of carbogen treatment further revealed that 30 min/day administration for 5 days did not show any significant difference in the cardiac function when compared to 1 h/day carbogen administration. However, 15 min/day of carbogen treatment did not yield similar functional improvement and it resulted in significantly less ejection fraction and fractional shortening when compared to 30 min/day treatment. An optimal condition of 30-min OxCy with 95% oxygen + 5% CO_2 under normobaric conditions was found to be optimal for cardioprotection. This study also showed that HBO administered at 1.5 or 2 ATA did not provide significantly better results with regard to recovery of cardiac function when compared to 100% oxygen treatment administered at ambient pressure [147]. For this reason, it was concluded that hyperbaric conditions, while beneficial overall, is

Fig. 9.9 (continued) breathing on myocardial oxygenation in healthy hearts. The myocardial pO_2 increases from ~15 mmHg baseline level to about 30 mmHg after 18 min of carbogen administration and returned to baseline up on withdrawal of carbogen. (**b**) Peak myocardial oxygenation after about 20 min of administration of hyperoxygen gas mixed with nitrogen ($^{\#}$mixed with 5% CO_2). Statistically significant increase in peak myocardial pO_2 is observed in all hyperoxygen groups when compared to baseline (21% O_2) group ($^{*}P < 0.01$). Carbogen administration shows the largest increase. Data represent Mean±SD ($N = 3$). (**c**) Myocardial pO_2 in rat hearts subjected myocardial infarction (MI) and infarcted group treated with OxCy using carbogen for 60 min/day for 5 days (MI+OxCy), measured at 1-week post-MI. MI was not induced in the control hearts. Data represent Mean±SD ($N = 3$). There were no statistically significant differences between the groups. (Reproduced from Prabhat et al. [147])

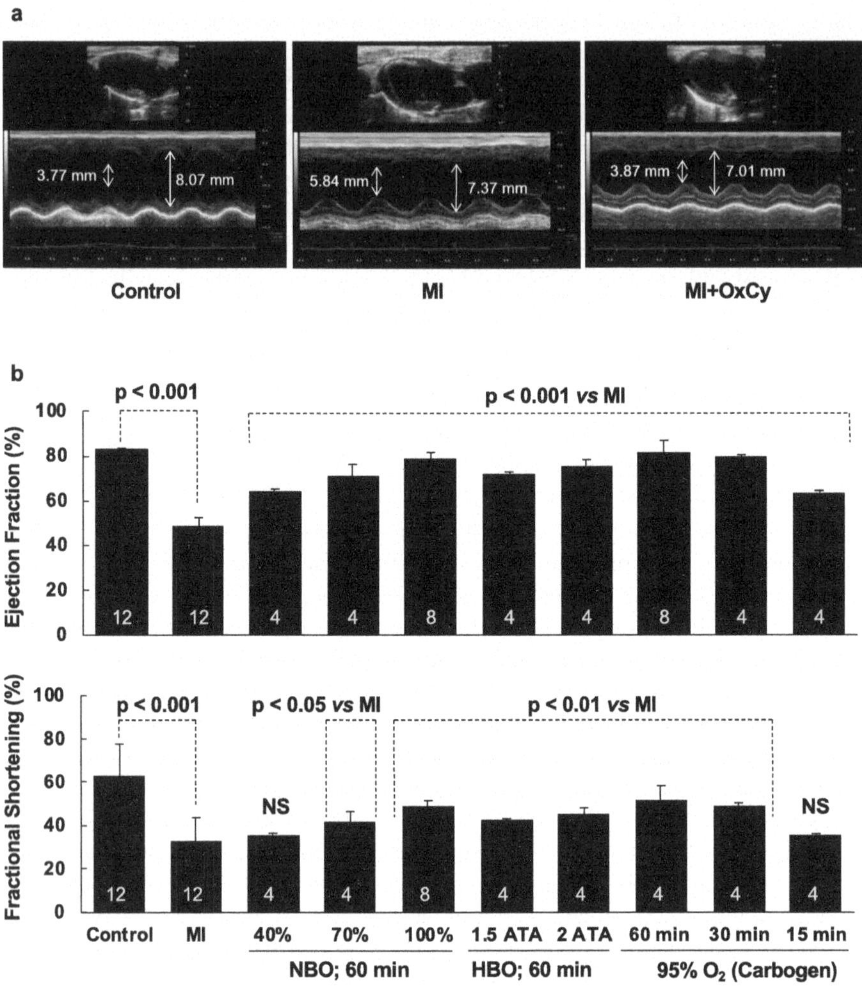

Fig. 9.10 Recovery of cardiac function of MI hearts following supplemental oxygen (oxygen cycling, OxCy) treatment. Cardiac function was measured using transthoracic M-mode echocardiography. (**a**) Representative echocardiogram of Control, MI, and MI hearts treated with OxCy using 95% oxygen (carbogen) for 60 min/day for 5 days. Systolic and diastolic diameters are shown on the images. (**b**) Left ventricular ejection fraction and fractional shortening are shown for non-MI hearts (Control), untreated MI hearts (MI), MI hearts treated with OxCy for 1 h/day with 40% O_2 (40%), 70% O_2 (70%), 100% O_2 (100%), 100% O_2 at 1.5 ATA (1.5 ATA), 100% O_2 at 2 ATA (2 ATA), 95% O_2/5% CO_2 (60 min), 30 min/day with 95% O_2/5% CO_2 (30 min), and 15 min/day with 95% O_2/5% CO_2 (15 min). Data represent Mean±SD. Number of hearts per group is shown on the bar. Statistical significance, as indicated; NS indicates not significant when compared to MI group. (Reproduced from Prabhat et al. [147])

not entirely necessary or even practical for oxygen therapy, especially considering a patient in a clinical setting would need to be placed inside a chamber to synthesize such an environment. Carbogen, a mixture of 95% O_2 and 5% CO_2, was found to enhance myocardial tissue oxygenation to the greatest extent, and this increase was notably greater than when 100% oxygen was used. This is most likely due to two well-documented hemodynamic responses: hyperoxygenation promotes vasoconstriction and CO_2 is a potent vasodilator. From these results, it became obvious that the vasodilatory effect of CO_2 outweighs the vasoconstrictive effect of hyperoxygenation under carbogen breathing. If the foremost objective of supplemental oxygen therapy as an adjuvant measure for MI patients is to increase myocardial tissue oxygenation, then the use of carbogen, instead of 100% oxygen, should be considered.

Testing various treatment conditions led to the conclusion that carbogen breathing at ambient pressure is arguably the ideal scenario for treating MI. Administration of carbogen for 1 h led to improved ejection fraction and fractional shortening. Furthermore, it is important to note that the 30- and 60-min carbogen-breathing groups did not show any significant differences in cardiac function, which suggests that a half-hour of carbogen treatment is sufficient to produce substantial improvements. In conclusion, the oxygen-cycling therapy serves as a very appealing option for myocardial infarction treatment because it yields some of the greatest benefits while also minimizing treatment time and inconvenience for the subject.

Oxygenation Enhances Transplanted Stem Cell Graft and Functional Recovery in the Infarct Heart

Stem cell therapy for treating myocardial infarction is a viable option but faced with several challenges. Hypoxia, which commonly occurs in ischemic/infarct myocardium, can lead to the production of oxygen free radicals. The hostile environment with persistent oxidative stress ultimately leads to apoptosis of the majority of the transplanted cells. Therefore, a straightforward physiological approach to reduce the severity of hypoxia is to oxygenate the infarct tissue following cell transplantation. Hyperbaric oxygenation (HBO) has been used as a primary therapy in patients with decompression sickness, arterial gas embolism and carbon-monoxide poisoning [165]. It is also used as an adjuvant therapy to promote wound healing [166], and for the treatment of various conditions, including ischemic injury [167]. HBO involves inhalation of 100% oxygen under greater-than-one atmospheric absolute (ATA) pressure. Such doses of oxygen have a number of beneficial biochemical, cellular, and physiologic effects [168]. HBO, administered at the onset of reperfusion in an open-chest rabbit model of myocardial ischemia-reperfusion injury, showed a significant reduction in infarct size [169]. Other studies have shown that HBO attenuates myocardial injury via nitric-oxide signaling [168], improves cardiac function in patients with acute myocardial infarction [170], and helps mobilization of stem cells by enhancing CXCR4 and VEGFR-2 in humans [171]. However, the efficacy of HBO as an adjuvant to cell therapy has not yet been studied.

The efficacy of HBO as an adjuvant to cell therapy has been evaluated in a rat model of MI. Myocardial infarction was induced in Fisher-344 rats by permanently ligating the left-anterior-descending coronary artery [150]. Bone-marrow-derived mesenchymal stem cells (MSCs) were transplanted in the infarct and peri-infarct regions of the MI hearts. HBO (100% oxygen at 2 ATA for 90 min) was administered daily for 2 weeks. Four MI groups were used: untreated (MI); HBO; MSC; MSC+HBO. Echocardiography, electro-vectorcardiography, and magnetic resonance imaging were used for functional evaluations. The engraftment of transplanted MSCs in the heart was confirmed by SPIO fluorescence and Prussian-blue staining. Immunohistochemical staining was used to identify key cellular and molecular markers including CD29, troponin-T, connexin-43, VEGF, α-smooth-muscle actin, and von-Willebrand factor in the tissue. Compared to MI and MSC groups, the MSC+HBO group showed a significantly increased recovery of cardiac function including left-ventricular (LV) ejection fraction, fraction-shortening , and QRS vector [150]. The effect of HBO and MSC transplantation on myocardial fibrosis and left-ventricular wall thickness (LV-WT) was assessed in hearts excised 2 weeks after stem-cell therapy. The results (Fig. 9.11) showed a significant reduction in tissue fibrosis and recovery of LV-WT in both the MSC and MSC+HBO groups when compared to MI group. Combined treatment of MSC and HBO resulted in a significant attenuation of fibrosis and restoration of LV-WT when compared to MSC alone treatment. Further, HBO treatment significantly increased the engraftment of CD29-positive cells, expression of connexin-43, troponin-T and VEGF, and angiogenesis in the infarct tissue. Immunostaining for VEGF expression was performed in heart sections at 2 weeks post-MSC transplantation. The images showed higher levels of VEGF expression in the infarct tissue of the MSC and MSC+HBO groups when compared to the MI group (Fig. 9.12). The VEGF expression was significantly higher in the MSC+HBO group when compared to MSC group. Since the increased VEGF expression group may enhance angiogenesis and new blood vessel formation in the HBO-treated hearts, blood vessels and capillary density in heart sections at 2 weeks post-MSC transplantation were evaluated. Cardiac tissue sections were stained using anti-α-SM-actin and anti-vWF antibodies to identify mature blood vessels and microcapillaries, respectively (Fig. 9.13). The results showed increased vasculogenesis and microcapillary formation in the MSC+HBO group, in both infarct and peri-infarct areas, when compared to the MI and MSC groups. Thus, HBO appears to be an effective and clinically applicable method to improve the survival and retention of stem cells used in the treatment of myocardial infarction and/or heart failure, thereby improving therapeutic efficacy and overall clinical outcome.

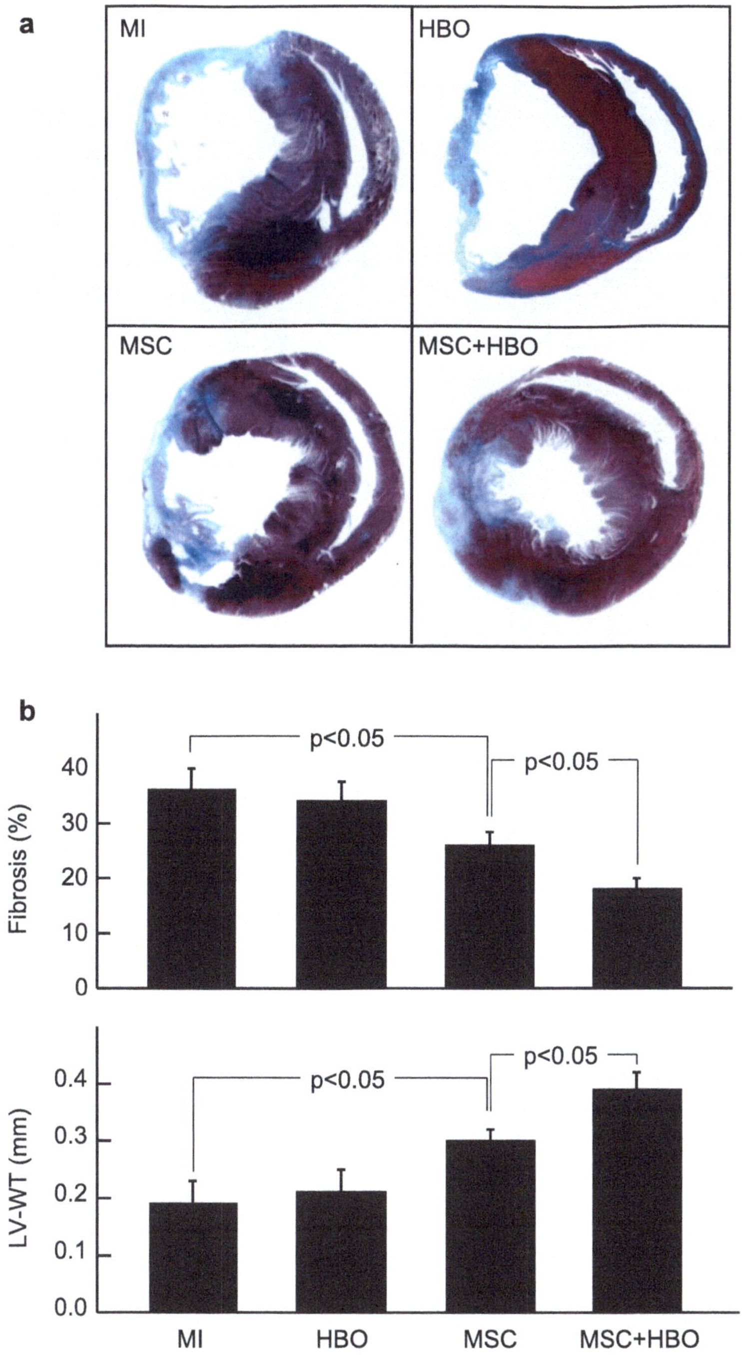

Fig. 9.11 Cardiac tissue fibrosis and remodeling in MI hearts treated with MSC and HBO. Masson-trichrome staining of heart sections was performed in infarcted hearts (**MI**), and infarcted hearts treated with HBO (**HBO**), MSC (**MSC**) or MSC followed by HBO (**MSC+HBO**) at 2 weeks after transplantation of MSCs. (**a**) Representative images of heart sections stained with Masson-trichrome. (**b**) Percentage of fibrosis and LV wall thickness (LV-WT), as determined by computer planimetry. Data represent mean ± SD obtained from six hearts per group. The **MSC+HBO** group exhibits a significant reduction in the fibrosis and improvement in LV-WT when compared to **MSC** group. (Reproduced from Khan et al. [150])

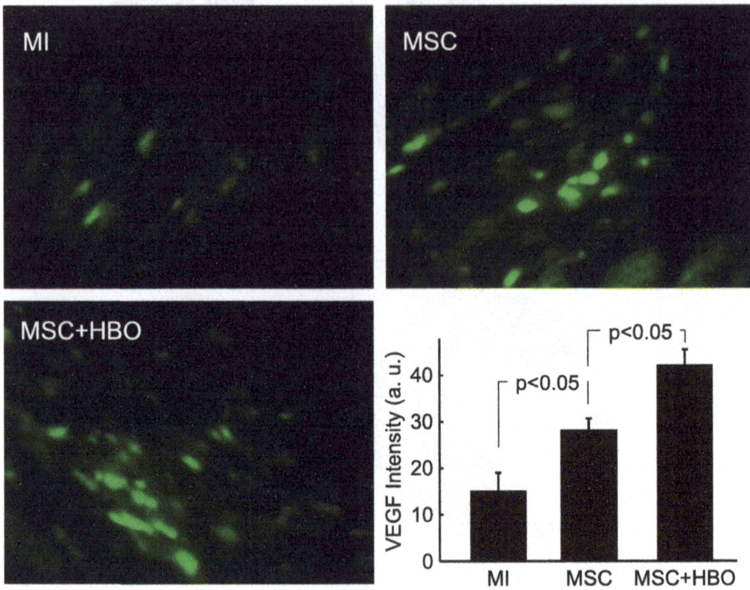

Fig. 9.12 VEGF expression in the infarct heart 2 weeks after MSC transplantation. Heart sections were stained with a VEGF antibody to identify the VEGF expression. Shown are VEGF immunofluorescence images (green) in sections from **MI**, **MSC**, and **MSC+HBO** hearts and quantitative assessment of the VEGF fluorescence intensity. Data represent mean±SD obtained from three hearts. The VEGF level is significantly higher in the **MSC+HBO** group when compared to the **MSC** group. (Reproduced from Khan et al. [150])

Oxygen Cycling Induces NOS3 Expression Leading to Attenuation of Fibrosis and Improved Cardiac Function

Nitric oxide (NO) plays a crucial role in the regulation of coronary vasodilatory tone [172], reduces post-ischemic hyperpermeability [173, 174], inhibits platelet adhesion and aggregation [175], neutrophil adherence, migration and associated injury[176], and impairs mast cell activation [177]. NO also exerts its beneficial effects by inhibiting neutrophilic generation of superoxide [178] through inhibition of membrane-bound nicotinamide adenine dinucleotide phosphate oxidase activity [179]. The administration of NO donors prior to ischemia has been shown to attenuate myocardial ischemia-reperfusion (I-R) injury, leading to reduced infarct size and endothelial dysfunction [180]. A study in a rat heart model of I-R injury demonstrated that hyperbaric oxygenation (HBO) is capable of stimulating NO release via increased NOS3 [181]. Buras et al. [182] reported that HBO, at 2.5 absolute atmospheres (ATA) for 90 min upon reperfusion, was capable of inducing NOS3, but not NOS2 in endothelial cells of an in vitro model of I-R injury.

We investigated the effect of oxygenation, stem cell transplantation, and the combination of both, on NOS3 gene expression in infarcted heart tissue [149]. The

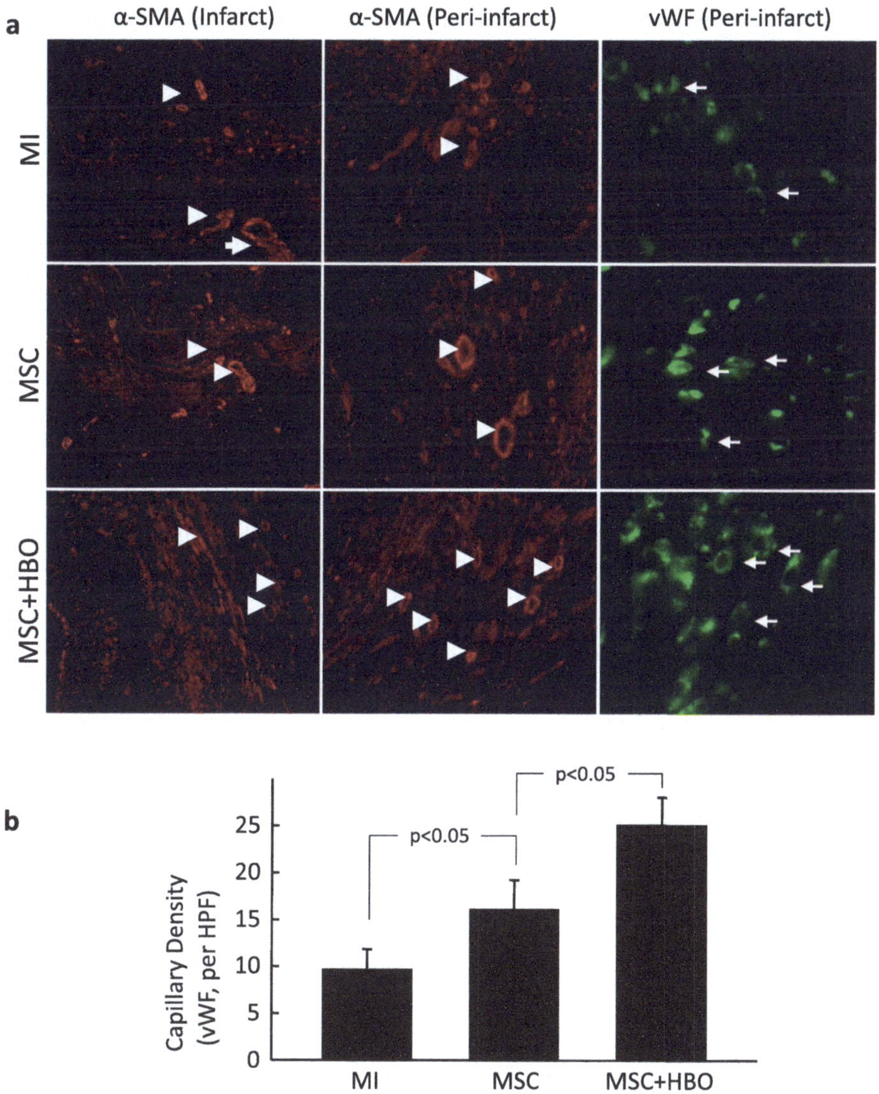

Fig. 9.13 Blood vessel and capillary densities in the infarct hearts 2 weeks after MSC transplantation. Blood vessels and capillaries in the tissue sections were imaged using antibodies for α-smooth-muscle actin (α-SMA) and von Willebrand Factor VIII (vWF), respectively. (**a**) Representative immunofluorescence images of tissue sections showing α-SMA (red, arrow heads) and vWF (green, arrows) in MI, MSC, and MSC+HBO groups. (**b**) Quantitative assessment of the total number of capillaries in the peri-infarct region of the heart. The images show a significantly increased number of capillaries in the peri-infarct region of MSC+HBO group when compared to MSC group. (Reproduced from Khan et al. [150])

expression of NOS3 mRNA, observed using RT-PCR, showed a basal level of NOS3 mRNA expression in the untreated control (MI) and infarcted hearts (MSC and Ox) (Fig. 9.14a). The expression of NOS3 mRNA was found to increase four-fold versus the basal level in rat heart tissues recovered from the MSC+Ox group. The results suggested that stem cell engraftment with hyperoxygenation induced NOS3 gene transcription, thus increasing NOS3 mRNA levels. Similar results were observed in NOS3 protein expression measured by Western blotting, where hyper oxygenation along with stem cell therapy were able to induce a four-fold increase in NOS3 protein expression as compared to other groups [149]. Oxygenation or stem cell therapy alone was not as effective in increasing NOS3 protein expression.

Immunoprecipitation of NOS3 protein from the heart tissue samples showed the combination of oxygenation and stem cell grafting was able to induce significant increases in NOS3 protein expression in the infarct heart. The expression of NOS3 enzyme is not the sole factor which determines NOS3 activation. NOS3 activity requires phosphorylation at Ser-1177, which is downstream of p53-dependent kinase signaling, and concomitant dephosphorylation of Thr-495 [183–186]. The effect of oxygenation- and stem cell therapy-based treatments on NOS3 phosphorylation at Ser-1177 and NOS3 dephosphorylation at Ser-495 was analyzed by immunoblotting. It was found that NOS3 was not phosphorylated at Ser-1177 in MI heart tissue. Oxygenation or stem cell therapy alone induced an increase in the phosphorylation of Ser-1177. The combined therapy of oxygenation and stem cell transplantation led to a 3.5-fold increase in the phosphorylation of the Ser-1177 residue of NOS3. In contrast, NOS3 was also found to be phosphorylated on the Ser-495 residue in infarcted hearts, indicative of an inactive state of NOS3 in the tissue. Oxygenation, stem cells, and the combination of both reversed NOS3 phosphorylation at the Ser-495 residue. Immunostaining for NOS3 in cardiac tissues treated with stem cells and oxygen-cycling showed a marked increase in NOS3 levels in MSC+Ox hearts (Fig. 9.14b), when compared to the MSC-alone treated group. Overall, the results established that myocardial infarction reduced NOS3 mRNA and gene expression in untreated infarcted heart tissue. Furthermore, the combination of oxygenation and stem cells led to a substantial increase in NOS3 gene and protein expression through regulation of NOS posttranslational modifications.

Supplemental Oxygen Protects Infarct Myocardium by Recruiting p53 on NOS3 Promoter

Administration of supplemental oxygen by oxygen cycling to a rat model of MI has been shown to protect cardiomyocytes from ischemia/reflow-induced death and is believed to function through upregulation of NOS3 expression [181]. Also, daily administration of oxygen (90 min/day for 4 weeks) to rats with experimentally induced MI resulted in a significant reduction of infarction and improvement of cardiac function [149]. The oxygen-cycling also improved engraftment of mesenchymal stem cells (MSC) transplanted in the infarcted myocardium. It was further

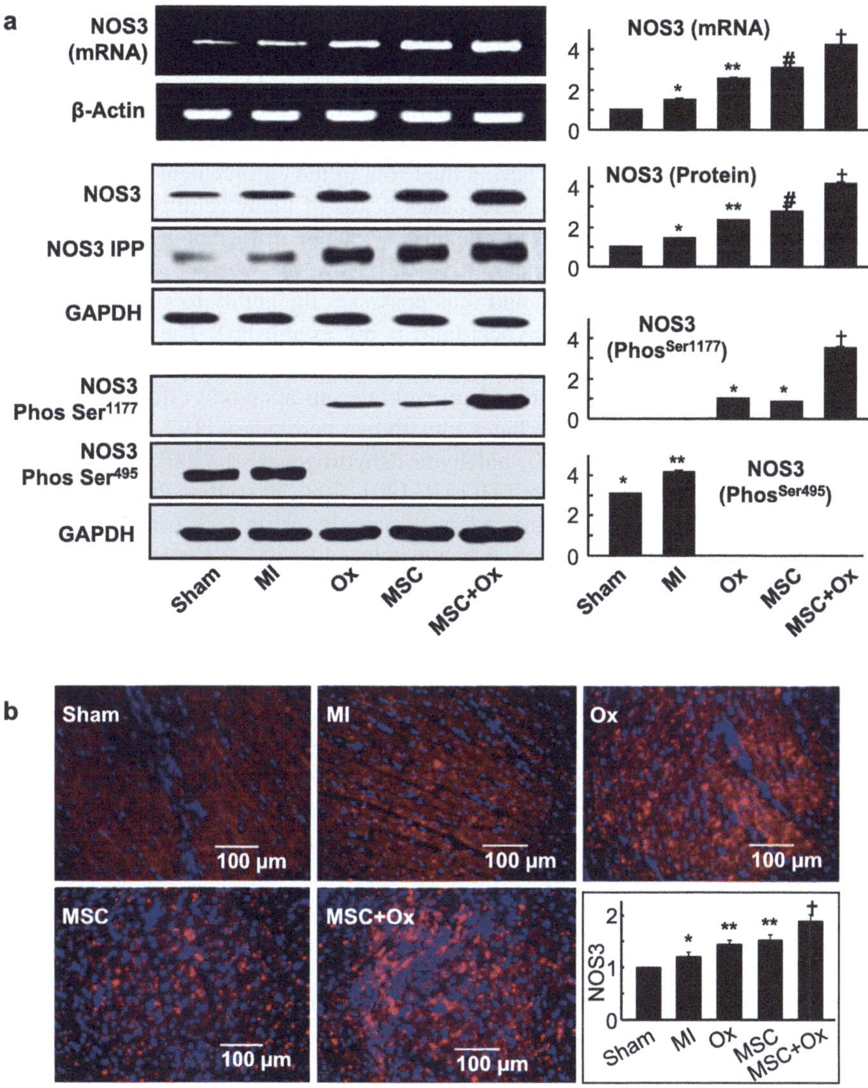

Fig. 9.14 NOS3 expression. The expressions of NOS3 gene and protein in the infarct hearts were measured. (**a**) NOS3 mRNA (by RT-PCR), NOS3 protein (by western blot and immunoprecipitation), NOS3 phosphorylation at Ser-1177 and NOS3 dephosphorylation at Ser-495 (by western blot). NOS3 was not phosphorylated at Ser-1177 in sham and MI hearts; however, NOS3 was phosphorylated in the treated groups. NOS3 was phosphorylated on the Ser-495 residue in sham and MI hearts, indicative of an inactive state of NOS3 in the tissue. $^{*}p < 0.05$ vs. Sham; $^{**}p < 0.05$ vs. MI; $^{\#}p < 0.05$ vs. Ox; $^{\dagger}p < 0.05$ vs. MSC). (**b**) Representative NOS3 immunostaining images and quantitative results expressed as a percent (mean±SD; $n = 6$) of Sham hearts. $^{*}p < 0.05$ vs. Sham; $^{**}p < 0.05$ vs. MI; $^{\dagger}p < 0.05$ vs. MSC). The results show a significant increase in NOS3 expression in the treated groups when compared to the MI group. (Reproduced from Khan et al. [149])

observed that endothelial nitric oxide synthase (eNOS or NOS3) was overexpressed both in the oxygenated and stem cell-treated hearts [149]. p53 is an established apoptotic effector in infarct heart [187, 188]. p53 and NOS3 have been shown to have clinical correlation [189]. p53 transcriptionally regulates other members of NOS family as well [190]. However, if p53 and NOS3 have a relation at the transcriptional level then p53 might have a dual role in the enforcement of "death" or "survival" in the infarct heart. p53 regulates nexus of many cellular pathways, it integrates abnormal signals and in response, induces arrest, apoptosis or DNA-repair in a context-dependent manner [191, 192]. However, whether p53 might have a role in cell-survival [193–195] and thus possesses the ability to support cardiac survival in oxygenated infarct myocardium is not established. The hypothesis is supported by the evidence that p53 positively regulates the expression of genes whose products are directly involved in evoking anti-apoptotic effects in cancer cells [196]. This list of genes includes glutathione peroxidase [197–199], manganese superoxide dismutase [197], aldehyde dehydrogenase-4 [200], p53-induced glycolysis and apoptosis regulator (TIGAR) [201, 202], as well as PA26 and Hi95 that encode two proteins of the sestrin family, namely sestrin 1, sestrin 2 [203–205] and Slug [206]. Another transcription factor induced by p53 is Krüppel-like factor 4 [207], which induces cell-cycle arrest at the G1/S and G2/M transition [208], thus participates in the cell-survival program. Similarly, Cop1 (constitutively photomorphogenic 1) and Pirh2 (p53-induced protein with a Ring-H2 domain) proteins [209–211], p53-induced R2 homolog gene with a p53-binding sequence in intron 1 [212–215], hematopoietic zinc finger [193, 216], heparin-binding epidermal growth factor-like growth factor [217], discoidin domain receptor 1 [218] and cyclooxygenase 2 [219] have been shown to be involved in the p53-mediated survival program of cancer cells.

We used a rat model of myocardial infarction induced by permanent ligation of left-anterior-descending (LAD) coronary artery. Rats were exposed to oxygen-cycling 90 min/day for 4 weeks [149]. Heart tissues harvested from the infarct region was used for analysis. The results showed that p53 exhibits a differential DNA binding, switching from BAX-p53RE in the infarct heart to NOS3-p53RE in the oxygenated heart, apparently regulated by oxygen-dependent TIP60 acetylase expression and posttranslational modification of p53 core domain at p53-Lys^{118} residue (Fig. 9.15). The study establishes a new role of p53 in cardioprotection. The present study provides a novel mechanistic insight and therapeutic strategy to target the infarction-induced myocyte apoptosis in the heart. The results have important biomedical and physiological relevance in the treatment of myocardial infarction. Oxygen therapy is expected to improve the oxygenation of the ischemic myocardium, reduce infarct size, and consequently morbidity and mortality. Although, the use of supplemental oxygen in the treatment of acute MI has been in practice for over 100 years, there is no conclusive data on its beneficial effect [162, 220]. Controversies continue to emerge regarding the applicability and efficacy of oxygen therapy for MI patients [221]. One-time administration of hyperoxygenation, intended as a preconditioning treatment before induction of myocardial injury, has been shown to be beneficial [168, 181, 222]. However, these studies lacked the

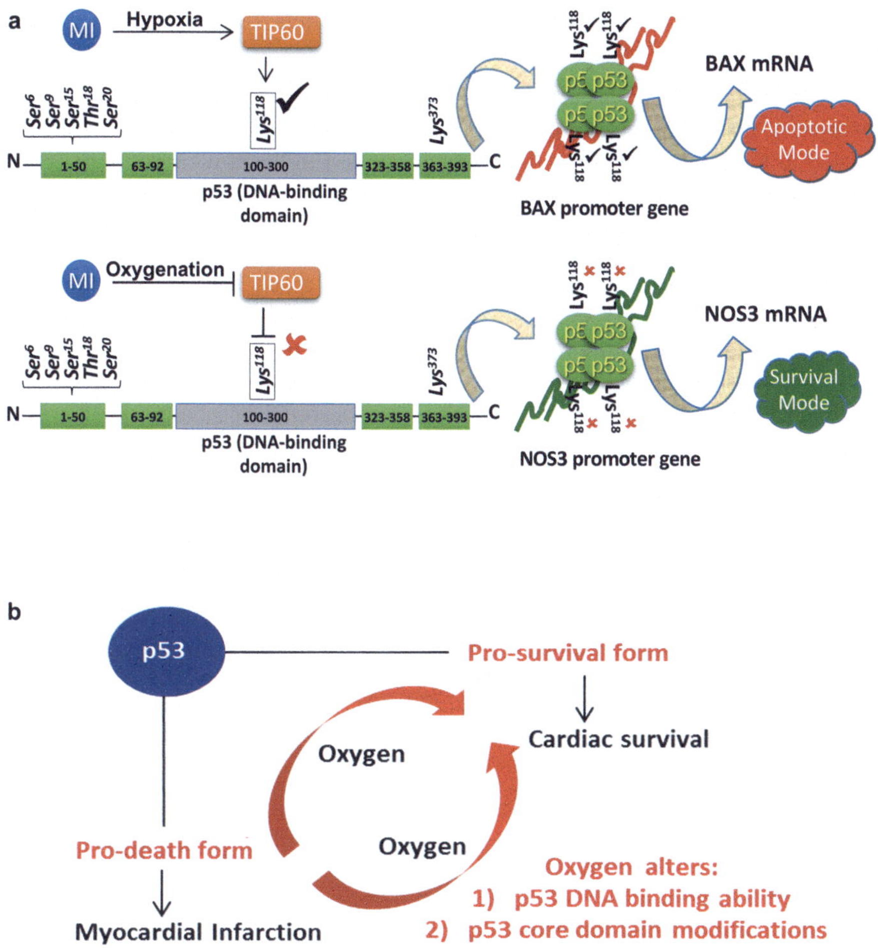

Fig. 9.15 Molecular mechanism of the dual role of p53 in the infarct myocardium subjected to supplemental oxygen therapy. (**a**) Schematic illustration of the molecular switch, which regulates the decision of p53 to activate BAX or NOS3 promoter. (**b**) Schematic illustration of the model explaining the physiological relevance of the research work. (Reproduced from Gogna et al. [256])

clinical relevance for treating post-MI patients. On the other hand, clinical protocols routinely use inhalation of high-flow oxygen in the first 24 h after acute MI. These clinical studies provided conflicting results, even detrimental effects, largely attributed to vasoconstrictive effect of oxygen [162, 221]. This study provides a post-MI approach with daily cycles of brief periods of oxygenation, which is more practical and clinically relevant. Furthermore, this study also provides the underlying molecular mechanism by which periodic administration of supplemental oxygenation results in pro-survival responses in the infarct heart.

Clinical Potential of EPR Oximetry

Although there are several methods to measure oxygen concentration or related parameters, a suitable method that can make direct and repeated measurements of oxygen in the same tissue over the treatment period in the clinical setting is currently not available. While electrode techniques have evolved as the standard methods for measurement of oxygen, they generate analytical artifacts and are not suitable for repeated measurements [16]. Near-infrared (NIR) and magnetic resonance techniques such as nuclear magnetic resonance (NMR), blood oxygen level-dependent (BOLD) magnetic resonance imaging (MRI), Overhauser-enhanced magnetic resonance imaging (OMRI) are noninvasive methods; however, they do not report absolute values of oxygen concentration [17–22]. EPR oximetry, closely related to the aforementioned magnetic resonance techniques, enables reliable and accurate measurements of the pO_2 [23]. EPR oximetry can provide direct and repeated measurements of absolute value of pO_2. The placement of the paramagnetic oxygen-sensing probe directly at the site of interest, where it remains and can be interrogated after any temporary perturbation has resolved, also avoids potential concerns about damaging the tissue at the time of measurement [25]. The ability of EPR oximetry to make repeated measurements from localized sites provides a very important capability that can enable critical aspects of a number of clinical applications. Our laboratory has pioneered the development of probes and instrumentation for pO_2 measurements in human subjects [223, 224]. The following sections provide a summary of the first-ever pO_2 study in the clinic using EPR oximetry.

Transcutaneous Oxygen Measurements in Healthy Human Subjects

Transcutaneous oxygen monitoring (TcOM) is a noninvasive, clinically approved device to obtain skin oxygen data [225, 226]. The method is quantitative, and measures oxygen delivery to the skin from underlying tissue. It has been used to monitor transcutaneous oxygen tension ($TcpO_2$) in the skin, especially for premature infants, but also for adults in the intensive care setting [227]. Physicians also use TcOM to determine whether or not adequate blood flow exists to heal lower extremity wounds, such as venous stasis or diabetic foot ulcers. Vascular surgeons use TcOM measurements to assess the need for amputation and determine the level where the amputation should take place. Unfortunately, the TcOM method has some significant limitations. The method requires heating of the skin, it is prone to calibration errors and the electrode consumes oxygen during measurement and is thus not capable of measuring low levels of oxygen that may be present in the skin at room temperature. TcOM uses a heated metal electrode placed directly on the skin. During measurement, the skin beneath the electrode is heated up to 44 °C, which dilates the vessels in the underlying tissue and increases oxygen permeability. Alternative methods based on optical sensors using fluorescence—[228], phosphorescence—[229–231] or luminescence-quenching [232, 233] are under development to overcome the limitations associated with TcOM.

We have performed a preliminary evaluation of the SPOT chip's capabilities for $TcpO_2$ measurements using a small ($N = 10$) cohort of healthy human subjects. The measurements in human subjects demonstrated that the SPOTChip can measure transcutaneous pO_2 under ambient (room-temperature) conditions [73]. To evaluate the reliability of SPOTChip oximetry, we determined intra- and inter-person variability of $TcpO_2$ in a total of 29 measurements made in the flexor side of the forearm of 5 healthy subjects. A new chip was applied at approximately the same location on the forearm for each measurement and subject. The data exhibited a variation of $TcpO_2$ between the subjects, from 7.8 ± 0.8 mmHg to 22.0 ± 1.0 mmHg. Similar measurements in the foot between the first and second metatarsal heads showed a substantial variation of $TcpO_2$ between the subjects, from 8.1 ± 0.3 mmHg to 23.4 ± 1.3 mmHg, while the variation within each subject was smaller (Fig. 9.16). The SPOT chip technology is a noninvasive method for measuring tissue oxygenation [72, 73]. It can provide clinicians with real-time information about tissue oxygen levels in wounded or injured tissue. This technology could potentially be used to decrease mortality in critically ill patients. Giving surgeons quantitative measurements on levels of tissue oxygenation can help guide their debridement decisions to avoid taking too much tissue (which may increase the difficulty of reconstruction or compromise function of the remaining tissue) or avoid taking too little, which may result in tissue necrosis and infection. It may reduce the number of times a patient must be taken to the operating room for a surgical debridement and decrease the incidence of surgical site infections by reducing the likelihood of leaving behind nonviable tissue. It can be used to monitor the viability of tissue flaps used to close defects due to trauma, cancer or congenital causes as well as to identify early changes in oxygenation to recognize a compromised flap. An earlier return to the operating room to revise compromised flaps will result in an increased flap salvage rate. The technique could also potentially be used to track changes in vascular supply, such as due to hyperbaric oxygen or angiogenic agents.

Tumor Oxygenation in Cancer Patients

The oxygen level in solid malignancies is a critical parameter affecting clinical outcomes, particularly in radiation therapy [118, 234–238]. The dependence of radiation therapy's efficacy on oxygen levels in tumors appears to be as significant as the tumor stage, tumor morphology, or tumor size [238, 239]. Oxygen level is also an important factor in tumor response to chemotherapeutic approaches [240, 241]. There are also several critical surgical oncology issues that are strongly affected by oxygen levels, such as the status of tissues when surgery and radiation are used in combination, wound healing, and reconstructive tissue transplantation, which can be better resolved if we can measure the oxygen level (pO_2) at critical sites [10–12, 242].

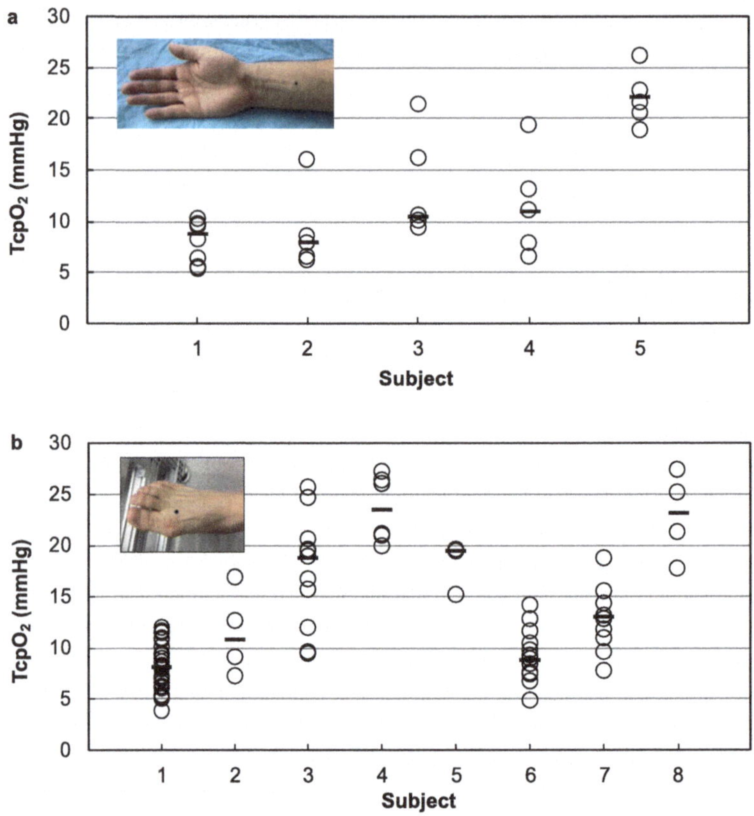

Fig. 9.16 $TcpO_2$ values measured in the arm and foot of healthy subjects. (**a**) Measurement in the arm. A total of 29 measurements were performed in the flexor forearm of 5 healthy subjects. The measurements were performed multiple times over a period of 24 weeks, by placing a new SPOT chip each time. $TcpO_2$ values in subjects 1–5 show the distribution within and among the subjects. The median value is indicated by the horizontal bar. (**b**) Measurement in the foot. A total of 86 measurements were performed in the foot of 8 healthy subjects. The measurements were performed multiple times over a period of 24 weeks, by placing a new SPOT chip each time. $TcpO_2$ values in subjects 1–8 show the distribution within and among the subjects. The median value is indicated by the horizontal bar. Overall, the results from the arm and foot demonstrate a substantial variation of $TcpO_2$ among the subjects measured. (Reproduced from Kmiec et al. [73])

Significance of Repeated Monitoring of pO_2 in Tumors

Because oxygen levels in tumors change with time [243–252], there is a great need to monitor and relate oxygen level changes to tumor progression and treatment. In addition, the pO_2 levels can also undergo dynamic variations during treatment. Therefore, it is highly desirable to monitor oxygen levels in tumors before, during, and after therapeutic intervention. The monitoring of pO_2 has been shown to have very significant implications for improving decision-making in cancer therapy at several timepoints, ranging from diagnosis, to primary treatment, to reconstructive procedures [118, 238, 253, 254]. This capability also would make it feasible to determine whether procedures designed to increase oxygenation are effective and, if so, indicate the time windows when they are most effectively administered. These applications require the ability to make repeated measurements, i.e., to have a method capable of measuring changes in overall tissue oxygenation over periods from a few minutes to many days or longer, and to ascertain tissue responsiveness to interventions to modify the level of oxygenation.

The promise of EPR oximetry technology to make repeated pO_2 measurements in human tumors under clinically applicable conditions would have a significant impact on the routine clinical decision-making process, by making previously unavailable information more easily accessible. The availability of an EPR clinical scanner will permit the evaluation of interventions designed to administer tumor oxygenation, which is crucial for improving treatment protocols currently used with patients. EPR oximetry can enable the evaluation of the effectiveness of a given tumor-oxygenation intervention in the subjects who potentially could benefit from it. Implementation of this technology may also help to improve the specificity of cancer therapy, e.g., reduce unnecessary courses of radiation therapy at nonoptimal times or in patients who are in an otherwise indistinguishable "grey area" of risk using standard clinical measures. EPR oximetry can provide information on both the presence of tumor hypoxia before the treatment, which helps in identifying those patients who can and cannot benefit from a proposed therapeutic procedure, and the changes in oxygenation induced by a hyperoxic protocol.

We have initiated the first-ever clinical trial for establishing the safety and efficacy of OxyChip for repeated measurements of tumor pO_2 in cancer patients [70]. To date, the trial has enrolled a total of 24 cancer patients with over 60 measurements conducted at the Dartmouth-Hitchcock Medical Center [71]. Out of the 24 patients with the OxyChip implanted, 18 patients did not receive any treatment prior to surgery, 5 patients received neoadjuvant chemotherapy, and 1 patient received neoadjuvant radiation therapy while the OxyChip was implanted. All patients underwent standard-of-care resection of their tumor with intent to remove the OxyChip with the tumor. The maximum duration of the OxyChip placement in a tumor prior to resection was 137 days. Most of the tumor pO_2 values in these patients were measured using a surface-loop resonator placed over the tumor. During each visit/session, pO_2 measurements were made continuously for 5–10 min. while the patient breathed room air (baseline pO_2), followed by 10 min of breathing 100% oxygen (using a nonrebreathing face mask), and then for 10 min period of room-air breathing. The tumor pO_2 values showed that varying levels of hypoxia and hyperoxygenation did lead to a detectable increase in tumor oxygenation, which has been linked to treatment efficacy.

Summary and Future Prospects for EPR Oximetry

EPR oximetry using implantable oxygen-sensors is a novel method that provides direct detection and absolute quantitation of tissue oxygen levels. The measurements are noninvasive and can be made repeatedly over long periods—several months or longer. The sensors are stable, biocompatible and provide oxygen measurement at any depth in the tissue, from topical to depths of several centimeters. The EPR oximetry has been established in preclinical models as a reliable and minimally invasive method for direct and repeated measurement of tissue oxygen levels (pO_2). Ongoing preliminary clinical studies in cancer patients at Dartmouth have confirmed that EPR oximetry can be extended to human subjects very effectively, meeting a currently important but unmet need—the capability to make repeated measurements of tissue oxygenation in the clinic. The unique potential of EPR for tissue oximetry has been recognized for nearly four decades. However, insufficient advancements in the hardware and procedures, and lack of development and availability of stable oxygen-sensing probes that are FDA-approved for human use, have been major impediments to translate the EPR technology into clinical use. Using innovative approaches, we have made significant progress on the development of novel probes for clinical oximetry. We have developed the necessary oxygen-sensing probes, procedures, and protocols for tumor oximetry. Our lab has further developed major components of the EPR instrumentation, including magnet designs, resonators, RF bridges, gradient coils and procedures, to enable clinical adaptation. We also have made an important technological leap by building the first-ever CW clinical EPR scanner (operating at 1.2 GHz), which we are currently using for establishing the safety and feasibility of making oxygen measurements in cancer patients. Our next goal is to develop clinical scanners with enhanced capabilities that will make it suitable for routine clinical use. The acquisition of this capability would be a new addition to clinical medicine and would significantly enhance treatment outcomes for multiple diseases.

Acknowledgments The studies presented in this chapter were largely supported by NIH grants R01 CA078886, R01 EB005004, R01 EB006153, R01 EB004031, and P01 CA190193.

References

1. Lane N. Oxygen : the molecule that made the world. Oxford: Oxford University Press; 2002. x, 374 p.
2. Kulkarni AC, Kuppusamy P, Parinandi N. Oxygen, the lead actor in the pathophysiologic drama: enactment of the trinity of normoxia, hypoxia, and hyperoxia in disease and therapy. Antioxid Redox Signal. 2007;9(10):1717–30.
3. Carreau A, El Hafny-Rahbi B, Matejuk A, Grillon C, Kieda C. Why is the partial oxygen pressure of human tissues a crucial parameter? Small molecules and hypoxia. J Cell Mol Med. 2011;15(6):1239–53.

4. Kutala VK, Khan M, Angelos MG, Kuppusamy P. Role of oxygen in postischemic myocardial injury. Antioxid Redox Signal. 2007;9(8):1193–206.
5. Bertout JA, Patel SA, Simon MC. The impact of O_2 availability on human cancer. Nat Rev Cancer. 2008;8(12):967–75.
6. Brown JM, Wilson WR. Exploiting tumour hypoxia in cancer treatment. Nat Rev Cancer. 2004;4(6):437–47.
7. Dewhirst MW, Cao Y, Moeller B. Cycling hypoxia and free radicals regulate angiogenesis and radiotherapy response. Nat Rev Cancer. 2008;8(6):425–37.
8. Harris AL. Hypoxia—a key regulatory factor in tumour growth. Nat Rev Cancer. 2002;2(1):38–47.
9. Davidson JD, Mustoe TA. Oxygen in wound healing: more than a nutrient. Wound Repair Regen. 2001;9(3):175–7.
10. Hopf HW, Holm J. Hyperoxia and infection. Best Pract Res Clin Anaesthesiol. 2008;22(3):553–69.
11. Hunt TK, Ellison EC, Sen CK. Oxygen: at the foundation of wound healing—introduction. World J Surg. 2004;28(3):291–3.
12. Sen CK. Wound healing essentials: let there be oxygen. Wound Repair Regen. 2009;17(1):1–18.
13. Hopf HW, Rollins MD. Wounds: an overview of the role of oxygen. Antioxid Redox Signal. 2007;9(8):1183–92.
14. Rockwell S, Moulder JE. Hypoxic fractions of human tumors xenografted into mice: a review. Int J Radiat Oncol Biol Phys. 1990;19(1):197–202.
15. Vikram DS, Zweier JL, Kuppusamy P. Methods for noninvasive imaging of tissue hypoxia. Antioxid Redox Signal. 2007;9(10):1745–56.
16. Vaupel P, Schlenger K, Knoop C, Hockel M. Oxygenation of human tumors: evaluation of tissue oxygen distribution in breast cancers by computerized O_2 tension measurements. Cancer Res. 1991;51(12):3316–22.
17. Aboagye EO, Maxwell RJ, Horsman MR, Lewis AD, Workman P, Tracy M, Griffiths JR. The relationship between tumour oxygenation determined by oxygen electrode measurements and magnetic resonance spectroscopy of the fluorinated 2-nitroimidazole SR-4554. Br J Cancer. 1998;77(1):65–70.
18. Kim JG, Zhao D, Song Y, Constantinescu A, Mason RP, Liu H. Interplay of tumor vascular oxygenation and tumor pO_2 observed using near-infrared spectroscopy, an oxygen needle electrode, and 19F MR pO_2 mapping. J Biomed Opt. 2003;8(1):53–62.
19. Laukemper-Ostendorf S, Scholz A, Burger K, Heussel CP, Schmittner M, Weiler N, Markstaller K, Eberle B, Kauczor HU, Quintel M, Thelen M, Schreiber WG. 19F-MRI of perflubron for measurement of oxygen partial pressure in porcine lungs during partial liquid ventilation. Magn Reson Med. 2002;47(1):82–9.
20. Mason RP, Hunjan S, Le D, Constantinescu A, Barker BR, Wong PS, Peschke P, Hahn EW, Antich PP. Regional tumor oxygen tension: fluorine echo planar imaging of hexafluorobenzene reveals heterogeneity of dynamics. Int J Radiat Oncol Biol Phys. 1998;42(4):747–50.
21. Mason RP, Rodbumrung W, Antich PP. Hexafluorobenzene: a sensitive 19F NMR indicator of tumor oxygenation. NMR Biomed. 1996;9(3):125–34.
22. Prasad PV, Edelman RR, Epstein FH. Noninvasive evaluation of intrarenal oxygenation with BOLD MRI. Circulation. 1996;94(12):3271–5.
23. Ahmad R, Kuppusamy P. Theory, instrumentation, and applications of electron paramagnetic resonance oximetry. Chem Rev. 2010;110(5):3212–36.
24. Swartz HM. Using EPR to measure a critical but often unmeasured component of oxidative damage: oxygen. Antioxid Redox Signal. 2004;6(3):677–86.
25. Swartz HM, Khan N, Buckey J, Comi R, Gould L, Grinberg O, Hartford A, Hopf H, Hou H, Hug E, Iwasaki A, Lesniewski P, Salikhov I, Walczak T. Clinical applications of EPR: overview and perspectives. NMR Biomed. 2004;17(5):335–51.
26. Smirnov AI, Norby SW, Walczak T, Liu KJ, Swartz HM. Physical and instrumental considerations in the use of lithium phthalocyanine for measurements of the concentration of the oxygen. J Magn Reson Ser B. 1994;103(2):95–102.

27. Elas M, Ahn KH, Parasca A, Barth ED, Lee D, Haney C, Halpern HJ. Electron paramagnetic resonance oxygen images correlate spatially and quantitatively with Oxylite oxygen measurements. Clin Cancer Res. 2006;12(14 Pt 1):4209–17.
28. Froncisz W, Lai CS, Hyde JS. Spin-label oximetry: kinetic study of cell respiration using a rapid-passage T1-sensitive electron spin resonance display. Proc Natl Acad Sci U S A. 1985;82(2):411–5.
29. Gallez B, Bacic G, Goda F, Jiang J, O'Hara JA, Dunn JF, Swartz HM. Use of nitroxides for assessing perfusion, oxygenation, and viability of tissues: in vivo EPR and MRI studies. Magn Reson Med. 1996;35(1):97–106.
30. Grinberg OY, Friedman BJ, Swartz HM. Intramyocardial pO_2 measured by EPR. Adv Exp Med Biol. 1997;428:261–8.
31. Halpern HJ, Yu C, Peric M, Barth ED, Karczmar GS, River JN, Grdina DJ, Teicher BA. Measurement of differences in pO_2 in response to perfluorocarbon/carbogen in FSa and NFSa murine fibrosarcomas with low-frequency electron paramagnetic resonance oximetry. Radiat Res. 1996;145(5):610–8.
32. Hou H, Dong R, Li H, Williams B, Lariviere JP, Hekmatyar SK, Kauppinen RA, Khan N, Swartz H. Dynamic changes in oxygenation of intracranial tumor and contralateral brain during tumor growth and carbogen breathing: a multisite EPR oximetry with implantable resonators. J Magn Reson. 2012;214(1):22–8.
33. Ilangovan G, Li H, Zweier JL, Kuppusamy P. Effect of carbogen-breathing on redox status of the RIF-1 tumor. Adv Exp Med Biol. 2003;510:13–7.
34. Ilangovan G, Liebgott T, Kutala VK, Petryakov S, Zweier JL, Kuppusamy P. EPR oximetry in the beating heart: myocardial oxygen consumption rate as an index of postischemic recovery. Magn Reson Med. 2004;51(4):835–42.
35. Khan M, Kutala VK, Wisel S, Chacko SM, Kuppusamy ML, Kwiatkowski P, Kuppusamy P. Measurement of oxygenation at the site of stem cell therapy in a murine model of myocardial infarction. Adv Exp Med Biol. 2008;614:45–52.
36. Khan M, Mohan IK, Kutala VK, Kotha SR, Parinandi NL, Hamlin RL, Kuppusamy P. Sulfaphenazole protects heart against ischemia-reperfusion injury and cardiac dysfunction by overexpression of iNOS, leading to enhancement of nitric oxide bioavailability and tissue oxygenation. Antioxid Redox Signal. 2009;11(4):725–38.
37. Kuppusamy P, Afeworki M, Shankar RA, Coffin D, Krishna MC, Hahn SM, Mitchell JB, Zweier JL. In vivo electron paramagnetic resonance imaging of tumor heterogeneity and oxygenation in a murine model. Cancer Res. 1998;58(7):1562–8.
38. Liu KJ, Gast P, Moussavi M, Norby SW, Vahidi N, Walczak T, Wu M, Swartz HM. Lithium phthalocyanine: a probe for electron paramagnetic resonance oximetry in viable biological systems. Proc Natl Acad Sci U S A. 1993;90(12):5438–42.
39. Krishna MC, English S, Yamada K, Yoo J, Murugesan R, Devasahayam N, Cook JA, Golman K, Ardenkjaer-Larsen JH, Subramanian S, Mitchell JB. Overhauser enhanced magnetic resonance imaging for tumor oximetry: coregistration of tumor anatomy and tissue oxygen concentration. Proc Natl Acad Sci U S A. 2002;99(4):2216–21.
40. Matsumoto A, Matsumoto K, Matsumoto S, Hyodo F, Sowers AL, Koscielniak JW, Devasahayam N, Subramanian S, Mitchell JB, Krishna MC. Intracellular hypoxia of tumor tissue estimated by noninvasive electron paramagnetic resonance oximetry technique using paramagnetic probes. Biol Pharm Bull. 2011;34(1):142–5.
41. Halpern HJ, Yu C, Peric M, Barth E, Grdina DJ, Teicher BA. Oxymetry deep in tissues with low-frequency electron paramagnetic resonance. Proc Natl Acad Sci U S A. 1994;91(26):13047–51.
42. Roschmann P. Radiofrequency penetration and absorption in the human body: limitations to high-field whole-body nuclear magnetic resonance imaging. Med Phys. 1987;14(6):922–31.
43. Pandian RP, Parinandi NL, Ilangovan G, Zweier JL, Kuppusamy P. Novel particulate spin probe for targeted determination of oxygen in cells and tissues. Free Radic Biol Med. 2003;35(9):1138–48.

44. Ilangovan G, Li H, Zweier JL, Krishna MC, Mitchell JB, Kuppusamy P. In vivo measurement of regional oxygenation and imaging of redox status in RIF-1 murine tumor: effect of carbogen-breathing. Magn Reson Med. 2002;48(4):723–30.
45. Ilangovan G, Bratasz A, Li H, Schmalbrock P, Zweier JL, Kuppusamy P. In vivo measurement and imaging of tumor oxygenation using coembedded paramagnetic particulates. Magn Reson Med. 2004;52(3):650–7.
46. Bratasz A, Pandian RP, Ilangovan G, Kuppusamy P. Monitoring oxygenation during the growth of a transplanted tumor. Adv Exp Med Biol. 2006;578:375–80.
47. Gallez B, Debuyst R, Dejehet F, Liu KJ, Walczak T, Goda F, Demeure R, Taper H, Swartz HM. Small particles of fusinite and carbohydrate chars coated with aqueous soluble polymers: preparation and applications for in vivo EPR oximetry. Magn Reson Med. 1998;40(1):152–9.
48. He J, Beghein N, Ceroke P, Clarkson RB, Swartz HM, Gallez B. Development of biocompatible oxygen-permeable films holding paramagnetic carbon particles: evaluation of their performance and stability in EPR oximetry. Magn Reson Med. 2001;46(3):610–4.
49. Meenakshisundaram G, Eteshola E, Pandian RP, Bratasz A, Lee SC, Kuppusamy P. Fabrication and physical evaluation of a polymer-encapsulated paramagnetic probe for biomedical oximetry. Biomed Microdevices. 2009;11(4):773–82.
50. Meenakshisundaram G, Pandian RP, Eteshola E, Lee SC, Kuppusamy P. A paramagnetic implant containing lithium naphthalocyanine microcrystals for high-resolution biological oximetry. J Magn Reson. 2010;203(1):185–9.
51. Belanger MC, Marois Y. Hemocompatibility, biocompatibility, inflammatory and in vivo studies of primary reference materials low-density polyethylene and polydimethylsiloxane: a review. J Biomed Mater Res. 2001;58(5):467–77.
52. Mata A, Fleischman AJ, Roy S. Characterization of polydimethylsiloxane (PDMS) properties for biomedical micro/nanosystems. Biomed Microdevices. 2005;7(4):281–93.
53. Shanbhag PV, Sirkar KK. Ozone and oxygen permeation behavior of silicon capillary membranes employed in membrane ozonators. J Appl Polym Sci. 1998;69(7):1263–73.
54. Zhang ZY. Facilitated oxygen transport in a novel silicone polymer membrane containing carboxylic cobalt groups. J Appl Polym Sci. 2003;90(4):1038–44.
55. Mitrovski SM, Nuzzo RG. An electrochemically driven poly(dimethylsiloxane) microfluidic actuator: oxygen sensing and programmable flows and pH gradients. Lab Chip. 2005;5(6):634–45.
56. Abbasi F, Mirzadeh H, Katbab AA. Modification of polysiloxane polymers for biomedical applications: a review. Polym Int. 2001;50(12):1279–87.
57. Kuncova-Kallio J, Kallio PJ. PDMS and its suitability for analytical microfluidic devices. Conf Proc IEEE Eng Med Biol Soc. 2006;2006:2486–9.
58. Eddings MA, Gale BK. A PDMS-based gas permeation pump for on-chip fluid handling in microfluidic devices. J Micromech Microeng. 2006;16(11):2396–402.
59. Mills CA, Martínez E, Errachid A, Gomila G, Samsó A, Samitier J. Small scale structures: the fabrication of polymeric nanostructures for biomedical applications using pattern replication techniques. Contrib Sci. 2005;3(1):47–56.
60. Quake SR, Scherer A. From micro- to nanofabrication with soft materials. Science. 2000;290(5496):1536–40.
61. Meenakshisundaram G, Eteshola E, Pandian RP, Bratasz A, Selvendiran K, Lee SC, Krishna MC, Swartz HM, Kuppusamy P. Oxygen sensitivity and biocompatibility of an implantable paramagnetic probe for repeated measurements of tissue oxygenation. Biomed Microdevices. 2009;11(4):817–26.
62. Tse D, Kuppusamy P. Biocompatibility of oxygen-sensing paramagnetic implants. Cell Biochem Biophys. 2019;77(3):197–202.
63. Hou H, Khan N, Gohain S, Kuppusamy ML, Kuppusamy P. Pre-clinical evaluation of OxyChip for long-term EPR oximetry. Biomed Microdevices. 2018;20(2):29.
64. Hou H, Khan N, Nagane M, Gohain S, Chen EY, Jarvis LA, Schaner PE, Williams BB, Flood AB, Swartz HM, Kuppusamy P. Skeletal muscle oxygenation measured by EPR oximetry using a highly sensitive polymer-encapsulated paramagnetic sensor. Adv Exp Med Biol. 2016;923:351–7.

65. Hou H, Khan N, Gohain S, Eskey CJ, Moodie KL, Maurer KJ, Swartz HM, Kuppusamy P. Dynamic EPR oximetry of changes in intracerebral oxygen tension during induced thromboembolism. Cell Biochem Biophys. 2017;75(3–4):285–94.
66. Hou H, Khan N, Lariviere J, Hodge S, Chen EY, Jarvis LA, Eastman A, Williams BB, Kuppusamy P, Swartz HM. Skeletal muscle and glioma oxygenation by carbogen inhalation in rats: a longitudinal study by EPR oximetry using single-probe implantable oxygen sensors. Adv Exp Med Biol. 2014;812:97–103.
67. Khan N, Hou H, Hodge S, Kuppusamy M, Chen EY, Eastman A, Kuppusamy P, Swartz HM. Recurrent low-dose chemotherapy to inhibit and oxygenate head and neck tumors. Adv Exp Med Biol. 2014;812:105–11.
68. Hou H, Khan N, Kuppusamy P. Measurement of pO_2 in a pre-clinical model of rabbit tumor using OxyChip, a paramagnetic oxygen sensor. Adv Exp Med Biol. 2017;977:313–8.
69. Polacco MA, Hou H, Kuppusamy P, Chen EY. Measuring flap oxygen using electron paramagnetic resonance oximetry. Laryngoscope. 2019;129(12):E415–9.
70. Jarvis LA, Williams BB, Schaner PE, Chen EY, Angeles CV, Hou H, Schreiber W, Wood VA, Flood AB, Swartz HM, Kuppusamy P. Phase 1 clinical trial of OxyChip, an implantable absolute pO(2) sensor for tumor oximetry. Int J Radiat Oncol Biol Phys. 2016;96(2):S109–10.
71. Schaner PE, Chen EY, Jarvis LA, Zuurbier RA, DiFlorio-Alexander RM, Paydarfar JA, Gosselin BJ, Pettus JR, Hou H, Schreiber WA, Kmiec MM, Petryakov SV, Demidenko E, Wood VA, Flood AB, Swartz HM, Kuppusamy P. Quantitation of pO_2 using the OxyChip in human tumors via electron paramagnetic resonance oximetry: baseline variability and response to hyperoxygenation. Int J Radiat Oncol Biol Phys. 2019;105(1):E675.
72. Kmiec MM, Hou H, Kuppusamy ML, Drews TM, Prabhat AM, Petryakov SV, Demidenko E, Schaner PE, Buckey JC, Blank A, Kuppusamy P. Application of SPOT chip for transcutaneous oximetry. Magn Reson Med. 2019;81(5):2837–40.
73. Kmiec MM, Hou H, Lakshmi Kuppusamy M, Drews TM, Prabhat AM, Petryakov SV, Demidenko E, Schaner PE, Buckey JC, Blank A, Kuppusamy P. Transcutaneous oxygen measurement in humans using a paramagnetic skin adhesive film. Magn Reson Med. 2019;81(2):781–94.
74. Kmiec MM, Tse D, Mast JM, Ahmad R, Kuppusamy P. Implantable microchip containing oxygen-sensing paramagnetic crystals for long-term, repeated, and multisite in vivo oximetry. Biomed Microdevices. 2019;21(3):71.
75. Caston RM, Schreiber WA, Hou H, Williams BB, Chen EY, Schaner PE, Jarvis LA, Flood AB, Petryakov SV, Kmiec MM, Kuppusamy P, Swartz HM. Development of the implantable resonator system for clinical EPR oximetry. Cell Biochem Biophys. 2017;75(3–4):275–83.
76. Pandian RP, Raju NP, Gallucci JC, Woodward PM, Epstein AJ, Kuppusamy P. A new tetragonal crystalline polymorph of lithium octa-n-butoxy-naphthalocyanine (LiNc-BuO) radical: structural, magnetic and oxygen-sensing properties. Chem Mater. 2010;22(23):6254–62.
77. Fuster V, Badimon L, Badimon JJ, Chesebro JH. The pathogenesis of coronary artery disease and the acute coronary syndromes (1). N Engl J Med. 1992;326(4):242–50.
78. Agbulut O, Vandervelde S, Al Attar N, Larghero J, Ghostine S, Leobon B, Robidel E, Borsani P, Le Lorc'h M, Bissery A, Chomienne C, Bruneval P, Marolleau JP, Vilquin JT, Hagege A, Samuel JL, Menasche P. Comparison of human skeletal myoblasts and bone marrow-derived CD133+ progenitors for the repair of infarcted myocardium. J Am Coll Cardiol. 2004;44(2):458–63.
79. Chiu RC, Zibaitis A, Kao RL. Cellular cardiomyoplasty: myocardial regeneration with satellite cell implantation. Ann Thorac Surg. 1995;60(1):12–8.
80. Ghostine S, Carrion C, Souza LC, Richard P, Bruneval P, Vilquin JT, Pouzet B, Schwartz K, Menasche P, Hagege AA. Long-term efficacy of myoblast transplantation on regional structure and function after myocardial infarction. Circulation. 2002;106(12 Suppl 1):I131–6.
81. Kamihata H, Matsubara H, Nishiue T, Fujiyama S, Tsutsumi Y, Ozono R, Masaki H, Mori Y, Iba O, Tateishi E, Kosaki A, Shintani S, Murohara T, Imaizumi T, Iwasaka T. Implantation of bone marrow cells into ischemic myocardium enhances collateral perfusion and regional function via side supply of angioblasts, angiogenic ligands, and cytokines. Circulation. 2001;104(9):1046–52.

82. Orlic D, Kajstura J, Chimenti S, Bodine DM, Leri A, Anversa P. Transplanted adult bone marrow cells repair myocardial infarcts in mice. Ann N Y Acad Sci. 2001;938:221–9; discussion 229–30.
83. Pagani FD, DerSimonian H, Zawadzka A, Wetzel K, Edge AS, Jacoby DB, Dinsmore JH, Wright S, Aretz TH, Eisen HJ, Aaronson KD. Autologous skeletal myoblasts transplanted to ischemia-damaged myocardium in humans. Histological analysis of cell survival and differentiation. J Am Coll Cardiol. 2003;41(5):879–88.
84. Retuerto MA, Schalch P, Patejunas G, Carbray J, Liu N, Esser K, Crystal RG, Rosengart TK. Angiogenic pretreatment improves the efficacy of cellular cardiomyoplasty performed with fetal cardiomyocyte implantation. J Thorac Cardiovasc Surg. 2004;127(4):1041–9; discussion 1049–51.
85. Tse HF, Kwong YL, Chan JK, Lo G, Ho CL, Lau CP. Angiogenesis in ischaemic myocardium by intramyocardial autologous bone marrow mononuclear cell implantation. Lancet. 2003;361(9351):47–9.
86. McConnell PI, del Rio CL, Jacoby DB, Pavlicova M, Kwiatkowski P, Zawadzka A, Dinsmore JH, Astra L, Wisel S, Michler RE. Correlation of autologous skeletal myoblast survival with changes in left ventricular remodeling in dilated ischemic heart failure. J Thorac Cardiovasc Surg. 2005;130(4):1001.
87. Ezashi T, Das P, Roberts RM. Low O_2 tensions and the prevention of differentiation of hES cells. Proc Natl Acad Sci U S A. 2005;102(13):4783–8.
88. Salim A, Giaccia AJ, Longaker MT. Stem cell differentiation. Nat Biotechnol. 2004;22(7):804–5; author reply 805–6.
89. Wang DW, Fermor B, Gimble JM, Awad HA, Guilak F. Influence of oxygen on the proliferation and metabolism of adipose derived adult stem cells. J Cell Physiol. 2005;204(1):184–91.
90. Silverman HS, Wei S, Haigney MC, Ocampo CJ, Stern MD. Myocyte adaptation to chronic hypoxia and development of tolerance to subsequent acute severe hypoxia. Circ Res. 1997;80(5):699–707.
91. Khan M, Kutala VK, Vikram DS, Wisel S, Chacko SM, Kuppusamy ML, Mohan IK, Zweier JL, Kwiatkowski P, Kuppusamy P. Skeletal myoblasts transplanted in the ischemic myocardium enhance in situ oxygenation and recovery of contractile function. Am J Physiol Heart Circ Physiol. 2007;293(4):H2129–39.
92. Hearse DJ, Bolli R. Reperfusion induced injury: manifestations, mechanisms, and clinical relevance. Cardiovasc Res. 1992;26(2):101–8.
93. Das DK, Maulik N. Antioxidant effectiveness in ischemia-reperfusion tissue injury. Methods Enzymol. 1994;233:601–10.
94. Chen Z, Siu B, Ho YS, Vincent R, Chua CC, Hamdy RC, Chua BH. Overexpression of MnSOD protects against myocardial ischemia/reperfusion injury in transgenic mice. J Mol Cell Cardiol. 1998;30(11):2281–9.
95. Chen Z, Oberley TD, Ho Y, Chua CC, Siu B, Hamdy RC, Epstein CJ, Chua BH. Overexpression of CuZnSOD in coronary vascular cells attenuates myocardial ischemia/reperfusion injury. Free Radic Biol Med. 2000;29(7):589–96.
96. Yoshida T, Watanabe M, Engelman DT, Engelman RM, Schley JA, Maulik N, Ho YS, Oberley TD, Das DK. Transgenic mice overexpressing glutathione peroxidase are resistant to myocardial ischemia reperfusion injury. J Mol Cell Cardiol. 1996;28(8):1759–67.
97. Moncada S, Higgs A. The L-arginine-nitric oxide pathway. N Engl J Med. 1993;329(27):2002–12.
98. Liaudet L, Soriano FG, Szabo C. Biology of nitric oxide signaling. Crit Care Med. 2000;28(4 Suppl):N37–52.
99. Zweier JL, Wang P, Kuppusamy P. Direct measurement of nitric oxide generation in the ischemic heart using electron paramagnetic resonance spectroscopy. J Biol Chem. 1995;270(1):304–7.
100. Depre C, Fierain L, Hue L. Activation of nitric oxide synthase by ischaemia in the perfused heart. Cardiovasc Res. 1997;33(1):82–7.

101. Woolfson RG, Patel VC, Neild GH, Yellon DM. Inhibition of nitric oxide synthesis reduces infarct size by an adenosine-dependent mechanism. Circulation. 1995;91(5):1545–51.
102. Naseem SA, Kontos MC, Rao PS, Jesse RL, Hess ML, Kukreja RC. Sustained inhibition of nitric oxide by NG-nitro-L-arginine improves myocardial function following ischemia/reperfusion in isolated perfused rat heart. J Mol Cell Cardiol. 1995;27(1):419–26.
103. Jones SP, Greer JJ, Kakkar AK, Ware PD, Turnage RH, Hicks M, van Haperen R, de Crom R, Kawashima S, Yokoyama M, Lefer DJ. Endothelial nitric oxide synthase overexpression attenuates myocardial reperfusion injury. Am J Physiol Heart Circ Physiol. 2004;286(1):H276–82.
104. Pernow J, Wang QD. The role of the L-arginine/nitric oxide pathway in myocardial ischaemic and reperfusion injury. Acta Physiol Scand. 1999;167(2):151–9.
105. Weyrich AS, Ma XL, Lefer AM. The role of L-arginine in ameliorating reperfusion injury after myocardial ischemia in the cat. Circulation. 1992;86(1):279–88.
106. Shinmura K, Tang XL, Takano H, Hill M, Bolli R. Nitric oxide donors attenuate myocardial stunning in conscious rabbits. Am J Phys. 1999;277(6 Pt 2):H2495–503.
107. Sato H, Zhao ZQ, Vinten-Johansen J. L-Arginine inhibits neutrophil adherence and coronary artery dysfunction. Cardiovasc Res. 1996;31(1):63–72.
108. Lefer AM. Attenuation of myocardial ischemia-reperfusion injury with nitric oxide replacement therapy. Ann Thorac Surg. 1995;60(3):847–51.
109. Hunter AL, Cruz RP, Cheyne BM, McManus BM, Granville DJ. Cytochrome p450 enzymes and cardiovascular disease. Can J Physiol Pharmacol. 2004;82(12):1053–60.
110. Granville DJ, Tashakkor B, Takeuchi C, Gustafsson AB, Huang C, Sayen MR, Wentworth P Jr, Yeager M, Gottlieb RA. Reduction of ischemia and reperfusion-induced myocardial damage by cytochrome P450 inhibitors. Proc Natl Acad Sci U S A. 2004;101(5):1321–6.
111. Nithipatikom K, Gross ER, Endsley MP, Moore JM, Isbell MA, Falck JR, Campbell WB, Gross GJ. Inhibition of cytochrome P450omega-hydroxylase: a novel endogenous cardioprotective pathway. Circ Res. 2004;95(8):e65–71.
112. Seubert J, Yang B, Bradbury JA, Graves J, Degraff LM, Gabel S, Gooch R, Foley J, Newman J, Mao L, Rockman HA, Hammock BD, Murphy E, Zeldin DC. Enhanced postischemic functional recovery in CYP2J2 transgenic hearts involves mitochondrial ATP-sensitive K+ channels and p42/p44 MAPK pathway. Circ Res. 2004;95(5):506–14.
113. Fleming I, Michaelis UR, Bredenkotter D, Fisslthaler B, Dehghani F, Brandes RP, Busse R. Endothelium-derived hyperpolarizing factor synthase (Cytochrome P450 2C9) is a functionally significant source of reactive oxygen species in coronary arteries. Circ Res. 2001;88(1):44–51.
114. Fichtlscherer S, Dimmeler S, Breuer S, Busse R, Zeiher AM, Fleming I. Inhibition of cytochrome P450 2C9 improves endothelium-dependent, nitric oxide-mediated vasodilatation in patients with coronary artery disease. Circulation. 2004;109(2):178–83.
115. Elmi S, Sallam NA, Rahman MM, Teng X, Hunter AL, Moien-Afshari F, Khazaei M, Granville DJ, Laher I. Sulfaphenazole treatment restores endothelium-dependent vasodilation in diabetic mice. Vasc Pharmacol. 2008;48(1):1–8.
116. Khan M, Mohan IK, Kutala VK, Kumbala D, Kuppusamy P. Cardioprotection by sulfaphenazole, a cytochrome p450 inhibitor: mitigation of ischemia-reperfusion injury by scavenging of reactive oxygen species. J Pharmacol Exp Ther. 2007;323(3):813–21.
117. Okunieff P, Hoeckel M, Dunphy EP, Schlenger K, Knoop C, Vaupel P. Oxygen tension distributions are sufficient to explain the local response of human breast tumors treated with radiation alone. Int J Radiat Oncol Biol Phys. 1993;26(4):631–6.
118. Hockel M, Knoop C, Schlenger K, Vorndran B, Baussmann E, Mitze M, Knapstein PG, Vaupel P. Intratumoral pO_2 predicts survival in advanced cancer of the uterine cervix. Radiother Oncol. 1993;26(1):45–50.
119. Brizel DM, Scully SP, Harrelson JM, Layfield LJ, Bean JM, Prosnitz LR, Dewhirst MW. Tumor oxygenation predicts for the likelihood of distant metastases in human soft tissue sarcoma. Cancer Res. 1996;56(5):941–3.

120. Rofstad EK. Microenvironment-induced cancer metastasis. Int J Radiat Biol. 2000;76(5):589–605.
121. Menon C, Fraker DL. Tumor oxygenation status as a prognostic marker. Cancer Lett. 2005;221(2):225–35.
122. Evans SM, Koch CJ. Prognostic significance of tumor oxygenation in humans. Cancer Lett. 2003;195(1):1–16.
123. Stone HB, Brown JM, Phillips TL, Sutherland RM. Oxygen in human tumors: correlations between methods of measurement and response to therapy. Summary of a workshop held November 19-20, 1992, at the National Cancer Institute, Bethesda, Maryland. Radiat Res. 1993;136(3):422–34.
124. Milosevic M, Fyles A, Hedley D, Hill R. The human tumor microenvironment: invasive (needle) measurement of oxygen and interstitial fluid pressure. Semin Radiat Oncol. 2004;14(3):249–58.
125. Olive PL, Aquino-Parsons C. Measurement of tumor hypoxia using single-cell methods. Semin Radiat Oncol. 2004;14(3):241–8.
126. Dorie MJ, Kallman RF. Reoxygenation in the RIF-1 tumor. Int J Radiat Oncol Biol Phys. 1984;10(5):687–93.
127. Olive PL, Vikse CM, Durand RE. Hypoxic fractions measured in murine tumors and normal tissues using the comet assay. Int J Radiat Oncol Biol Phys. 1994;29(3):487–91.
128. Mueller-Klieser W, Vaupel P, Manz R, Schmidseder R. Intracapillary oxyhemoglobin saturation of malignant tumors in humans. Int J Radiat Oncol Biol Phys. 1981;7(10):1397–404.
129. Koch CJ, Evans SM. Non-invasive PET and SPECT imaging of tissue hypoxia using isotopically labeled 2-nitroimidazoles. Adv Exp Med Biol. 2003;510:285–92.
130. Koch CJ. Measurement of absolute oxygen levels in cells and tissues using oxygen sensors and 2-nitroimidazole EF5. Methods Enzymol. 2002;352:3–31.
131. Hockel M, Schlenger K, Knoop C, Vaupel P. Oxygenation of carcinomas of the uterine cervix: evaluation by computerized O_2 tension measurements. Cancer Res. 1991;51(22):6098–102.
132. Vinogradov SA, Lo LW, Jenkins WT, Evans SM, Koch C, Wilson DF. Noninvasive imaging of the distribution in oxygen in tissue in vivo using near-infrared phosphors. Biophys J. 1996;70(4):1609–17.
133. Braun RD, Lanzen JL, Snyder SA, Dewhirst MW. Comparison of tumor and normal tissue oxygen tension measurements using OxyLite or microelectrodes in rodents. Am J Physiol Heart Circ Physiol. 2001;280(6):H2533–44.
134. Zhao D, Jiang L, Mason RP. Measuring changes in tumor oxygenation. Methods Enzymol. 2004;386:378–418.
135. al-Hallaq HA, Zamora MA, Fish BL, Halpern HJ, Moulder JE, Karczmar GS. Using high spectral and spatial resolution bold MRI to choose the optimal oxygenating treatment for individual cancer patients. Adv Exp Med Biol. 2003;530:433–40.
136. Elas M, Williams BB, Parasca A, Mailer C, Pelizzari CA, Lewis MA, River JN, Karczmar GS, Barth ED, Halpern HJ. Quantitative tumor oxymetric images from 4D electron paramagnetic resonance imaging (EPRI): methodology and comparison with blood oxygen level-dependent (BOLD) MRI. Magn Reson Med. 2003;49(4):682–91.
137. Gallez B, Baudelet C, Jordan BF. Assessment of tumor oxygenation by electron paramagnetic resonance: principles and applications. NMR Biomed. 2004;17(5):240–62.
138. Fan X, River JN, Zamora M, Al-Hallaq HA, Karczmar GS. Effect of carbogen on tumor oxygenation: combined fluorine-19 and proton MRI measurements. Int J Radiat Oncol Biol Phys. 2002;54(4):1202–9.
139. Dunn JF, O'Hara JA, Zaim-Wadghiri Y, Lei H, Meyerand ME, Grinberg OY, Hou H, Hoopes PJ, Demidenko E, Swartz HM. Changes in oxygenation of intracranial tumors with carbogen: a BOLD MRI and EPR oximetry study. J Magn Reson Imaging. 2002;16(5):511–21.
140. Goda F, O'Hara JA, Rhodes ES, Liu KJ, Dunn JF, Bacic G, Swartz HM. Changes of oxygen tension in experimental tumors after a single dose of X-ray irradiation. Cancer Res. 1995;55(11):2249–52.

141. Goda F, Bacic G, O'Hara JA, Gallez B, Swartz HM, Dunn JF. The relationship between partial pressure of oxygen and perfusion in two murine tumors after X-ray irradiation: a combined gadopentetate dimeglumine dynamic magnetic resonance imaging and in vivo electron paramagnetic resonance oximetry study. Cancer Res. 1996;56(14):3344–9.
142. Sonveaux P, Dessy C, Brouet A, Jordan BF, Gregoire V, Gallez B, Balligand JL, Feron O. Modulation of the tumor vasculature functionality by ionizing radiation accounts for tumor radiosensitization and promotes gene delivery. FASEB J. 2002;16(14):1979–81.
143. Olive PL. Radiation-induced reoxygenation in the SCCVII murine tumour: evidence for a decrease in oxygen consumption and an increase in tumour perfusion. Radiother Oncol. 1994;32(1):37–46.
144. Mast JM, Kuppusamy P. Hyperoxygenation as a therapeutic supplement for treatment of triple negative breast cancer. Front Oncol. 2018;8:527.
145. Mast JM, Hinds JW, Tse D, Axelrod K, Kuppusamy ML, Kmiec MM, Bognar B, Kalai T, Kuppusamy P. Selective induction of cellular toxicity and anti-tumor efficacy by N-methylpiperazinyl diarylidenylpiperidone and its pro-nitroxide conjugate through ROS-mediated mitochondrial dysfunction and G2/M cell-cycle arrest in human pancreatic cancer. Cell Biochem Biophys. 2020;78(2):191–202.
146. Rivera BK, Naidu SK, Subramanian K, Joseph M, Hou H, Khan N, Swartz HM, Kuppusamy P. Real-time, in vivo determination of dynamic changes in lung and heart tissue oxygenation using EPR oximetry. Adv Exp Med Biol. 2014;812:81–6.
147. Prabhat AM, Kuppusamy ML, Naidu SK, Meduru S, Reddy PT, Dominic A, Khan M, Rivera BK, Kuppusamy P. Supplemental oxygen protects heart against acute myocardial infarction. Front Cardiovasc Med. 2018;5:114.
148. Khan N, Hou H, Eskey CJ, Moodie K, Gohain S, Du G, Hodge S, Culp WC, Kuppusamy P, Swartz HM. Deep-tissue oxygen monitoring in the brain of rabbits for stroke research. Stroke. 2015;46(3):e62–6.
149. Khan M, Meduru S, Gogna R, Madan E, Citro L, Kuppusamy ML, Sayyid M, Mostafa M, Hamlin RL, Kuppusamy P. Oxygen cycling in conjunction with stem cell transplantation induces NOS3 expression leading to attenuation of fibrosis and improved cardiac function. Cardiovasc Res. 2012;93(1):89–99.
150. Khan M, Meduru S, Mohan IK, Kuppusamy ML, Wisel S, Kulkarni A, Rivera BK, Hamlin RL, Kuppusamy P. Hyperbaric oxygenation enhances transplanted cell graft and functional recovery in the infarct heart. J Mol Cell Cardiol. 2009;47(2):275–87.
151. Khan M, Meduru S, Pandian RP, Rivera BK, Kuppusamy P. Effect of oxygenation on stem-cell therapy for myocardial infarction. Adv Exp Med Biol. 2011;701:175–81.
152. O'Connor RE, Brady W, Brooks SC, Diercks D, Egan J, Ghaemmaghami C, Menon V, O'Neil BJ, Travers AH, Yannopoulos D. Part 10: acute coronary syndromes: 2010 American Heart Association Guidelines for Cardiopulmonary Resuscitation and Emergency Cardiovascular Care. Circulation. 2010;122(18 Suppl 3):S787–817.
153. Antman EM, Anbe DT, Armstrong PW, Bates ER, Green LA, Hand M, Hochman JS, Krumholz HM, Kushner FG, Lamas GA, Mullany CJ, Ornato JP, Pearle DL, Sloan MA, Smith SC Jr, American College of Cardiology, American Heart Association, Canadian Cardiovascular Society. ACC/AHA guidelines for the management of patients with ST-elevation myocardial infarction—executive summary. A report of the American College of Cardiology/American Heart Association Task Force on Practice Guidelines (Writing Committee to revise the 1999 guidelines for the management of patients with acute myocardial infarction). J Am Coll Cardiol. 2004;44(3):671–719.
154. Van de Werf F, Ardissino D, Betriu A, Cokkinos DV, Falk E, Fox KA, Julian D, Lengyel M, Neumann FJ, Ruzyllo W, Thygesen C, Underwood SR, Vahanian A, Verheugt FW, Wijns W, Task Force on the Management of Acute Myocardial Infarction of the European Society of Cardiology. Management of acute myocardial infarction in patients presenting with ST-segment elevation. The Task Force on the Management of Acute Myocardial Infarction of the European Society of Cardiology. Eur Heart J. 2003;24(1):28–66.

155. O'Driscoll BR, Howard LS, Davison AG, British Thoracic Society. BTS guideline for emergency oxygen use in adult patients. Thorax. 2008;63(Suppl 6):vi1–68.
156. Kenmure AC, Murdoch WR, Beattie AD, Marshall JC, Cameron AJ. Circulatory and metabolic effects of oxygen in myocardial infarction. Br Med J. 1968;4(5627):360–4.
157. Thomas M, Malmcrona R, Shillingford J. Haemodynamic effects of oxygen in patients with acute myocardial infarction. Br Heart J. 1965;27:401–7.
158. Shillingford JP, Thomas M. Cardiovascular and pulmonary changes in patients with myocardial infarction treated in an intensive care and research unit. Am J Cardiol. 1967;20(4):484–93.
159. Cameron AJ, Hutton I, Kenmure AC, Murdoch WR. Haemodynamic and metabolic effects of hyperbaric oxygen in myocardial infarction. Lancet. 1966;2(7468):833–7.
160. Foster GL, Casten GG, Reeves TJ. The effects of oxygen breathing in patients with acute myocardial infarction. Cardiovasc Res. 1969;3(2):179–89.
161. Mackenzie GJ, Flenley DC, Taylor SH, McDonald AH, Staunton HP, Donald KW. Circulatory and respiratory studies in myocardial infarction and cardiogenic shock. Lancet. 1964;2(7364):825–32.
162. Wijesinghe M, Perrin K, Ranchord A, Simmonds M, Weatherall M, Beasley R. Routine use of oxygen in the treatment of myocardial infarction: systematic review. Heart. 2009;95(3):198–202.
163. Stavitsky Y, Shandling AH, Ellestad MH, Hart GB, Van Natta B, Messenger JC, Strauss M, Dekleva MN, Alexander JM, Mattice M, Clarke D. Hyperbaric oxygen and thrombolysis in myocardial infarction: the 'HOT MI' randomized multicenter study. Cardiology. 1998;90(2):131–6.
164. Ashfield R, Gavey CJ. Severe acute myocardial infarction treated with hyperbaric oxygen. Report on forty patients. Postgrad Med J. 1969;45(528):648–54.
165. Tibbles PM, Edelsberg JS. Hyperbaric-oxygen therapy. N Engl J Med. 1996;334(25):1642–8.
166. Thackham JA, McElwain DL, Long RJ. The use of hyperbaric oxygen therapy to treat chronic wounds: a review. Wound Repair Regen. 2008;16(3):321–30.
167. Yogaratnam JZ, Laden G, Madden LA, Seymour AM, Guvendik L, Cowen M, Greenman J, Cale A, Griffin S. Hyperbaric oxygen: a new drug in myocardial revascularization and protection? Cardiovasc Revasc Med. 2006;7(3):146–54.
168. Yogaratnam JZ, Laden G, Guvendik L, Cowen M, Cale A, Griffin S. Pharmacological preconditioning with hyperbaric oxygen: can this therapy attenuate myocardial ischemic reperfusion injury and induce myocardial protection via nitric oxide? J Surg Res. 2008;149(1):155–64.
169. Sterling DL, Thornton JD, Swafford A, Gottlieb SF, Bishop SP, Stanley AW, Downey JM. Hyperbaric oxygen limits infarct size in ischemic rabbit myocardium in vivo. Circulation. 1993;88(4 Pt 1):1931–6.
170. Dekleva M, Neskovic A, Vlahovic A, Putnikovic B, Beleslin B, Ostojic M. Adjunctive effect of hyperbaric oxygen treatment after thrombolysis on left ventricular function in patients with acute myocardial infarction. Am Heart J. 2004;148(4):E14.
171. Thom SR, Bhopale VM, Velazquez OC, Goldstein LJ, Thom LH, Buerk DG. Stem cell mobilization by hyperbaric oxygen. Am J Physiol Heart Circ Physiol. 2006;290(4):H1378–86.
172. McGowan FX Jr, Davis PJ, del Nido PJ, Sobek M, Allen JW, Downing SE. Endothelium-dependent regulation of coronary tone in the neonatal pig. Anesth Analg. 1994;79(6):1094–101.
173. Kubes P, Granger DN. Nitric oxide modulates microvascular permeability. Am J Phys. 1992;262(2 Pt 2):H611–5.
174. Noel AA, Fallek SR, Hobson RW 2nd, Duran WN. Inhibition of nitric oxide synthase attenuates primed microvascular permeability in the in vivo microcirculation. J Vasc Surg. 1995;22(6):661–9; discussion 669–70.
175. Radomski MW, Palmer RM, Moncada S. Comparative pharmacology of endothelium-derived relaxing factor, nitric oxide and prostacyclin in platelets. Br J Pharmacol. 1987;92(1):181–7.
176. Jordan JE, Zhao ZQ, Vinten-Johansen J. The role of neutrophils in myocardial ischemia-reperfusion injury. Cardiovasc Res. 1999;43(4):860–78.
177. Johnson G III, Tsao PS, Mulloy D, Lefer AM. Cardioprotective effects of acidified sodium nitrite in myocardial ischemia with reperfusion. J Pharmacol Exp Ther. 1990;252(1):35–41.

178. Kubes P, Suzuki M, Granger DN. Nitric oxide: an endogenous modulator of leukocyte adhesion. Proc Natl Acad Sci U S A. 1991;88(11):4651–5.
179. Clancy RM, Leszczynska-Piziak J, Abramson SB. Nitric oxide, an endothelial cell relaxation factor, inhibits neutrophil superoxide anion production via a direct action on the NADPH oxidase. J Clin Invest. 1992;90(3):1116–21.
180. Bolli R. Cardioprotective function of inducible nitric oxide synthase and role of nitric oxide in myocardial ischemia and preconditioning: an overview of a decade of research. J Mol Cell Cardiol. 2001;33(11):1897–918.
181. Cabigas BP, Su J, Hutchins W, Shi Y, Schaefer RB, Recinos RF, Nilakantan V, Kindwall E, Niezgoda JA, Baker JE. Hyperoxic and hyperbaric-induced cardioprotection: role of nitric oxide synthase 3. Cardiovasc Res. 2006;72(1):143–51.
182. Buras JA, Stahl GL, Svoboda KK, Reenstra WR. Hyperbaric oxygen downregulates ICAM-1 expression induced by hypoxia and hypoglycemia: the role of NOS. Am J Physiol Cell Physiol. 2000;278(2):C292–302.
183. Cai H, Li Z, Davis ME, Kanner W, Harrison DG, Dudley SC Jr. Akt-dependent phosphorylation of serine 1179 and mitogen-activated protein kinase kinase/extracellular signal-regulated kinase 1/2 cooperatively mediate activation of the endothelial nitric-oxide synthase by hydrogen peroxide. Mol Pharmacol. 2003;63(2):325–31.
184. Mount PF, Kemp BE, Power DA. Regulation of endothelial and myocardial NO synthesis by multi-site eNOS phosphorylation. J Mol Cell Cardiol. 2007;42(2):271–9.
185. Tanaka T, Nakamura H, Yodoi J, Bloom ET. Redox regulation of the signaling pathways leading to eNOS phosphorylation. Free Radic Biol Med. 2005;38(9):1231–42.
186. Thomas SR, Chen K, Keaney JF Jr. Hydrogen peroxide activates endothelial nitric-oxide synthase through coordinated phosphorylation and dephosphorylation via a phosphoinositide 3-kinase-dependent signaling pathway. J Biol Chem. 2002;277(8):6017–24.
187. Bialik S, Geenen DL, Sasson IE, Cheng R, Horner JW, Evans SM, Lord EM, Koch CJ, Kitsis RN. Myocyte apoptosis during acute myocardial infarction in the mouse localizes to hypoxic regions but occurs independently of p53. J Clin Invest. 1997;100(6):1363–72.
188. Crow MT, Mani K, Nam YJ, Kitsis RN. The mitochondrial death pathway and cardiac myocyte apoptosis. Circ Res. 2004;95(10):957–70.
189. Alvarado-Vasquez N, Zapata E, Alcazar-Leyva S, Masso F, Montano LF. Reduced NO synthesis and eNOS mRNA expression in endothelial cells from newborns with a strong family history of type 2 diabetes. Diabetes Metab Res Rev. 2007;23(7):559–66.
190. Chen YK, Hsue SS, Lin LM. Correlation between inducible nitric oxide synthase and p53 expression for DMBA-induced hamster buccal-pouch carcinomas. Oral Dis. 2003;9(5):227–34.
191. Ko LJ, Prives C. p53: puzzle and paradigm. Genes Dev. 1996;10(9):1054–72.
192. Levine AJ, Hu W, Feng Z. The P53 pathway: what questions remain to be explored? Cell Death Differ. 2006;13(6):1027–36.
193. Vousden KH. Outcomes of p53 activation--spoilt for choice. J Cell Sci. 2006;119(Pt 24):5015–20.
194. Gogna R, Madan E, Keppler B, Pati U. Gallium compound GaQ(3)-induced Ca(2+) signalling triggers p53-dependent and -independent apoptosis in cancer cells. Br J Pharmacol. 2012;166(2):617–36.
195. Madan E, Gogna R, Bhatt M, Pati U, Kuppusamy P, Mahdi AA. Regulation of glucose metabolism by p53: emerging new roles for the tumor suppressor. Oncotarget. 2011;2(12):948–57.
196. Janicke RU, Sohn D, Schulze-Osthoff K. The dark side of a tumor suppressor: anti-apoptotic p53. Cell Death Differ. 2008;15(6):959–76.
197. Hussain SP, Amstad P, He P, Robles A, Lupold S, Kaneko I, Ichimiya M, Sengupta S, Mechanic L, Okamura S, Hofseth LJ, Moake M, Nagashima M, Forrester KS, Harris CC. p53-induced up-regulation of MnSOD and GPx but not catalase increases oxidative stress and apoptosis. Cancer Res. 2004;64(7):2350–6.
198. Tan M, Li S, Swaroop M, Guan K, Oberley LW, Sun Y. Transcriptional activation of the human glutathione peroxidase promoter by p53. J Biol Chem. 1999;274(17):12061–6.

199. Yan W, Chen X. GPX2, a direct target of p63, inhibits oxidative stress-induced apoptosis in a p53-dependent manner. J Biol Chem. 2006;281(12):7856–62.
200. Donald SP, Sun XY, Hu CA, Yu J, Mei JM, Valle D, Phang JM. Proline oxidase, encoded by p53-induced gene-6, catalyzes the generation of proline-dependent reactive oxygen species. Cancer Res. 2001;61(5):1810–5.
201. Bensaad K, Tsuruta A, Selak MA, Vidal MN, Nakano K, Bartrons R, Gottlieb E, Vousden KH. TIGAR, a p53-inducible regulator of glycolysis and apoptosis. Cell. 2006;126(1):107–20.
202. Madan E, Gogna R, Kuppusamy P, Bhatt M, Pati U, Mahdi AA. TIGAR induces p53-mediated cell-cycle arrest by regulation of RB–E2F1 complex. Br J Cancer. 2012;107(3):516–26.
203. Masutani H, Ueda S, Yodoi J. The thioredoxin system in retroviral infection and apoptosis. Cell Death Differ. 2005;12(Suppl 1):991–8.
204. Velasco-Miguel S, Buckbinder L, Jean P, Gelbert L, Talbott R, Laidlaw J, Seizinger B, Kley N. PA26, a novel target of the p53 tumor suppressor and member of the GADD family of DNA damage and growth arrest inducible genes. Oncogene. 1999;18(1):127–37.
205. Budanov AV, Shoshani T, Faerman A, Zelin E, Kamer I, Kalinski H, Gorodin S, Fishman A, Chajut A, Einat P, Skaliter R, Gudkov AV, Chumakov PM, Feinstein E. Identification of a novel stress-responsive gene Hi95 involved in regulation of cell viability. Oncogene. 2002;21(39):6017–31.
206. Wu WS, Heinrichs S, Xu D, Garrison SP, Zambetti GP, Adams JM, Look AT. Slug antagonizes p53-mediated apoptosis of hematopoietic progenitors by repressing puma. Cell. 2005;123(4):641–53.
207. Zhang W, Geiman DE, Shields JM, Dang DT, Mahatan CS, Kaestner KH, Biggs JR, Kraft AS, Yang VW. The gut-enriched Kruppel-like factor (Kruppel-like factor 4) mediates the transactivating effect of p53 on the p21WAF1/Cip1 promoter. J Biol Chem. 2000;275(24):18391–8.
208. Rowland BD, Bernards R, Peeper DS. The KLF4 tumour suppressor is a transcriptional repressor of p53 that acts as a context-dependent oncogene. Nat Cell Biol. 2005;7(11):1074–82.
209. Fuchs SY, Adler V, Buschmann T, Yin Z, Wu X, Jones SN, Ronai Z. JNK targets p53 ubiquitination and degradation in nonstressed cells. Genes Dev. 1998;12(17):2658–63.
210. Dornan D, Wertz I, Shimizu H, Arnott D, Frantz GD, Dowd P, O'Rourke K, Koeppen H, Dixit VM. The ubiquitin ligase COP1 is a critical negative regulator of p53. Nature. 2004;429(6987):86–92.
211. Leng RP, Lin Y, Ma W, Wu H, Lemmers B, Chung S, Parant JM, Lozano G, Hakem R, Benchimol S. Pirh2, a p53-induced ubiquitin-protein ligase, promotes p53 degradation. Cell. 2003;112(6):779–91.
212. Kimura T, Takeda S, Sagiya Y, Gotoh M, Nakamura Y, Arakawa H. Impaired function of p53R2 in Rrm2b-null mice causes severe renal failure through attenuation of dNTP pools. Nat Genet. 2003;34(4):440–5.
213. Tanaka H, Arakawa H, Yamaguchi T, Shiraishi K, Fukuda S, Matsui K, Takei Y, Nakamura Y. A ribonucleotide reductase gene involved in a p53-dependent cell-cycle checkpoint for DNA damage. Nature. 2000;404(6773):42–9.
214. Utrera R, Collavin L, Lazarevic D, Delia D, Schneider C. A novel p53-inducible gene coding for a microtubule-localized protein with G2-phase-specific expression. EMBO J. 1998;17(17):5015–25.
215. Monte M, Benetti R, Buscemi G, Sandy P, Del Sal G, Schneider C. The cell cycle-regulated protein human GTSE-1 controls DNA damage-induced apoptosis by affecting p53 function. J Biol Chem. 2003;278(32):30356–64.
216. Braithwaite AW, Del Sal G, Lu X. Some p53-binding proteins that can function as arbiters of life and death. Cell Death Differ. 2006;13(6):984–93.
217. Fang L, Li G, Liu G, Lee SW, Aaronson SA. p53 induction of heparin-binding EGF-like growth factor counteracts p53 growth suppression through activation of MAPK and PI3K/Akt signaling cascades. EMBO J. 2001;20(8):1931–9.
218. Ongusaha PP, Kim JI, Fang L, Wong TW, Yancopoulos GD, Aaronson SA, Lee SW. p53 induction and activation of DDR1 kinase counteract p53-mediated apoptosis and influence p53 regulation through a positive feedback loop. EMBO J. 2003;22(6):1289–301.

219. Han JA, Kim JI, Ongusaha PP, Hwang DH, Ballou LR, Mahale A, Aaronson SA, Lee SW. P53-mediated induction of Cox-2 counteracts p53- or genotoxic stress-induced apoptosis. EMBO J. 2002;21(21):5635–44.
220. Beasley R, Aldington S, Weatherall M, Robinson G, McHaffie D. Oxygen therapy in myocardial infarction: an historical perspective. J R Soc Med. 2007;100(3):130–3.
221. Kones R. Oxygen therapy for acute myocardial infarction-then and now. A century of uncertainty. Am J Med. 2011;124(11):1000–5.
222. Yogaratnam JZ, Laden G, Guvendik L, Cowen M, Cale A, Griffin S. Hyperbaric oxygen preconditioning improves myocardial function, reduces length of intensive care stay, and limits complications post coronary artery bypass graft surgery. Cardiovasc Revasc Med. 2010;11(1):8–19.
223. Swartz HM, Hou H, Khan N, Jarvis LA, Chen EY, Williams BB, Kuppusamy P. Advances in probes and methods for clinical EPR oximetry. Adv Exp Med Biol. 2014;812:73–9.
224. Swartz HM, Williams BB, Zaki BI, Hartford AC, Jarvis LA, Chen EY, Comi RJ, Ernstoff MS, Hou H, Khan N, Swarts SG, Flood AB, Kuppusamy P. Clinical EPR: unique opportunities and some challenges. Acad Radiol. 2014;21(2):197–206.
225. Huch R, Lubbers DW, Huch A. Quantitative continuous measurement of partial oxygen pressure on the skin of adults and newborn babies. Pflügers Arch ges Physiol. 1972;337:185–98.
226. Huch R, Huch A, Lubbers DW. Transcutaneous measurement of blood Po2 (tcPo2)—method and application in perinatal medicine. J Perinat Med. 1973;1(3):183–91.
227. Kaur S, Pawar M, Banerjee N, Garg R. Evaluation of the efficacy of hyperbaric oxygen therapy in the management of chronic nonhealing ulcer and role of periwound transcutaneous oximetry as a predictor of wound healing response: a randomized prospective controlled trial. J Anaesthesiol Clin Pharmacol. 2012;28(1):70–5.
228. Stucker M, Struk A, Altmeyer P, Herde M, Baumgartl H, Lubbers DW. The cutaneous uptake of atmospheric oxygen contributes significantly to the oxygen supply of human dermis and epidermis. J Physiol. 2002;538(Pt 3):985–94.
229. Geis S, Babilas P, Schreml S, Angele P, Nerlich M, Jung EM, Prantl L. Transcutaneous pO_2 measurement during tourniquet-induced venous occlusion using dynamic phosphorescence imaging. Clin Hemorheol Microcirc. 2008;40(4):249–58.
230. Li Z, Navarro-Alvarez N, Keeley EJ, Nowell NH, Goncalves BMM, Huang CA, Evans CL. Non-invasive monitoring of skin inflammation using an oxygen-sensing paint-on bandage. Biomed Opt Express. 2017;8(10):4640–51.
231. Li Z, Roussakis E, Koolen PG, Ibrahim AMS, Kim K, Rose LF, Wu J, Nichols AJ, Baek Y, Birngruber R, Apiou-Sbirlea G, Matyal R, Huang T, Chan R, Lin SJ, Evans CL. Non-invasive transdermal two-dimensional mapping of cutaneous oxygenation with a rapid-drying liquid bandage. Biomed Opt Express. 2014;5(11):3748–64.
232. Babilas P, Lamby P, Prantl L, Schreml S, Jung EM, Liebsch G, Wolfbeis OS, Landthaler M, Szeimies RM, Abels C. Transcutaneous pO_2 imaging during tourniquet-induced forearm ischemia using planar optical oxygen sensors. Skin Res Technol. 2008;14(3):304–11.
233. Schreml S, Meier RJ, Kirschbaum M, Kong SC, Gehmert S, Felthaus O, Kuchler S, Sharpe JR, Woltje K, Weiss KT, Albert M, Seidl U, Schroder J, Morsczeck C, Prantl L, Duschl C, Pedersen SF, Gosau M, Berneburg M, Wolfbeis OS, Landthaler M, Babilas P. Luminescent dual sensors reveal extracellular pH-gradients and hypoxia on chronic wounds that disrupt epidermal repair. Theranostics. 2014;4(7):721–35.
234. Doll CM, Milosevic M, Pintilie M, Hill RP, Fyles AW. Estimating hypoxic status in human tumors: a simulation using Eppendorf oxygen probe data in cervical cancer patients. Int J Radiat Oncol Biol Phys. 2003;55(5):1239–46.
235. Evans SM, Judy KD, Dunphy I, Jenkins WT, Nelson PT, Collins R, Wileyto EP, Jenkins K, Hahn SM, Stevens CW, Judkins AR, Phillips P, Geoerger B, Koch CJ. Comparative measurements of hypoxia in human brain tumors using needle electrodes and EF5 binding. Cancer Res. 2004;64(5):1886–92.
236. Gagel B, Piroth M, Pinkawa M, Reinartz P, Zimny M, Kaiser HJ, Stanzel S, Asadpour B, Demirel C, Hamacher K, Coenen HH, Scholbach T, Maneschi P, DiMartino E, Eble MJ. pO polarography, contrast enhanced color duplex sonography (CDS), [18F] fluoromisoni-

dazole and [18F] fluorodeoxyglucose positron emission tomography: validated methods for the evaluation of therapy-relevant tumor oxygenation or only bricks in the puzzle of tumor hypoxia? BMC Cancer. 2007;7:113.
237. Vaupel P, Frinak S, O'Hara M. Direct measurement of reoxygenation in malignant mammary tumors after a single large dose of irradiation. Adv Exp Med Biol. 1984;180:773–82.
238. Vaupel P, Mayer A. Hypoxia in cancer: significance and impact on clinical outcome. Cancer Metastasis Rev. 2007;26(2):225–39.
239. Vaupel P, Thews O, Hoeckel M. Treatment resistance of solid tumors: role of hypoxia and anemia. Med Oncol. 2001;18(4):243–59.
240. Cosse JP, Michiels C. Tumour hypoxia affects the responsiveness of cancer cells to chemotherapy and promotes cancer progression. Anti Cancer Agents Med Chem. 2008;8(7):790–7.
241. Shannon AM, Bouchier-Hayes DJ, Condron CM, Toomey D. Tumour hypoxia, chemotherapeutic resistance and hypoxia-related therapies. Cancer Treat Rev. 2003;29(4):297–307.
242. Biswas S, Roy S, Banerjee J, Hussain SR, Khanna S, Meenakshisundaram G, Kuppusamy P, Friedman A, Sen CK. Hypoxia inducible microRNA 210 attenuates keratinocyte proliferation and impairs closure in a murine model of ischemic wounds. Proc Natl Acad Sci U S A. 2010;107(15):6976–81.
243. Baudelet C, Ansiaux R, Jordan BF, Havaux X, Macq B, Gallez B. Physiological noise in murine solid tumours using T2*-weighted gradient-echo imaging: a marker of tumour acute hypoxia? Phys Med Biol. 2004;49(15):3389–411.
244. Brurberg KG, Graff BA, Rofstad EK. Temporal heterogeneity in oxygen tension in human melanoma xenografts. Br J Cancer. 2003;89(2):350–6.
245. Brurberg KG, Skogmo HK, Graff BA, Olsen DR, Rofstad EK. Fluctuations in pO_2 in poorly and well-oxygenated spontaneous canine tumors before and during fractionated radiation therapy. Radiother Oncol. 2005;77(2):220–6.
246. Cardenas-Navia LI, Yu D, Braun RD, Brizel DM, Secomb TW, Dewhirst MW. Tumor-dependent kinetics of partial pressure of oxygen fluctuations during air and oxygen breathing. Cancer Res. 2004;64(17):6010–7.
247. Cardenas-Navia LI, Mace D, Richardson RA, Wilson DF, Shan S, Dewhirst MW. The pervasive presence of fluctuating oxygenation in tumors. Cancer Res. 2008;68(14):5812–9.
248. Matsumoto S, Yasui H, Mitchell JB, Krishna MC. Imaging cycling tumor hypoxia. Cancer Res. 2010;70(24):10019–23.
249. Pogue BW, Braun RD, Lanzen JL, Erickson C, Dewhirst MW. Analysis of the heterogeneity of pO_2 dynamics during photodynamic therapy with verteporfin. Photochem Photobiol. 2001;74(5):700–6.
250. Pogue BW, O'Hara JA, Goodwin IA, Wilmot CJ, Fournier GP, Akay AR, Swartz H. Tumor PO(2) changes during photodynamic therapy depend upon photosensitizer type and time after injection. Comp Biochem Physiol A Mol Integr Physiol. 2002;132(1):177–84.
251. Yasui H, Matsumoto S, Devasahayam N, Munasinghe JP, Choudhuri R, Saito K, Subramanian S, Mitchell JB, Krishna MC. Low-field magnetic resonance imaging to visualize chronic and cycling hypoxia in tumor-bearing mice. Cancer Res. 2010;70(16):6427–36.
252. Magat J, Jordan BF, Cron GO, Gallez B. Noninvasive mapping of spontaneous fluctuations in tumor oxygenation using 19F MRI. Med Phys. 2010;37(10):5434–41.
253. Hockel M, Schlenger K, Aral B, Mitze M, Schaffer U, Vaupel P. Association between tumor hypoxia and malignant progression in advanced cancer of the uterine cervix. Cancer Res. 1996;56(19):4509–15.
254. Vaupel P. Hypoxia and aggressive tumor phenotype: implications for therapy and prognosis. Oncologist. 2008;13(Suppl 3):21–6.
255. Bratasz A, Pandian RP, Deng Y, Petryakov S, Grecula JC, Gupta N, Kuppusamy P. In vivo imaging of changes in tumor oxygenation during growth and after treatment. Magn Reson Med. 2007;57(5):950–9.
256. Gogna R, Madan E, Khan M, Pati U, Kuppusamy P. p53's choice of myocardial death or survival: oxygen protects infarct myocardium by recruiting p53 on NOS3 promoter through regulation of p53-Lys(118) acetylation. EMBO Mol Med. 2013;5(11):1662–83.

Chapter 10
Resonators for Clinical Electron Paramagnetic Resonance (EPR)

Hiroshi Hirata, Sergey Petryakov, and Wilson Schreiber

Abstract In pulsed electron paramagnetic resonance (EPR), free-induction decay (FID) or spin echo (SE) signals of unpaired electrons are recorded in the time-domain. In both methods, electromagnetic waves play an important role in the detection of unpaired electrons in EPR spectroscopy. The resonator generates and senses electromagnetic waves and therefore serves as a critical interface between unpaired electrons and the transmit/receive systems of an EPR spectrometer. Since a resonator is a sensitive electrical circuit that can amplify voltages and currents when the electrical circuit of the resonator is on resonance, the resonator is an essential component for EPR detection in continuous wave and pulsed EPR. Without the resonator, EPR signals cannot be detected with sufficient sensitivity. In this chapter, the basics of resonators and some examples of resonators used in preclinical studies with small animals and human subjects are explained.

Keywords Resonators for clinical EPR · Clinical electron paramagnetic resonance

Introduction

For EPR detection, a radiofrequency (RF) or microwave resonator is commonly used to apply electromagnetic waves to unpaired electrons in a sample. The frequency of electromagnetic waves is determined by the static magnetic field B_0 applied to the sample. In continuous-wave (CW) electron paramagnetic resonance

H. Hirata (✉)
Division of Bioengineering and Bioinformatics, Graduate School of Information Science and Technology, Hokkaido University, Sapporo, Hokkaido, Japan
e-mail: hhirata@ist.hokudai.ac.jp

S. Petryakov · W. Schreiber
EPR Center for the Study of Viable Systems, Geisel School of Medicine at Dartmouth, Hanover, NH, USA

L. J. Berliner, N. L. Parinandi (eds.), *Measuring Oxidants and Oxidative Stress in Biological Systems*, Biological Magnetic Resonance 34,
https://doi.org/10.1007/978-3-030-47318-1_10

(EPR), the energy absorption in the resonator is measured as a function of the magnetic field B_0. In pulsed EPR, free-induction decay (FID) or spin echo (SE) signals of unpaired electrons are recorded in the time-domain. In both methods, electromagnetic waves play an important role in the detection of unpaired electrons in EPR spectroscopy. The resonator both generates and senses electromagnetic waves and therefore serves as a critical interface between unpaired electrons and the transmit/receive systems of an EPR spectrometer.

In CW-EPR, the energy absorption of unpaired electrons due to the EPR phenomenon is detected through the reflection of incident electromagnetic waves at the resonator. When the energy absorption of unpaired electrons occurs at the resonator, the input impedance of the resonator changes; this results in an impedance mismatch between the resonator and the transmission line, which is usually 50 Ω. To detect EPR absorption, the resonator should be sensitive to the energy absorption due to electron spins by having sufficient quality and filling factors, and by permitting sufficient microwave power to be transmitted to the measured sample [1]. Since a resonator is a sensitive electrical circuit that can amplify voltages and currents when the electrical circuit of the resonator is on resonance, the resonator is an essential component for EPR detection in CW and pulsed EPR. Without the resonator, EPR signals cannot be detected with sufficient sensitivity. This is similar to the detection of nuclear magnetic resonance (NMR).

The requirements of the resonator differ between CW-EPR and pulsed EPR because the methods of detection involve different experimental setups, which measure different aspects of the resonance characteristics at different points in time. Therefore, the resonator is designed to meet specific needs in applications and measurements. In this chapter, the basics of resonators and some examples of resonators used in preclinical studies with small animals and human subjects are explained.

Basics of Resonators for EPR

Resonant Circuit

An RF resonator is an electrical circuit that stores electrical and magnetic energy. The simplest resonant circuit is a series LC circuit and a parallel LC circuit. Figure 10.1 illustrates the RLC series resonant circuit, which is driven by a sinusoidal voltage source $V(t)$. In this circuit, a resistor R satisfies Ohm's law, a capacitor C stores electric charges, and an inductor L stores magnetic fields.

Let us consider a sinusoidal signal of frequency f. The impedance $Z_{a-a'}$ of the RLC circuit at port $a-a'$ is given by

$$Z_{a-a'} = R + j\left(\omega L - \frac{1}{\omega C}\right), \tag{10.1}$$

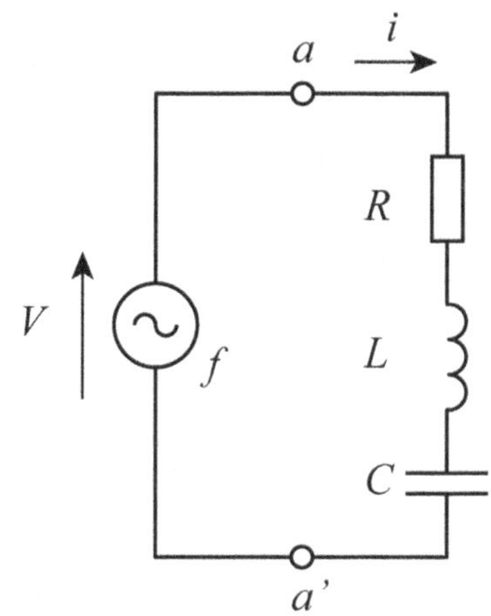

Fig. 10.1 An RLC series resonant circuit with a sinusoidal voltage source $V(t)$

where w is the angular frequency, $\omega = 2\,pf$, and j is the imaginary unit, $j = \sqrt{-1}$. If you are not familiar with sinusoidal steady-state analysis, textbooks of basic circuit theory will be helpful. By Ohm's law, current $i(t)$ is expressed as

$$i(t) = \frac{V(t)}{Z_{a-a'}} = \frac{V(t)}{R + j\left(\omega L - \dfrac{1}{\omega C}\right)}. \tag{10.2}$$

When the imaginary part in Eq. (10.1) vanishes, the impedance becomes minimal and consists of only the real part R. At this point, current $i(t)$ is expressed as

$$i(t) = \frac{V(t)}{R}. \tag{10.3}$$

In Eq. (10.2), current $i(t)$ is frequency-dependent, and is maximal at angular frequency

$$\omega = \frac{1}{\sqrt{LC}}. \tag{10.4}$$

From Eq. (10.4), the resonant frequency f_r is given by

$$f_r = \frac{1}{2\pi\sqrt{LC}}. \tag{10.5}$$

At the resonant frequency, current i flowing in the circuit is maximized, and magnetic flux in the inductor also becomes maximized. Therefore, RF magnetic fields are efficiently generated at the resonant frequency f_r.

Sensitivity of EPR Detection

The sensitivity of EPR signal detection is well documented [2]. In a reflection-type EPR bridge, as illustrated in Fig. 10.2, the RF resonator is connected to the transmission line to apply the electromagnetic waves to a sample. Figure 10.3 illustrates a resonator connected to the transmission line.

The signal intensity V_S of an EPR absorption spectrum in CW-EPR is given by

$$V_S = \chi'' \eta Q \sqrt{P Z_0}, \tag{10.6}$$

where η is the filling factor, Q is the quality factor of the resonator loaded with a sample, P is the incident RF power to the resonator, and Z_0 is the characteristic impedance of the transmission line connected to the resonator [1]. In Eq. (10.6), χ'' is the imaginary component of the effective RF susceptibility and depends on the sample. From the viewpoint of the electrical circuit, the product $\eta Q \sqrt{P Z_0}$ should be maximized to obtain the maximum signal intensity. The incident RF power P has a limitation in EPR spectroscopy that is sample-specific, because saturation effects of EPR signals are observed when a spin system being measured is saturated with a higher RF magnetic field. The RF power P also depends on the sample and the conversion efficiency of RF magnetic fields in the resonator. The characteristic impedance Z_0 depends on the selection of the transmission lines of the EPR spectrometer. The most commonly used characteristic impedance in RF and microwave commercial products is 50 Ω. Thus, the filling factor η and the quality factor Q are key factors for optimizing the sensitivity of the resonator in EPR spectroscopy. Further descriptions of the filling factor and quality factor are given below.

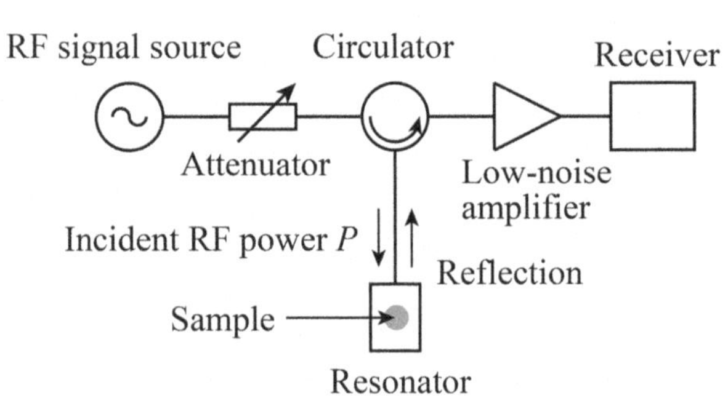

Fig. 10.2 General schematic of a simplified reflection-type bridge for CW-EPR

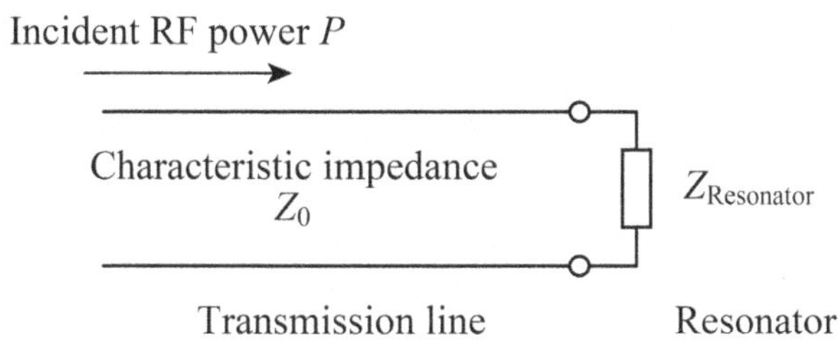

Fig. 10.3 Transmission line and a resonator

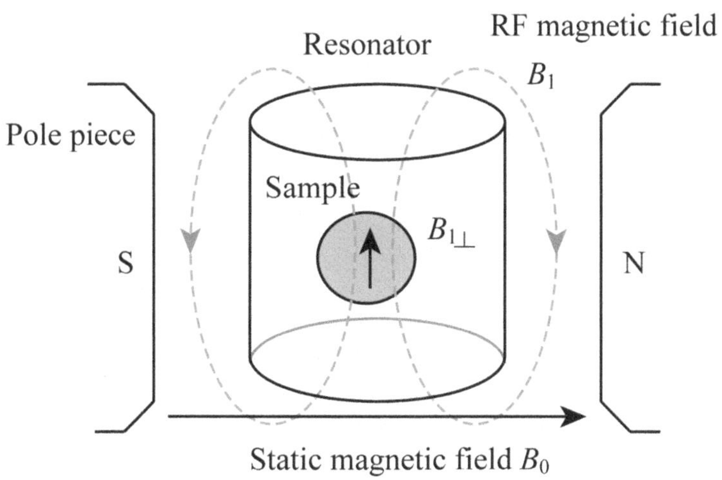

Fig. 10.4 Geometry of magnetic fields, a resonator, and a sample regarding the filling factor

Filling Factor and Quality Factor

The filling factor is generally defined as the ratio between the stored magnetic energy in the resonator and the stored magnetic energy in the sample that can contribute to EPR phenomena. Figure 10.4 illustrates the resonator and the sample, as well as the magnetic fields.

The RF magnetic field $B_{1\perp}$ perpendicular to the static magnetic field B_0 is taken into account in EPR phenomenon. The density of magnetic energy w_m in the space can be calculated as

$$w_{\mathrm{m}} = \frac{1}{2}\boldsymbol{H} \bullet \boldsymbol{B} = \frac{\mu}{2}\boldsymbol{H} \bullet \boldsymbol{B} = \frac{1}{2\mu}\boldsymbol{B} \bullet \boldsymbol{B}, \tag{10.7}$$

where $\boldsymbol{H}$ is a magnetic field in A/m, $\boldsymbol{B}$ is a magnetic flux density ($\mu\boldsymbol{H}$) in T, and μ is the magnetic permeability of the space and is written as $\mu_0 = 4\pi \times 10^{-7}$ (N/A^2) in free space. When the magnetic field distribution of RF magnetic field B_1 in the resonator is known, the stored magnetic energies in the sample and the resonator can be calculated as

$$\eta = \frac{\int_{\text{sample}} B_{1\perp}^2 dV}{\int_{\text{resonator}} B_1^2 dV}. \tag{10.8}$$

The quality factor is a dimensionless value that quantifies a resonator's ability to store and dissipate energy, and can be evaluated as

$$Q = \frac{2\pi \times (\text{energy stored})}{\text{energy dissipated per one cycle}}. \tag{10.9}$$

However, direct measurements of the stored energy in the resonator and the energy dissipated per cycle are not practical. Another equation for the quality factor is often used:

$$Q = \frac{f_r}{\Delta f} = \frac{f_r}{f_1 - f_2}, \tag{10.10}$$

where δf is the 3-dB bandwidth of the resonator that is the difference between the frequencies, f_1 and f_2, in which the stored energy becomes half of the maximum stored energy in the resonator. Presently, frequencies can be measured very precisely through the use of a vector network analyzer. Equation (10.9) is a better definition for understanding the physical meaning of the quality factor. However, Eq. (10.10) is commonly used in practice by measuring the RF resonance characteristics of the resonator.

For CW-EPR, a high quality factor and a high filling factor are both required to ensure sufficient EPR signal intensity according to Eq. (10.6). A high filling factor with a biological sample, which has a loss in terms of electromagnetic waves, results in an increase in energy dissipation in the resonator, which results in a low quality factor. In contrast, when a small part of a biological sample is placed in the resonator, the losses due to the resonator decrease, which results in a low filling factor and a high quality factor. In biological EPR spectroscopy and imaging, the balance between the quality factor and the filling factor should be optimized so that the biological sample can be measured adequately. Therefore, the resonator should be designed for a specific application to obtain the maximal EPR signal intensity with the optimal conditions for the quality factor and the filling factor. For pulsed EPR, the quality factor is significantly lowered to reduce the ring-down time after the RF pulses and to have a broader bandwidth. This aspect of pulsed EPR is explained in another chapter.

Frequency Selection of Preclinical and Clinical EPR

The RF frequency η for EPR spectroscopy and imaging is determined by the static magnetic field B_0 applied to a subject, according to the fundamental equation for the EPR phenomenon, $h\nu = g\beta_B B_0$, where h is Plank constant (6.626×10^{-34} Js), g is the g-factor of the species to be measured (2.002 for the free electron), and β_B is Bohr magneton (9.274×10^{-24} JT^{-1}) [3]. In laboratories, an X-band EPR spectrometer is commonly used as an analytical tool for various samples. However, the penetration of microwaves into biological tissues is critical in biomedical applications for small animals and even human subjects. As a measure of the penetration of electromagnetic waves in a subject, the skin depth is considered. The skin depth defines the distance at which the amplitude of electromagnetic waves becomes $1/e = 0.3679$, where e is Napier's constant. At angular frequency w, the skin depth d can be calculated as

$$d = \left[\frac{\varepsilon \mu \omega^2}{2} \sqrt{1 + \left(\frac{\sigma}{\varepsilon \omega} \right)^2} - 1 \right]^{-1/2}, \tag{10.11}$$

where $\mathcal{E}$ is the permittivity (dielectric constant) of the medium, ε is the conductivity of the medium (biological tissues), and μ is the permeability of the medium [4]. As you can see, the skin depth depends on the angular frequency. The dielectric constant and the conductivity are also frequency-dependent. Only permeability μ in biological tissues is usually considered to be the permeability in free space.

In vivo EPR measurements with mice were initially conducted in the X-band ($\nu = 8.5$ GHz, $B_0 = 0.3$ T) [5], but soon thereafter the frequency of electromagnetic waves was reduced to a lower frequency to allow increased penetration of electromagnetic waves in biological tissues [6]. The EPR studies below the X-band are well documented by Eaton and Eaton [7]. From the 1990s to the present, most in vivo small animal studies have been performed at frequencies from 250 to 1200 MHz (VHF/UHF to L-band). EPR imaging on human skin has been conducted in the S-band (2 GHz) [8, 9], since it does not require a significant penetration of electromagnetic waves in tissues and provides better sensitivity.

Frequency selection is important for biomedical applications of EPR spectroscopy and imaging, since it may limit the detection volume and the overall sensitivity of the EPR measurements. Several factors that affect EPR measurements should be taken into account. The criteria for frequency selection are as follows:

(a) The penetration depth of electromagnetic waves

A higher frequency may limit the penetration depth (see Eq. (10.11)).

(b) The sensitivity of EPR signals

A higher magnetic field and a corresponding frequency give stronger EPR signals. If there is no problem regarding the penetration depth, a frequency as high as possible is beneficial.

(c) Size of the species to be measured

How large is the subject? A mouse, rabbit, pig, or even a human?

(d) What will be measured, and where will it be measured?

The region of interest, such as superficial tissues or a deep region, determines the required penetration depth.

(e) Specific absorption rate (SAR) and safety regulations

To avoid unwanted effects in a subject, energy absorption in biological tissues should be below certain levels, which are promulgated in safety regulations by national governments or international organizations.

Types of Resonators

The resonators used for biological EPR applications can be divided into two major categories: volume resonators and surface resonators (similar to clinical magnetic resonance imaging (MRI)).

Samples are typically placed inside a volume resonator, which can accommodate a whole body or part of a subject animal. These types of resonators, such as a loop-gap resonator, benefit from a higher filling factor [10]. Since the sample is placed in the volume of the resonator where the magnetic flux generated by the resonator is the highest, the spins in the sample are stimulated more efficiently. The wavelength of electromagnetic waves may limit the dimensions of the resonator. When a large sample space is needed and the sample is not small in comparison to the wavelength, the resonator structure should be considered to obtain homogeneous RF magnetic fields in the sample space.

Surface resonators are typically placed on the surface of a sample. A surface resonator can measure EPR signals from localized regions of a subject animal, such as the skin or tissues close to the external surface of the animal [11]. As mentioned above, the region of interest is a key factor when choosing the type of resonator. In addition to these major categories, implantable resonators have been developed for small animal experiments, which address the problem of the penetration depth. Surgical procedures may be required for the use of an implantable resonator in animals. Specific examples of EPR resonators are introduced in section "Brief History of Technical Developments for In Vivo EPR Resonators".

Other Technical Aspects for RF Resonators in EPR

Previous sections addressed several technical aspects (filling factor, quality factor, and penetration depth). This section briefly explains other important aspects of the resonators used in EPR spectroscopy and imaging.

Conversion Efficiency of RF Magnetic Fields

The conversion efficiency of an RF magnetic field is the ratio of RF magnetic flux density B_1 and the square root of the incident RF power to the resonator. This conversion efficiency directly affects the sensitivity of the resonator because the density of magnetic energy in the sample is proportional to the square of the RF magnetic field B_1 generated by the resonator. When the input RF power is constant, a resonator with a high conversion efficiency will provide a stronger B_1 field than a resonator with a lower conversion efficiency. As a result, the signal intensity is proportional to the square of the conversion efficiency [12]. A resonator with a high conversion efficiency is desirable for all EPR experiments.

Frequency Tuning Adjustment

The microwave frequency of the RF source can be controlled and locked to the resonant frequency of the resonator, which is commonly referred to as automatic frequency control (AFC). With AFC, there is no need to adjust the frequency of the resonator. However, experiments that measure the narrow line-width of a first-derivative EPR absorption spectrum (such as EPR oximetry) require a very stable source frequency that is not always attainable. Another approach to maintaining equal frequency between the resonator and the source is called automatic tuning control (ATC). With ATC, the resonator is designed to accommodate its own frequency adjustment. There are several approaches to adjust the resonant frequency, including the use of an (1) induction motor [13], (2) piezoelectric actuator [14], and (3) varactor diodes [15, 16].

Impedance Matching (Coupling) Adjustment

For the bridge to adequately transfer the reflections of electromagnetic waves due to EPR from the resonator to the detector, the impedance of the loaded resonator should be adjusted to the characteristic impedance of the bridge (usually 50 Ω). Inductive coupling or capacitive coupling networks have been used for this purpose [17]. Traditionally, in X-band cavity resonators, the iris of a waveguide was adjusted by hand. In modern X-band cavity resonators, motorized iris control is available. When samples are measured in vivo where motion of the subject is expected, the resonator should automatically tune and couple to the perturbation due to motion of the subject to ensure high-quality recorded spectra. As with frequency adjustment, there are several approaches to adjusting the impedance of the resonator: (1) induction motor [13], (2) piezoelectric actuator [18], (3) varactor diodes [15], and (4) photo-resistor [16]. This technique is commonly referred to as automatic coupling control (ACC) or automatic matching control (AMC).

Homogeneity of RF Magnetic Fields

As mentioned previously, the EPR signal intensity depends on the square of RF magnetic field B_1. To achieve a uniform sensitivity of EPR detection within the resonator, a homogeneous B_1 field needs to be established. Three factors alter the RF magnetic field:

(a) The distance from the currents. In the radial direction, the RF magnetic field is weakest at the center of the resonator. However, as the distance from the current source to the observation point decreases, the RF magnetic field increases.
(b) In many cases, the physical length of the resonator is not negligibly small in comparison to the wavelength of electromagnetic waves. In this case, the RF current flowing in the conductors is no longer uniform.
(c) Also, the electrical properties of biological tissues in animals are not the same. These properties affect the RF electromagnetic fields.

Figure 10.5 shows the calculated RF magnetic field for a single-turn loop driven at 300 MHz (55 mm in diameter) [19]. Figure 10.5a, b show the loop segments and the distribution of the RF current flowing in the loop. Since the length of the loop circumference is not negligibly small in comparison to the wavelength at 300 MHz,

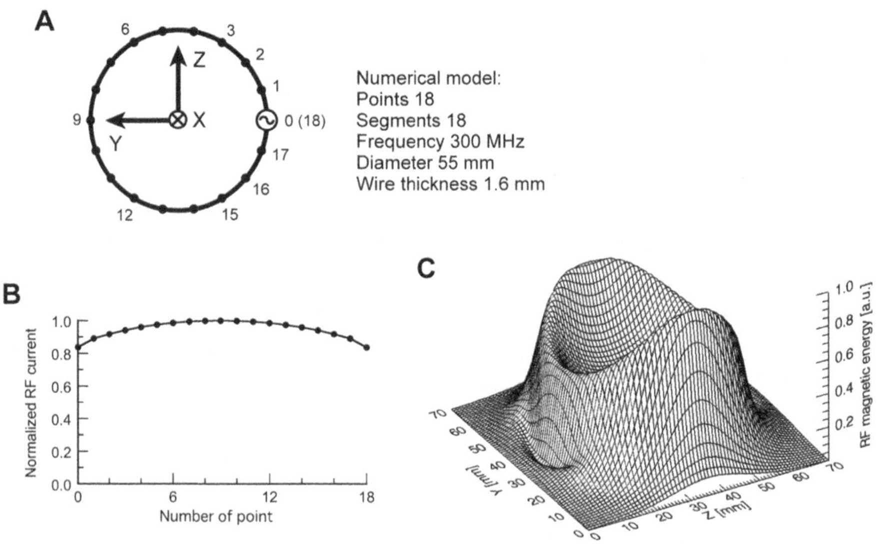

Fig. 10.5 Calculated RF magnetic field for a single-turn loop driven at 300 MHz (55 mm in diameter) [19]. (**a**) Modeling and segments of the loop for the calculation of the current and the magnetic field, (**b**) RF current distribution on the loop, and (**c**) the distribution of RF magnetic energy in the *YZ*-plane with an offset of 5 mm from the loop in the *X*-direction. In panel **b**, the point number on the abscissa refers to the location on the circumference of the loop in panel **a**. (Reprinted and adapted from Journal of Magnetic Resonance, Vol. 190, Hirata H, He G, Deng Y, Salikhov I, Petryakov S, Zweier JL, A loop resonator for slice-selective in vivo EPR imaging in rats, pp. 124–134, Copyright (2008), with permission from Elsevier)

the RF current in the loop is no longer uniform. Figure 10.5c shows the calculated RF magnetic energy that contributes to EPR. As mentioned above, the RF magnetic energy (square of the RF magnetic field B_1, see Eq. (10.7)) at the center of the loop is less than that in the proximity of the loop. To investigate the RF magnetic fields in the resonator, the finite-element method (FEM) or the finite-difference time-domain (FDTD) method can be used, and commercially available simulators for electromagnetic waves have been used to investigate the resonator and its RF electromagnetic fields. These simulations can then guide the manufacture, geometric dimensions and tolerance of resonators and their components to ensure that a given resonator design delivers maximal performance.

Ease of Operation

In animal experiments or a clinical setting, sample/subject handling is an important practical issue. An animal should be properly placed in the resonator or a person must be placed comfortably to allow for proper resonator placement. If ease of operation of the resonator is not well-considered, measurements may be confounded by several factors, such as the resonator not being placed properly in or relative to the magnetic field, inability to tune/couple, etc. A good example is a pop-up retractable resonator for use with mouse tumor-bearing legs [20].

Specific applications have different needs and technical considerations. For example, in EPR-based tooth dosimetry, the external loop resonator should be placed on the surface of the incisor of the human subject. Since placement of the loop is critical for good reproducibility of the measurements, the mechanical holder of the resonator should be stable and the other mechanical elements in the measurement setup should be considered in the context of ease of operation. In such a case, the resonator used has to be considered in terms of operation and sample handling.

Brief History of Technical Developments for In Vivo EPR Resonators

In vivo EPR spectroscopy and imaging of small animals have a history of almost four decades. This section will briefly look back at the development of resonators for in vivo EPR with small animals.

First In Vivo EPR Experiment Using a Helix Coil

In a report by Feldman et al. in 1975, a helix coil was implanted in the liver of a rat, and EPR spectra from exogenously injected free radical spin probes were observed [5]. Helices were made of gold-coated brass wire and enclosed in thin-wall Teflon tubing. The diameter of helices was in the range 1.6–2.4 mm. The helix coil was connected to a Varian X-band spectrometer via a waveguide-to-coax adapter. EPR spectra were recorded at a magnetic field of 0.3 T, which corresponds to a microwave frequency of 8.5 GHz.

Surface Coils

A helix coil and a flat coil were used as resonators at 1.86 GHz for animals in vivo [11]. A single-turn flat loop coil was used for in vivo EPR imaging of a living murine tumor (Cloudman S-91 melanoma in the tail of a mouse) [6]. An electronically tunable surface coil operating at 3 GHz was used for in vivo skin measurements in a human forearm [21]. A surface coil (4 mm in diameter) at 2.4 GHz was used for EPR spectroscopy and imaging of the surface domain of a large subject [22]. In that report, Herrling et al. explained two imaging approaches using a surface coil. One approach uses mechanical scanning of the surface coil and the other uses a magnetic field gradient as a conventional spatial imaging. Skin measurements are a good application for EPR spectroscopy and imaging using surface coils. Another skin study was performed by Takeshita et al. with a surface coil at 1.1 GHz [23, 24]. Another surface coil resonator, called an external loop resonator, was used for small animal experiments [15]. The single-turn loop was placed on the surface of tissue within which the oxygen-sensitive crystal LiPc was implanted.

Loop-Gap Resonators

In 1982, Froncisz and Hyde reported a loop-gap resonator (LGR) for EPR in the range of 1–10 GHz. They demonstrated EPR spectroscopy with LGRs at 3 and 9 GHz [10]. Figure 10.6 shows the structure of the LGR in the literature [10]. It is considered to be a lumped circuit when the dimensions of the LGR are smaller than 1/4 wavelength. Also, Nardy and Whitehead reported a split-ring resonator in the context of NMR in a frequency range from 200 to 2000 MHz [25]. These resonators are lumped circuits and are considered to be an LC resonant circuit. The LGR has the advantage of a high filling factor, which results in a greater EPR signal intensity. In the context of in vivo small animal EPR, a part of or the whole body of a subject animal can be placed in the loop. As previously mentioned, the resonant frequency

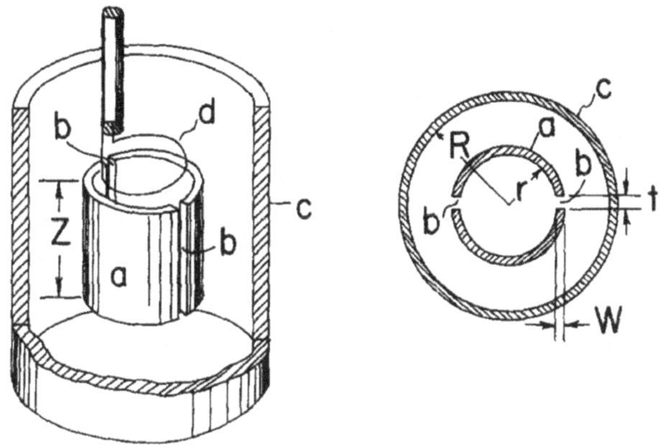

Fig. 10.6 Structure of the loop-gap resonator (two-gap one-loop) reported in 1982 [10]. (Reprinted from Journal of Magnetic Resonance, Vol. 47, Froncisz W, Hyde JS, The loop-gap resonator: A new lumped circuit ESR sample structure, pp. 515–521, Copyright (1982), with permission from Elsevier)

should be appropriately selected to give enough penetration depth for electromagnetic waves in biological tissues.

A modified structure of the LGR, called a bridged loop-gap resonator (BLGR), was reported in 1988 [26]. Curved conductive plates (or thin foils) called bridges are placed near the gap(s). In Pfenninger's BLGR, the bridge was located outside of the loops, and they intended to use BLGR for pulsed EPR at an X-band frequency. In the context of in vivo EPR, Ono et al. reported another structure for BLGR at an L-band frequency [27], in which electric shields are located inside the gaps to prevent the heating of biological tissues due to the electric field. LGR and BLGR have been used in small animal EPR spectroscopy and imaging [28, 29].

The design of an LGR is important for specific applications. Diodato et al. reported optimization of the axial RF field distribution in an LGR [30]. They showed that the coupling loop influenced the axial RF magnetic field and proposed the combination of two LGRs, with the coupling loop between the two LGRs. Ono et al. reported that electric shields influenced the RF magnetic field distribution in a BLGR [31]. The RF magnetic and electric fields inside the BLGR were measured. Their experiments clarified that appropriate angles of the electric shields can improve the homogeneity of the RF magnetic field.

Reentrant Resonators

Another type of resonator is a reentrant resonator. In 1983, Giordano et al. reported a reentrant cavity with a gap and a path for magnetic flux, which operated at 2 GHz [32]. They intended to use their reentrant resonator in studies on anisotropic line broadening. Later, Sotgiu and coworkers reported the designs of reentrant resona-

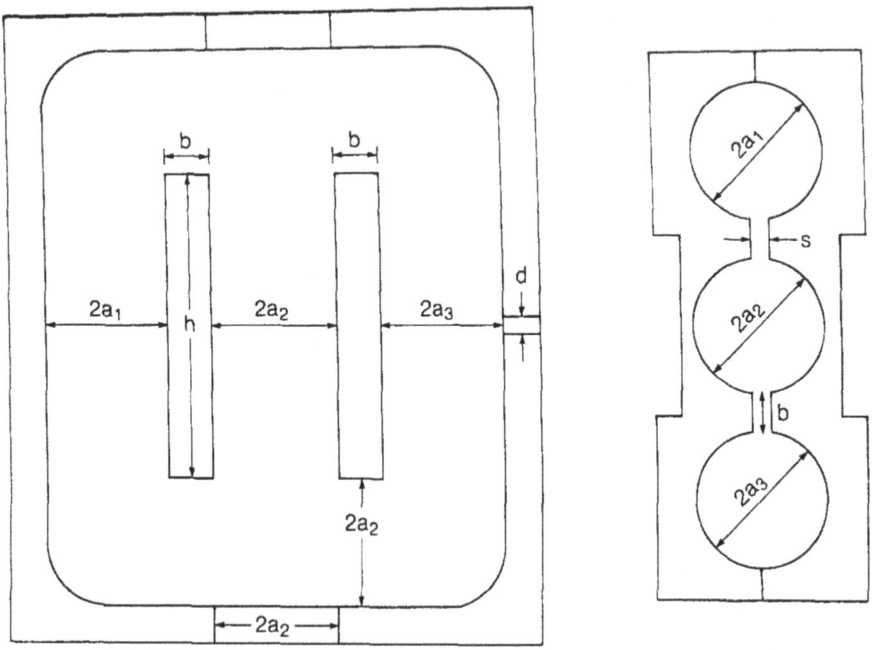

Fig. 10.7 Cross-section of a reentrant resonator reported in 1985 [35]. (Reprinted from Journal of Magnetic Resonance, Vol. 65, Sotgiu A, Resonator design for in vivo ESR spectroscopy, pp. 206–214, Copyright (1985), with permission from Elsevier)

tors that operated in the range of 2–10 GHz [33, 34]. Since reentrant resonators can operate at a low frequency, they have been used in small animal experiments. Figure 10.7 shows a cross-section of the reentrant resonator reported by Sotgiu in 1985 [35]. Chzhan et al. also developed reentrant resonators that operated at 1.2 GHz [36]. Chzhan et al. improved their reentrant resonator by making it capable of frequency tuning using piezoelectric actuators [14]. After that, Zweier and colleagues continued to develop reentrant resonators for EPR imaging in small animals at 750 MHz and 1.2 GHz [37, 38].

Surface Coils for Teeth or Fingernails

In human subjects, resonators can be used for the retrospective measurement of radiation dose, called EPR dosimetry. Stable free radicals in enamel in teeth and keratin in fingernails can be measured by EPR [39]. In measurements of the teeth, the surface coil is placed on an incisor or a molar.

Regarding EPR tooth dosimetry, Hochi et al. reported an X-band EPR spectrometer that used a cavity resonator with a small aperture [40]. In their approach, a human subject bit the cavity resonator and the electromagnetic waves from the aperture of the cavity resonator excited free radicals in the enamel of a tooth [41]. While a cavity resonator in the X-band can provide highly sensitive EPR signals because of greater Zeeman energy splitting, it requires a significantly larger magnet to establish the necessary fields, as compared to a magnet for low-field EPR, such as for L-band measurements. A lower magnetic field generated by an air-core magnet is suitable in a portable EPR spectrometer for triage dosimetry [42]. A 1.1-GHz surface loop resonator was used for EPR-based human tooth dosimetry. This application is explained in another subsection of this chapter.

For EPR dosimetry with fingernails, EPR measurements require the shallow penetration of electromagnetic waves in the fingernail. This is because fingernails are approximately 1 mm thick, and deeper penetration leads to a loss of electromagnetic waves in tissue in the fingertips. To satisfy this requirement, surface coil resonators in the X-band (9.5 GHz) were developed for in vivo EPR fingernail dosimetry. Sidabras et al. developed a microwave surface resonator array (SRA) in the X-band that is suitable for EPR-based fingernail dosimetry [43]. Grinberg et al. also reported a dielectric-backed aperture resonator in the X-band for in vivo nail dosimetry [44].

Resonators for Preclinical (Small Animal) Studies

For EPR spectroscopy and imaging in small animals such as mice and rats, a variety of resonators have been reported. This section introduces some of the more commonly used resonators in preclinical studies involving small animals.

Volume Resonators

Loop-Gap Resonator

LGRs can accommodate the whole body or part of a mouse or a rat. An early work on an LGR in 1986 involved the measurement of the partial pressure of oxygen in the peritoneal cavity of a mouse. Subczynski et al. used a 1 GHz LGR (25 mm in diameter and 30 mm long) to accommodate the whole body of a mouse [45]. A BLGR (43 mm in diameter and 30 mm long) operating at 800 MHz was used for measurement of the rat head. Ishida et al. reported the time-course of EPR signal intensities for the nitroxyl radical spin probes CTPO and TEMPOL [28]. A whole-body resonator based on an LGR (30 mm in diameter) at 1200 MHz was used for EPR spectroscopy in mice and rats [16]. For cancer studies using mouse tumor models, the LGR has been intensively used for 250 MHz CW and pulsed EPR imaging. For example, an LGR (16 mm in diameter and 15 mm long) was used for four-dimensional (4D) spectral-spatial EPR imaging to visualize the oxygen partial pressure in tumor-bearing mouse legs [46–48]. In addition to the CW protocol, the LGR has been used in pulsed EPR at 250 MHz for imaging of oxygen in tumor-bearing mouse legs [49, 50]. This LGR operating at 250 MHz was specifically designed for tumor-bearing mouse legs. Therefore, the sample space is rather small in comparison to other LGRs.

Reentrant Resonator

Many biomedical EPR studies have used reentrant resonators. Alecci et al. visualized nitroxyl radical probes in the rat tail by using three-dimensional EPR imaging and a 1.2-GHz reentrant resonator (sample space of 12 mm in diameter and 24 mm in length) [51]. He et al. reported EPR imaging in the beating heart of a mouse with a reentrant resonator operating at 1.2 GHz [37]. This resonator has varactor-based tuning and impedance-matching capabilities to compensate for the motion caused by the heartbeat of the mouse. A reentrant resonator operating at 750 MHz was used for EPR imaging of the whole-body of a mouse [52]. A 1.3-GHz ceramic three-loop two-gap reentrant resonator (for a sample 20 mm in diameter) [36] was used for EPR spectroscopy and imaging of nitric oxide generation in the mouse [53]. The same reentrant resonator was also used for oxygen mapping in the rat tail [54]. The gastrointestinal tract of a living mouse was visualized using EPR imaging and a 750-MHz tunable reentrant resonator [55].

Parallel Coil Resonator

In low-field pulsed EPR, the ring-down time of the resonator is critical [56]. After an RF pulse is transmitted to the resonator and a subject, the RF energy in the resonator remains stored for a period and disturbs signal detection at the receiver of a pulsed EPR instrument. This signal disturbance after the RF pulse is called dead time and is related to the ring-down time of the resonator. To overcome the problem regarding ringing after an RF pulse, the quality factor of the resonator is significantly reduced. A parallel coil resonator at 300 MHz has been reported by Devasahayam et al. [57]. The parallel coil resonator has been used in mouse whole-body imaging [58] and pO_2 mapping for tumor-bearing legs [59] at 300 MHz. A low quality factor (20–25) was achieved by over-coupling of the resonator [58]. The parallel coil resonator operating at 300 MHz could be used for EPR and 7 T proton MRI, which is important for EPR/NMR co-registration imaging of an animal subject.

Figure 10.8 shows a schematic of the parallel coil resonator [57]. The coils are connected to the coupling and tuning capacitors. Trimmer capacitors are adjusted to obtain the desired resonance frequency and a degree of coupling for the resonator. Since pulsed EPR does not require magnetic field modulation, a modulation coil is no longer needed. Thus, the structure of the parallel coil resonator is simple in comparison to a resonator with a modulation coil.

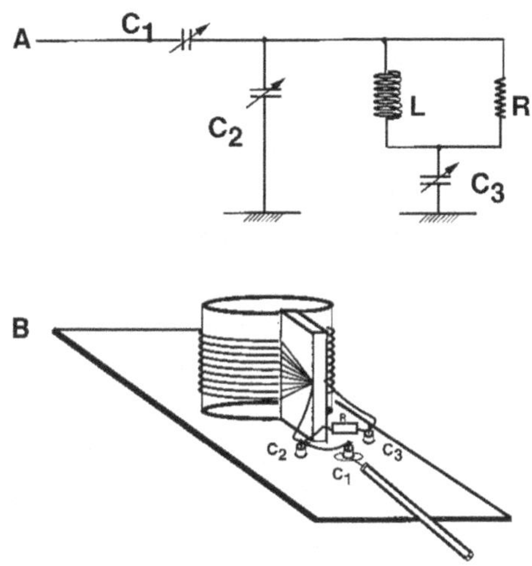

Fig. 10.8 Circuitry and schematic of a parallel coil resonator used for 300-MHz pulsed EPR [57]. (Reprinted from Journal of Magnetic Resonance, Vol. 142, Devasahayam N, Subramanian S, Murugesan R, Cook JA, Afeworki M, Tschudin RG, Mitchell JB, Krishna MC, Parallel coil resonators for time-domain radiofrequency electron paramagnetic resonance imaging of biological objects, pp. 168–176, Copyright (2000), with permission from Elsevier)

Alderman-Grant Resonator

The Alderman-Grant resonator (AGR) has been used in EPR and NMR. An important characteristic of the AGR is that the RF magnetic field is perpendicular to the axis of the AGR. The AGR was first reported by Alderman and Grant in 1979 [60]. In EPR-related studies, AGR has been used to excite electron spins in proton-electron double resonance imaging (PEDRI), which is also called Overhauser-enhanced magnetic resonance imaging (OMRI). Petryakov et al. developed an AGR at 590 MHz for exciting electron spins and demonstrated experiments with a high input power (up to 60 W for 3 min and 10 W for 10 min) [61]. Figure 10.9 illustrates the structure of the original AGR. This resonator can be used with other types of resonator such as a solenoid-type resonator because the RF magnetic fields are orthogonal to each other. This is necessary for double-resonance techniques (PEDRI/OMRI) [62].

Surface Resonator

Surface coil resonators have been used for measurements of the superficial regions of subject animals. A surface coil can be located at the region of interest for a measurement, even if a volume coil such as an LGR cannot accommodate a large subject. The simplest surface coil is just a combination of a single-turn loop and a capacitor to form a resonant circuit. However, additional circuitry is usually added to the coil to make it both possible and practical to adjust the resonant frequency and enable impedance matching. This section introduces several concepts regarding surface coil resonators.

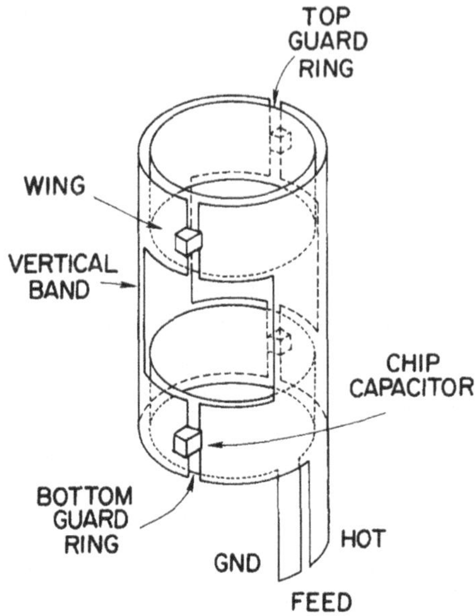

Fig. 10.9 Schematic of the Alderman-Grant resonator reported in 1979 [60]. (Reprinted from Journal of Magnetic Resonance, Vol. 36, Alderman DW, Grant DM, An efficient decoupler coil design which reduces heating in conductive samples in superconducting spectrometers, pp. 447–451, Copyright (1979), with permission from Elsevier)

Surface Coil/Loop Resonators

An electronically tunable resonator is useful for low-field EPR because perturbation due to motion of the subject should be accommodated. Figure 10.10 shows a circuit diagram of an electronically tunable surface coil resonator at 1.1 GHz [63]. This resonator uses varactor diodes for impedance matching and frequency adjustment. Also, a half-wavelength balun was used to ensure that the resonant circuit was properly balanced. A pair of varactor diodes that are connected in the opposite direction can increase the ability of the resonator to tolerate an incident RF power. This reduces the generation of the harmonics of incident RF signals. This type of surface coil resonator can be used for in vivo EPR spectroscopy.

A surface coil resonator is also called an external loop resonator (ELR). Figure 10.11 shows a circuit diagram of an ELR at 1.1 GHz [16, 64]. Varactor diodes are used to adjust the resonant frequency, and a photo-resistor and a light-emitting diode (LED) are used to control the quality factor. A shift in the quality factor can control impedance-matching between the 50-Ω transmission line and the resonator. Thus, the voltage applied to the LED controls the impedance of the ELR. An inductive coupling scheme is also used in the ELR. The length of a twisted wire connected to the inductive coupling loop is adjusted to minimize the shift of the resonant frequency due to the adjustment of impedance-matching [65].

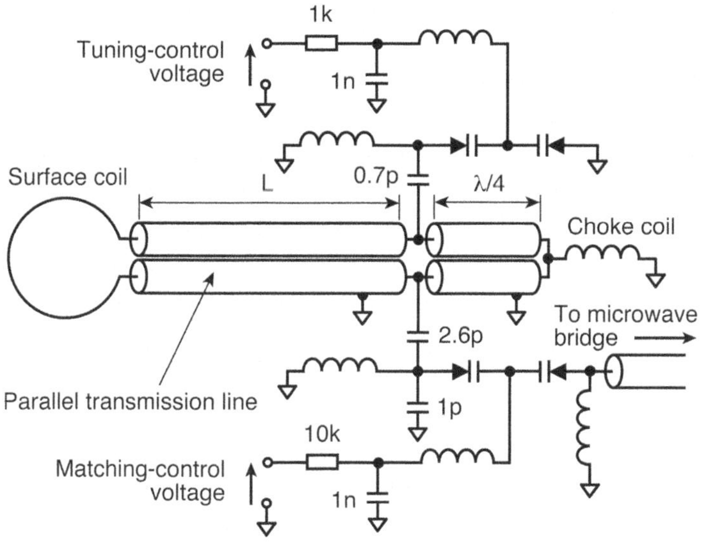

Fig. 10.10 Schematic of a tunable surface coil resonator reported in 2003 [63]. (Reprinted from Journal of Magnetic Resonance, Vol. 164, Hirata H, Kuyama T, Ono M, Shimoyama Y, Detection of electron paramagnetic resonance absorption using frequency modulation, pp. 233–241, Copyright (2003), with permission from Elsevier)

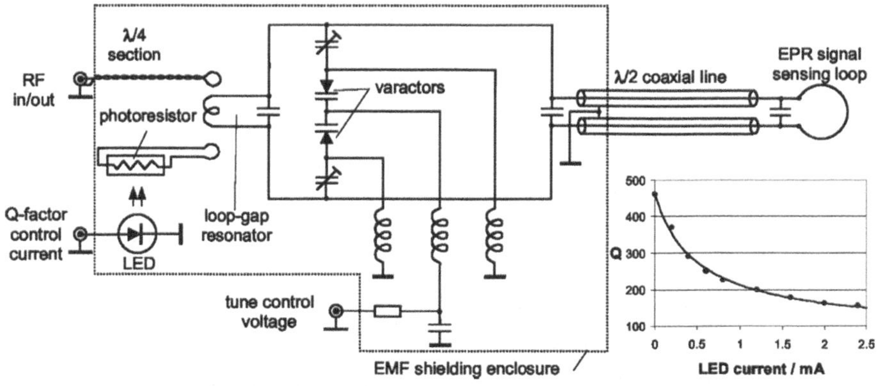

Fig. 10.11 Circuit diagram of the external loop resonator (ELR) reported in 2005 [16]. (Reprinted from Walczak T, Lesniewski P, Salikhov I, Sucheta A, Szybinski K, Swartz HM. 2005. L-band electron paramagnetic resonance spectrometer for use in vivo and in studies of aqueous biological samples. *Rev Sci Instrum* **76**:013107, with the permission of AIP Publishing)

An ELR has been used for in vivo L-band EPR spectroscopy. However, the details of the ELR were published in 2005. More recently, the ELR was modeled and analyzed by the FEM [66]. Sugawara et al. reported a FEM analysis and optimization of a 1.1-GHz surface coil resonator for EPR tooth dosimetry [67].

Surface Coil Array

To overcome the limited sensitive region of the surface coil resonator, an array of surface coil resonators can be used. This approach is also used in clinical MRI. Due to the higher RF frequency in EPR than in NMR/MRI, the size of the surface coil is usually limited. If the circumference of the loop is not negligibly short in comparison to the wavelength, i.e., <1/4 wavelength, of the RF electromagnetic waves, the fields generated by the loop are no longer in the correct phase relative to each other. This leads to an inhomogeneous RF magnetic field in the loop. This section briefly introduces a surface coil array for in vivo EPR imaging.

There are two approaches to image acquisition using a surface coil array: (1) sequential acquisition and (2) simultaneous (parallel) acquisition. Sequential acquisition refers to when one of the surface coil resonators within the array is selected for EPR signal recording. This is a simple approach to EPR imaging. However, the interaction between the neighboring coils should be suppressed to avoid interference while an image is being acquired. This is because inductive coupling between coils that are located in close proximity affects the resonance characteristics of the surface coil resonator to which the RF electromagnetic waves are fed. Enomoto et al. used PIN-diode switches to suppress the interaction between the neighboring resonators in a CW-EPR detection protocol [68, 69]. Figure 10.12 shows a four-channel surface coil array operating at 750 MHz. This array was tested with a living mouse [69]. The array was placed on the back of a subject mouse. After the mouse was injected with spin probes, EPR image-acquisition was performed with each surface coil in sequence. The reconstructed images were then combined. In pulsed EPR, the quality factor of the resonator is significantly decreased. This can reduce the influence of the interaction between the neighboring coils. Enomoto et al. demonstrated a surface coil array in pulsed EPR imaging at 300 MHz. As in CW-EPR, the resonant frequencies of unexcited resonators in the array were shifted to reduce the mutual inductive coupling of the coils [70]. In addition, passive decoupling of the coils was also tested for a four-channel surface coil array [71]. The feasibility of

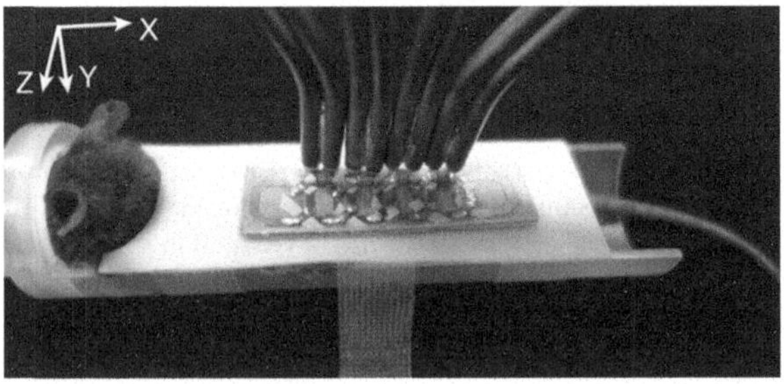

Fig. 10.12 Photograph of a four-channel surface coil array operating at 750 MHz for EPR imaging in mice [69]. (Reprinted from Journal of Magnetic Resonance, Vol. 234, Enomoto A, Emoto M, Fujii H, Hirata H, Four-channel surface coil array for sequential CW-EPR image acquisition, pp. 21–29, Copyright (2013), with permission from Elsevier)

the simultaneous acquisition of multiple coils has also been reported [72]. Development of a surface coil array for EPR imaging is an important direction for preclinical and clinical applications.

Surface Detection Using LGR

Another approach to detecting EPR signals from a superficial region in biological subjects uses the fringe magnetic field of a volume coil resonator such as the LGR instead of a surface coil. A surface probe using a double split-ring resonator operating at 1 GHz was reported by Nilges et al. [73]. In a more clinically related application, EPR imaging for human skin was performed with a surface probe using a BLGR in the S-band (2.2 GHz) [8, 9]. With a higher frequency in the S-band, nitroxide probes in human skin were successfully detected and the distribution of nitroxide radicals was visualized. While the frequency of 2 GHz limits the penetration depth of electromagnetic waves in biological tissues, measurements in human skin do not require deep penetration by electromagnetic waves.

Implantable Resonator

As mentioned above, surface coils have a limited sensitive volume, although there are some advantages of a comparatively high sensitivity and no limitation to the subject size. When a region of interest is deep in tissue that is a part of a large object, neither a surface coil nor a volume coil, e.g., the LGR, can be applied. In particular, the internal organs of human subjects are difficult to measure with either a surface coil or an LGR with a commonly used low-field EPR spectrometer/imager. One possible solution is an implantable resonator.

Li et al. demonstrated an implantable resonator for an EPR oximetry study in rat brain [74]. The implantable resonator has oxygen-sensitive materials in a small loop or multiple small loops that are located at the end of a twisted pair of transmission lines [75]. Figure 10.13 shows a photograph of an implantable resonator for rabbit brain [76]. The length of the twisted wires can be adjusted to reach the region of interest. The other end of the twisted wires has a larger loop that can be coupled with the loop located on the surface of the object.

RF signals can be delivered to the oxygen-sensitive probe in a small loop (or multiple loops) of the implantable device. After the implantable device is placed in the subject, EPR spectroscopy can be repeatedly conducted noninvasively. The small loop can increase the sensitivity of EPR detection because a strong RF magnetic field is generated in a small loop. This phenomenon is a primary reason why an implantable resonator can work effectively in deep tissues. With an implantable resonator with multiple loops, the partial pressure of oxygen at multiple sites can be measured, and, for example, it is possible to compare oxygenation at several regions

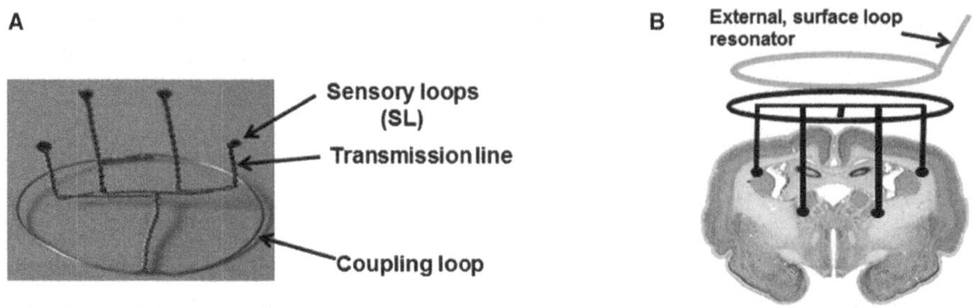

Fig. 10.13 Schematic of implantable resonators for rabbit brain [76]. (Reprinted from Stroke, Vol. 46, Khan N, Hou H, Eskey CJ, Moodie K, Gohain S, Du G, Hodge S, Culp WC, Kuppusamy P, Swartz HM, Deep-tissue oxygen monitoring in the brain of rabbits for stroke research, pp. e62–e66, copyright (2015), with permission from Wolters Kluwer)

of interest in rat brain. These measurements could be useful for studies in an animal model of ischemic stroke [75, 76] or in a tumor xenograft model [77, 78].

From an engineering point of view, the materials used to make an implantable resonator should be biocompatible if the resonator is to be placed in biological tissues for a long time. An implantable resonator made of copper can be coated with a biocompatible material. For oxygen measurements, the coating materials should be oxygen-permeable.

Resonators for In Vivo EPR in Human Subjects

Surface Resonators for Human Skin/Surface

EPR spectroscopy and imaging on human skin have been reported by He et al. [8]. They reported the distribution and metabolism of nitroxide radicals in human skin using a loop-gap surface resonator operating at 2.2 GHz [9]. A penetration depth of electromagnetic waves of approximately 2 mm is enough for skin EPR measurements. Reports of in vivo EPR measurements in human skin are still limited. For in vivo studies in human skin, the microwave frequency can be higher than conventional in vivo frequencies (L-band or lower RF frequencies), since the limited penetration depth is not a problem, but rather a desirable feature, for studies in skin. Successful applications of EPR spectroscopy and imaging in human skin help foster the further development of surface resonators. Wolfson et al. developed an LGR-based surface-detection resonator operating at 2.3 GHz [79]. This resonator was intended for transcutaneous oxygen-monitoring using the oxygen-sensitive crystal LiNc-BuO. The dimensions of this resonator are 13 mm in outer diameter, 3.6 mm in inner diameter, and 19 mm in height. The gap in the loop is 0.1 mm. This LGR was installed in an array of small magnets that generate a static magnetic field.

L-Band Surface Loop Resonators for Tooth Dosimetry

When teeth receive ionizing radiation in medical treatments, or because of other reasons such as an accident in a nuclear power plant, very stable CO_2^- radical is generated in the enamel [39]. EPR spectroscopy can detect these radiation-induced free radicals in human teeth after they have been exposed to ionizing radiation. Surface coil resonators have been developed for EPR-based in vivo tooth dosimetry.

An ELR at 1.15 GHz was developed and used for human EPR tooth dosimetry [80]. To improve the detection capabilities to more accurately estimate the radiation dose, surface coil resonators were further developed [66, 67, 81]. The loop was placed on an incisor to detect free radicals in the enamel. Recently, a unique EPR resonator was developed for tooth dosimetry. It consisted of a combination of an inductively coupled ELR printed on a flexible substrate which was driven by a non-resonant quarter-wavelength-loop antenna [82]. The printed ELR can be affixed to an incisor for an in vivo measurement. The advantages of this wireless flexible surface coil resonator are as follows:

(a) A printed ELR can be mass-produced to have a low unit cost, and therefore be disposable. This is important for biomedical applications and high throughput if a large number of people are to be measured in a short period of time, e.g., a triage operation in the event of a nuclear incident.
(b) A printed ELR can be easily affixed to a tooth via suitable medical double-sided tape outside of the magnet, reducing the need for cumbersome placement inside the restricted space of the magnet.
(c) The field generated by the antenna coupler encompasses both of the front incisor teeth, and therefore precise placement of the coupler inside the magnet during the measurement is not necessary—the coupler can simply be placed universally for all measurement subjects, reducing the burden on the operator and increasing the potential for automated actions of the EPR dosimeter.

Resonators for tooth dosimetry were tested in human subjects [83]. Since tooth dosimetry is a topical EPR measurement, the small surface loop operating at 1.1–1.2 GHz is not limited by the penetration depth of electromagnetic waves, and loss of the tooth itself is not so significant in comparison to the loss of other soft biological tissues [84].

X-Band Resonators for Fingernail Dosimetry

Several types of resonators have also been developed for another EPR-based dosimetry application: in vivo fingernail dosimetry. Fingernail dosimetry is commonly performed in the X-band, as radiation-induced signals in the fingernails are significantly weaker than those in tooth enamel. This increased frequency provides a high sensitivity of EPR detection due to a greater energy gap of Zeeman splitting. Since the region of EPR spectroscopy detection in the X-band is superficial and is suitable only for nails.

The radiation-induced EPR signal in human fingernails has been reported [85]. Radiation-induced signals in chipped fingernails were initially measured with a conventional X-band cavity resonator to qualify and generally characterize the dosimetry technique as a suitable candidate for physically based dosimetry [86]. This approach of using a microwave cavity is not suitable for in vivo human fingernails, as the conventional cavity resonator is designed for glass capillary tubes. Therefore, X-band resonators have been developed for in vivo studies of fingernails.

A microwave surface resonator array (SRA) has been developed for in vivo EPR nail dosimetry as a practical in vivo replacement for a conventional microwave cavity [43]. Figure 10.14 shows a photograph of an SRA for EPR fingernail dosimetry. This metallic arrayed structure was designed to limit the depth of penetration of the stimulating magnetic field so as to focus the field generated by the SRA into the fingernail itself rather than the underlying tissue. The intent of this is to help ensure that no energy applied to the resonator is then deposited into tissues where there is not expected to be any signal, i.e., the tissue under the nail. For a given nail dosimetry measurement, the resonator is simply placed on the fingernail within (or outside) the magnet, and then the finger and resonator assembly are then mated to an ergonomic receptacle within the EPR dosimeter.

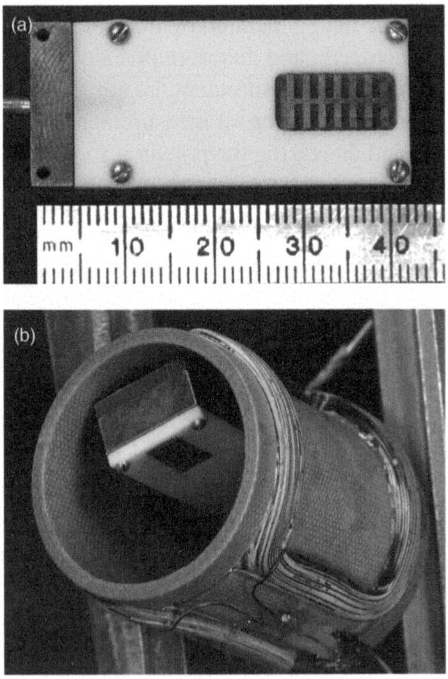

Fig. 10.14 Photograph of an SRA for EPR fingernail dosimetry [43]. (Reprinted from Sidabras JW, Varanasi SK, Mett RR, Swarts SG, Swartz HM, Hyde JS. 2014. A microwave resonator for limiting depth sensitivity for electron paramagnetic resonance spectroscopy of surfaces. *Rev Sci Instrum*, **85**:104707, licensed under a Creative Commons Attribution (CC BY) license)

Several resonators that use dielectric resonant structures have also been developed for the purpose of surface measurements at an X-band frequency, e.g., in vivo nail dosimetry. These dielectric resonator assemblies seek to more closely match the dielectric of the resonant structure to the dielectric of the tissue being measured to maximize the amount of RF energy transferred into the tissue of interest, i.e., the fingernail. Proper high-dielectric substrates also provide the opportunity for a higher quality factor for a resonator, given a suitable selection of dielectric resonator for the application. For the dielectric resonator in Ref. [44], RF power is driven to the dielectric resonator via the use of a modified resonant microwave cavity, and Ref. [87] describes a resonator topology where RF power is driven to the dielectric resonator via the use of a non-resonant antenna.

Implantable Resonators

Another direction for resonator development in human subjects is to measure EPR signals from tissues that are not on the surface of the body or superficial to the surface, i.e., at depths >10 mm below the skin. This is primarily of concern when attempting to measure oxygen levels in a given organ in vivo in a noninvasive fashion. While resonator implantation is invasive, repeated measurements can be performed by placing an ELR over the coupling loop of the implanted resonator [88] and collecting EPR data as in a usual surface EPR oximetry measurement.

A possible solution is the use of an implantable resonator to detect EPR signals deep in tissue. A clinically acceptable implantable EPR resonator is currently under development [89], and an implantable resonator for a rat brain [75, 76, 78] is a similar model of such an implantable resonator for human subjects. A rabbit brain was also measured with an implantable resonator [77]. Implantable coils have been investigated in proton magnetic resonance imaging (MRI) [90, 91]. The implantable resonator being developed in Ref. [89] seeks to accommodate and address aspects of clinical acceptance, including insertion/removal procedures, biocompatibility, MRI compatibility, and other aspects associated with obtaining an investigation device exemption from the U.S. Food and Drug Administration (FDA) with the goal of bringing implantable resonators closer to reality in the clinic. The development of implantable resonators for use in EPR spectroscopy and imaging in human subjects is a potential direction for future studies.

Future Perspective

From the viewpoint of clinical EPR, the development of resonators has several directions that can contribute to the future clinical use of EPR spectroscopy and imaging.

A Resonator for Superficial Tissues

As mentioned above, this is the most practical target for present EPR resonators. For example, skin, teeth, and fingernails are easy to access. When the RF frequency for EPR decreases to an RF frequency similar to that in clinical MRI, the problem of the penetration depth of electromagnetic waves may be partly solved; however, the absolute EPR amplitude sensitivity will suffer. A small resonator used for endoscope-like EPR measurement may also be considered for measurements of superficial regions.

An Implantable Resonator

Surgery is required to place implantable resonators deep in tissue. This is a drawback in clinical EPR. However, once placed, implantable resonators enable repeated, noninvasive measurements in EPR spectroscopy and imaging. The risks and benefits of implantable resonators should be considered, as with any other clinical treatment. However, technologies regarding implantable resonators should be explored further to achieve clinical EPR, as each implantable resonator should meet the specific needs of the clinical application of EPR spectroscopy and imaging.

A Volume Coil Resonator for Further Low-Frequency Clinical EPR

To overcome the limited penetration depth of electromagnetic waves in tissues, a further low RF frequency might be used. MRIs with magnetic fields of 1.5 T and 3 T operate at 64 MHz and 128 MHz, respectively. By analogy with MRI, such a lower RF frequency provides better penetration of electromagnetic waves in tissue as well as a smaller energy gap in Zeeman splitting, thereby resulting in a decrease in the sensitivity of EPR detection. In contrast to a smaller energy gap, a volume coil operating at a frequency similar to that in clinical MRI can accommodate a larger subject. This might compensate for the weaker EPR signals at a lower magnetic field and a corresponding RF frequency. Both the resonator and other parts of EPR instruments should be further improved to achieve a reasonable sensitivity of EPR detection at a low RF frequency to approach that of clinical MRI. Whole-body human EPR imaging, like proton MRI, is a dream in the application of EPR spectroscopy and imaging.

The future development of resonators for clinical EPR is not limited to these directions. Innovative technologies in RF/microwave electronics and advanced manufacturing techniques such as 3D printing may provide breakthroughs for EPR resonators in the future.

References

1. Rinard GA, Eaton SS, Eaton GR. Sensitivity (Chap. 1). In: Poole Jr CP, Farach HA, editors. Handbook of electron spin resonance, vol. 2. New York: Springer; 1999.
2. Feher G. Sensitivity considerations in microwave paramagnetic resonance absorption techniques. Bell System Technol J. 1957;36:449–84.
3. Weil JA, Bolton JR. Electron paramagnetic resonance. 2nd ed. Hoboken, NY: Wiley; 2007. p. 1–35.
4. Griffiths DJ. Introduction to electrodynamics. 2nd ed. Englewood Cliffs, NJ: Prentice Hall; 1989. p. 369–72.
5. Feldman A, Wildman E, Bartolinini G, Piette LH. In vivo electron spin resonance in rats. Phys Med Biol. 1975;20:602–12.
6. Berliner LJ, Fujii H, Wan X, Lukiewicz SJ. Feasibility study of imaging a living murine tumor by electron paramagnetic resonance. Magn Reson Med. 1987;4:380–4.
7. Eaton GR, Eaton SS. EPR spectrometers at frequencies below X-band (Chap. 2). In: Bender CJ, Beriner LJ, editors. EPR instrumental methods. Biological magnetic resonance, vol. 21. New York: Kluwer Academic/Plenum; 2004.
8. He G, Samouilov A, Kuppusamy P, Zweier JL. In vivo EPR imaging of the distribution and metabolism of nitroxide radicals in human skin. J Magn Reson. 2001;148:155–64.
9. Petryakov S, Chzhan M, Samouilov A, He G, Kuppusamy P, Zweier JL. A bridged loop-gap S-band surface resonator for topical EPR spectroscopy. J Magn Reson. 2001;151:124–8.
10. Froncisz W, Hyde JS. The loop-gap resonator: a new lumped circuit ESR sample structure. J Magn Reson. 1982;47:515–21.
11. Nishikawa H, Fujii H, Beriliner LJ. Helices and surface coils for low-field *in vivo* ESR and EPR imaging applications. J Magn Reson. 1985;62:79–86.
12. Eaton GR, Eaton SS, Barr DP, Weber RT. Quantitative EPR (Chap. 7). Wien: Springer; 2010.
13. Brivati JA, Stevens AD, Symons MCR. A radiofrequency ESR spectrometer for *in vivo* imaging. J Magn Reson. 1991;92:480–9.
14. Chzhan M, Kuppusamy P, Zweier JL. Development of an electronically tunable L-band resonator for EPR spectroscopy and imaging of biological samples. J Magn Reson Ser B. 1995;108:67–72.
15. Hirata H, Walczak T, Swartz HM. Electronically tunable surface-coil-type resonator for L-band EPR spectroscopy. J Magn Reson. 2000;142:159–67.
16. Walczak T, Lesniewski P, Salikhov I, Sucheta A, Szybinski K, Swartz HM. L-band electron paramagnetic resonance spectrometer for use *in vivo* and in studies of aqueous biological samples. Rev Sci Instrum. 2005;76:013107.
17. Rinard GA, Quine RW, Eaton SS, Eaton GR. Microwave coupling structures for spectroscopy. J Magn Reson Ser A. 1993;105:137–44.
18. McCallum S, Resmer F. Automatic coupling control system for radio frequency *in vivo* electron paramagnetic resonance based on a piezoelectric controlled capacitor. Rev Sci Instrum. 1999;70:4706–10.
19. Hirata H, He G, Deng Y, Salikhov I, Petryakov S, Zweier JL. A loop resonator for slice-selective in vivo EPR imaging in rats. J Magn Reson. 2008;190:124–34.
20. Epel B, Subramanian VS, Halpern HJ. Retractable loop-gap resonator for electron paramagnetic resonance imaging with in situ irradiation capabilities. Concepts Magn Reson B. 2011;39B:167–72.
21. Herrling TE, Groth NK, Fuchs J. Biochemical EPR imaging of skin. Appl Magn Reson. 1996;11:471–86.
22. Herrling T, Rehberg J, Jung K, Groth N. SURF_ER–surface electron spin resonance (ESR) of surface domain of large subjects. Spectrochim Acta A. 2002;58:1337–44.
23. Takeshita K, Takajo T, Hirata H, Ono M, Utsumi H. In vivo oxygen radical generation in the skin of the protoporphyria model mouse with visible light exposure: an L-band ESR study. J Invest Dermatol. 2004;122:1463–70.

24. Takeshita K, Chi C, Hirata H, Ono M, Ozawa T. In vivo generation of free radical in the skin of living mouse under ultraviolet light measured by L-band EPR spectroscopy. Free Radic Biol Med. 2006;40:876–85.
25. Nardy WN, Whitehead LA. Split-ring resonator for use in magnetic resonance from 200–2000 MHz. Rev Sci Instrum. 1981;52:213–6.
26. Pfenninger S, Forrer J, Schweiger A, Weiland T. Bridged loop-gap resonator: a resonant structure for pulsed ESR transparent to high-frequency radiation. Rev Sci Instrum. 1988;59:752–60.
27. Ono M, Ogata T, Hsieh K-C, Suzuki M, Yoshida E, Kamada H. L-band ESR spectrometer using a loop-gap resonator for *in vivo* analysis. Chem Lett. 1986;15:491–4.
28. Ishida S, Kumashiro H, Tsuchihashi N, Ogata T, Ono M, Kamada H, Yoshida E. *In vivo* analysis of nitroxide radicals injected into small animals by L-band ESR technique. Phys Med Biol. 1989;34:1317–23.
29. Ishida S, Matsumoto S, Yokoyama H, Mori N, Kumashiro H, Tsuchihashi N, Ogata T, Yamaga M, Ono M, Kitajima T, Kamada H, Yoshida E. An ESR-CT imaging of the head of a living rat receiving an administration of a nitroxide radical. Magn Reson Imaging. 1992;10:109–14.
30. Diodato R, Alecci M, Brivati JA, Varoli V, Sotgiu A. Optimization of axial RF field distribution in low-frequency EPR loop-gap resonators. Phys Med Biol. 1999;44:N69–75.
31. Ono M, Suenaga A, Hirata H. Experimental investigation of RF magnetic field homogeneity in a bridged loop-gap resonator. Magn Reson Med. 2002;47:415–9.
32. Giordano M, Momo F, Sotgiu A. On the design of a reentrant square cavity as resonator for low-frequency ESR spectroscopy. J Phys E Sci Instrum. 1983;16:774–9.
33. Momo F, Sotgiu A. Re-entrant resonators for ESR spectroscopy between 2 and 10 GHz. J Phys E Sci Instrum. 1984;17:556–8.
34. Sotgiu A, Gualtieri G. Cavity resonator for *in vivo* electron-spin-resonance spectroscopy. J Phys E Sci Instrum. 1985;18:899–901.
35. Sotgiu A. Resonator design for *in vivo* ESR spectroscopy. J Magn Reson. 1985;65:206–14.
36. Chzhan M, Shteynbuk M, Kuppusamy P, Zweier JL. An optimized L-band ceramic resonator for EPR imaging of biological samples. J Magn Reson Ser A. 1993;105:49–53.
37. He G, Petryakov S, Samoulilov A, Chzhan M, Kuppusamy P, Zweier JL. Development of a resonator with automatic tuning and coupling capability to minimize sample motion noise for *in vivo* EPR spectroscopy. J Magn Reson. 2001;149:218–27.
38. He G, Dumitrescu C, Petryakov S, Deng Y, Kesselring E, Zweier JL. Transverse oriented electric field reentrant resonator (TERR) with automatic tuning and coupling control for EPR spectroscopy and imaging of the beating heat. J Magn Reson. 2007;187:57–65.
39. Swartz HM, Williams BB, Flood AB. Overview of the principles and practice of biodosimetry. Radiat Environ Biophys. 2014;53:221–32.
40. Hochi A, Furusawa M, Ikeya M. Applications of microwave scanning ESR microscope: human tooth with metal. Appl Radiat Isot. 1993;44:401–5.
41. Ishii H, Ikeya M. An electron spin resonance system for in-vivo human tooth dosimetry. Jpn J Appl Phys. 1990;29:871–5.
42. Williams BB, Flood AB, Salikhov I, Kobayashi K, Dong R, Rychert K, Du G, Schreiber W, Swartz HM. In vivo EPR tooth dosimetry for triage after a radiation event involving large populations. Radiat Environ Biophys. 2014;53:335–46.
43. Sidabras JW, Varanasi SK, Mett RR, Swarts SG, Swartz HM, Hyde JS. A microwave resonator for limiting depth sensitivity for electron paramagnetic resonance spectroscopy of surfaces. Rev Sci Instrum. 2014;85:104707.
44. Grinberg O, Sidabras JW, Tipikin DS, Krymov V, Mariani M, Feldman MM, Kmiec MM, Petryakov SV, Brugger S, Carr B, Schreiber W, Swarts SG, Swartz HM. Dielectric-backed aperture resonators for X-band *in vivo* EPR nail dosimetry. Radiat Prot Dosim. 2016;172:121–6.
45. Subczynski WK, Lukiewicz S, Hyde JS. Murine *in vivo* L-band ESR spin-label oximetry with a loop-gap resonator. Magn Reson Med. 1986;3:747–54.
46. Elas M, Williams BB, Parasca A, Malier C, Pelizzari CA, Lewis MA, River JN, Karczmar GS, Barth ED, Halpern HJ. Quantitative tumor oxymetric images from 4D electron paramagnetic

resonance imaging (EPRI): methodology and comparison with blood oxygen level-dependent (BOLD) MRI. Magn Reson Med. 2003;49:682–91.
47. Elas M, Ahn KH, Parasca A, Barth ED, Lee D, Haney C, Halpern HJ. Electron paramagnetic resonance oxygen images correlate spatially and quantitatively with oxylite oxygen measurements. Clin Cancer Res. 2006;12:4209–17.
48. Elas M, Bell R, Hleihel D, Barth ED, McFaul C, Haney CR, Bielanska J, Pustelny K, Ahn KH, Pelizzari CA, Kocherginsky M, Halpern HJ. Electron paramagnetic resonance oxygen image hypoxic fraction plus radiation dose strongly correlates with tumor cure in FSa fibrosarcomas. Int J Radiat Oncol Biol Phys. 2008;71:542–9.
49. Epel B, Sundramoorthy SV, Mailer C, Halpern HJ. A versatile high speed 250-MHz pulse imager for biomedical applications. Concepts Magn Reson B. 2008;33B:163–76.
50. Elas M, Magwood JM, Butler B, Li C, Wardak R, Barth ED, Epel B, Rubinstein S, Pelizzari CA, Weichselbaum RR, Halpern HJ. EPR oxygen images predict tumor control by a 50% tumor control radiation dose. Cancer Res. 2013;73:5328–35.
51. Alecci M, Colacicchi S, Indovina PL, Momo F, Pavone P, Sotgiu A. Three-dimensional *in vivo* ESR imaging in rats. Magn Reson Imaging. 1990;8:59–63.
52. He G, Deng Y, Li H, Kuppusamy P, Zweier JL. EPR/NMR co-imaging for anatomic registration of free-radical images. Magn Reson Med. 2002;47:571–8.
53. Kuppusamy P, Shankar RA, Roubaud VM, Zweier JL. Whole body detection and imaging of nitric oxide generation in mice following cardiopulmonary arrest: detection of intrinsic nitrosoheme complexes. Magn Reson Med. 2001;45:700–7.
54. Velan SS, Spencer RGS, Zweier JL, Kuppusamy P. Electron paramagnetic resonance oxygen mapping (EPROM): direct visualization of oxygen concentration in tissue. Magn Reson Med. 2000;43:804–9.
55. He G, Shankar RA, Chzhan M, Samouilov A, Kuppusamy P, Zweier JL. Noninvasive measurement of anatomic structure and intraluminal oxygenation in the gastrointestinal tract of living mice with spatial and spectral EPR imaging. Proc Natl Acad Sci U S A. 1999;96:4586–91.
56. Subramanian S, Matsumoto K, Mitchell JB, Krishna MC. Radio frequency continuous-wave and time-domain EPR imaging and Overhauser-enhanced magnetic resonance imaging of small animals: instrumental developments and comparison of relative merits for functional imaging. NMR Biomed. 2004;17:263–94.
57. Devasahayam N, Subramanian S, Murugesan R, Cook JA, Afeworki M, Tschudin RG, Mitchell JB, Krishna MC. Parallel coil resonators for time-domain radiofrequency electron paramagnetic resonance imaging of biological objects. J Magn Reson. 2000;142:168–76.
58. Subramanian S, Devasahayam N, Murgesan R, Yamada K, Cook J, Taube A, Mitchell JB, Lohman JAB, Krishna MC. Single-point (constant-time) imaging in radiofrequency Fourier transform electron paramagnetic resonance. Magn Reson Med. 2002;48:370–9.
59. Matsumoto S, Hyodo F, Subramanian S, Devasahayam N, Munasinghe J, Hyodo E, Gadisetti C, Cook JA, Mitchekk JB, Krishna MC. Low-field paramagnetic resonance imaging of tumor oxygenation and glycolytic activity in mice. J Clin Invest. 2008;118:1965–73.
60. Alderman DW, Grant DM. An efficient decoupler coil design which reduces heating in conductive samples in superconducting spectrometers. J Magn Reson. 1979;36:447–51.
61. Petryakov S, Samouilov A, Roytenberg M, Li H, Zweier JL. Modified Alderman-Grant resonator with high-power stability for proton electron double resonance imaging. Magn Reson Med. 2006;56:654–9.
62. Lurie DJ, Davis GR, Foster MA, Hutchison JMS. Field-cycled PEDRI imaging of free radicals with detection at 450 mT. Magn Reson Imaging. 2005;23:175–81.
63. Hirata H, Kuyama T, Ono M, Shimoyama Y. Detection of electron paramagnetic resonance absorption using frequency modulation. J Magn Reson. 2003;164:233–41.
64. Salikhov I, Hirata H, Walczak T, Swartz HM. An improved external loop resonator for *in vivo* L-band EPR spectroscopy. J Magn Reson. 2003;164:54–9.

65. Hirata H, Walczak T, Swartz HM. An improved inductive coupler for suppressing a shift in the resonance frequency of electron paramagnetic resonance resonators. Rev Sci Instrum. 1997;68:3187–91.
66. Pollock JD, Williams BB, Sidabras JW, Grinberg O, Salikhov I, Lesniewski P, Kmiec M, Swartz HM. Surface loop resonator design for *in vivo* EPR tooth dosimetry using finite element analysis. Health Phys. 2010;98:339–44.
67. Sugawara H, Hirata H, Petryakov S, Lesniewski P, Williams BB, Flood AB, Swartz HM. Design and evaluation of a 1.1-GHz surface coil resonator for electron paramagnetic resonance-based tooth dosimetry. IEEE Trans Biomed Eng. 2014;61:1894–901.
68. Enomoto A, Hirata H. Sequential CW-EPR image acquisition with 760-MHz surface coil array. J Magn Reson. 2011;209:244–9.
69. Enomoto A, Emoto M, Fujii H, Hirata H. Four-channel surface coil array for sequential CW-EPR image acquisition. J Magn Reson. 2013;234:21–9.
70. Enomoto A, Hirata H, Matsumoto S, Saito K, Subramanian S, Krishna MC, Devasahayam N. Four-channel surface coil array for 300-MHz pulsed EPR imaging: proof-of-concept experiments. Magn Reson Med. 2014;71:853–8.
71. Enomoto A, Saito K, Subramanian S, Krishna MC, Hirata H, Devasahayam N. Passive decoupling due to low Q-factors of four-channel coils in 300-MHz pulsed EPR imaging. Appl Magn Reson. 2015;46:671–83.
72. Enomoto A, Hirata H. Parallel image-acquisition in continuous-wave electron paramagnetic resonance imaging with a surface coil array: proof-of-concept experiments. J Magn Reson. 2014;239:29–33.
73. Nilges MJ, Walczak T, Swartz HM. 1 GHz in-vivo EPR spectrometer operating with a surface probe. Phys Med. 1989;5:195–201.
74. Li H, Hou H, Sucheta A, Williams BB, Lariviere JP, Khan MN, Lesniewski PN, Gallez B, Swartz HM. Implantable resonators—a technique for repeated measurement of oxygen at multiple deep sites with *in vivo* EPR. In: Takahashi E, Bruley D, editors. Oxygen transport to tissue XXXI. Advances in experimental medicine and biology, vol. 662. Boston, MA: Springer; 2010.
75. Hou H, Li H, Dong R, Khan N, Swartz H. Real-time monitoring of ischemic and contralateral brain pO_2 during stroke by variable length multisite resonators. Magn Reson Imaging. 2014;32:563–9.
76. Khan N, Hou H, Eskey CJ, Moodie K, Gohain S, Du G, Hodge S, Culp WC, Kuppusamy P, Swartz HM. Deep-tissue oxygen monitoring in the brain of rabbits for stroke research. Stroke. 2015;46:e62–6.
77. Hou H, Dong R, Li H, Williams B, Lariviere JP, Hekmatyar SK, Kauppinen RA, Khan N, Swartz H. Dynamic changes in oxygenation of intracranial tumor and contralateral brain during tumor growth and carbogen breathing: a multisite EPR oximetry with implantable resonators. J Magn Reson. 2012;214:22–8.
78. Hou H, Nemani VK, Du G, Montano R, Song R, Gimi B, Swartz HM, Eastman A, Khan N. Monitoring oxygen levels in orthotopic human glioma xenograft following carbogen inhalation and chemotherapy by implantable resonator-based oximetry. Int J Cancer. 2015;136:1688–96.
79. Wolfson H, Ahmad R, Twig Y, Kuppusamy P, Blank A. A miniature electron spin resonance probehead for transcutaneous oxygen monitoring. Appl Magn Reson. 2014;45:955–67.
80. Miyake M, Liu KJ, Walczak TM, Swartz HM. In vivo EPR dosimetry of accidental exposures to radiation: experimental results indicating the feasibility of practical use in human subjects. Appl Radiat Isot. 2000;52:1031–8.
81. Haga T, Hirata H, Lesniwski P, Rychert KM, Williams BB, Flood AB, Swartz HM. L-band surface-coil resonator with voltage-control impedance-matching for EPR tooth dosimetry. Concepts Magn Reson B. 2013;43B:32–40.

82. Schreiber W, Petryakov SV, Kmiec MM, Feldman MA, Meaney PM, Wood VA, Boyle HK, Flood AB, Williams BB, Swartz HM. Flexible, wireless, inductively coupled surface coil resonator for EPR tooth dosimetry. Radiat Prot Dosim. 2016;172:87–95.
83. Williams BB, Dong R, Nicolalde RJ, Matthews TP, Gladstone DJ, Demidenko E, Zaki BI, Salikhov IK, Lesniewski PN, Swartz HM. Physically-based biodosimetry using *in vivo* EPR of teeth in patients undergoing total body irradiation. Int J Radiat Biol. 2011;87:766–75.
84. Hoshi N, Nikawa Y, Kawai K, Ebisu S. Application of microwaves and millimeter waves for the characterization of teeth for dental diagnosis and treatment. IEEE Trans Microwave Theory Tech. 1998;46:834–8.
85. Trompier F, Romanyuka A, Kornak L, Calas C, LeBlanc B, Mitchell C, Swartz H, Clairand I. Electron paramagnetic resonance radiation dosimetry in fingernails. Radiat Meas. 2009;44:6–10.
86. Black PJ, Swarts SG. Ex vivo analysis of irradiated fingernails: chemical yields and properties of radiation-induced and mechanically-induced radicals. Health Phys. 2010;98:301–8.
87. Petryakov SV, Schreiber W, Kmiec MM, Williams BB, Swartz HM. Surface dielectric resonators for X-band EPR spectroscopy. Radiat Prot Dosim. 2016;172:127–32.
88. Swartz HM, Williams BB, Zaki BI, Hartford AC, Jarvis LA, Chen EY, Comi RJ, Ernstoff MS, Hou H, Khan N, Swarts SG, Flood AB, Kuppusamy P. Clinical EPR: unique opportunities and some challenges. Acad Radiol. 2014;21:197–206.
89. Caston RM, Schreiber W, Hou H, Williams BB, Chen EY, Schaner PE, Jarvis LA, Flood AB, Petryakov SV, Kmiec MM, Kuppusamy P, Swartz HM. Development of the implantable resonator system for clinical EPR oximetry. Cell Biochem Biophys. 2017;75:275–83.
90. Bilgen M, Elshafiey I, Narayana PA. In vivo magnetic resonance microscopy of rat spinal cord at 7 T using implantable RF coils. Magn Reson Med. 2001;46:1250–3.
91. Rivera DS, Cohen MS, Clark WG, Chu AC, Nunnally RL, Smith J, Mills D, Judy JW. An implantable RF solenoid for magnetic resonance microscopy and microspectroscopy. IEEE Trans Biomed Eng. 2012;59:2118–25.

Chapter 11
Biomedical Overhauser Magnetic Resonance Imaging (OMRI): Noninvasive Imaging of Redox Processes

Kazuhiro Ichikawa, Mayumi Yamato, and Tatsuya Naganuma

Abstract Overhauser effect is an energy transfer phenomenon between different spin systems. It is known as NOE (nuclear Overhauser effect) in NMR and widely used in molecular structure analysis. It is generally recognized as the dynamic nuclear polarization (DNP) process to establish a hyperpolarized spin state to increase the sensitivity of magnetic resonance detection of nuclei. This type of DNP approach is also regarded as a promising technique for sensitive detection of NMR/MRI of chemical probes in combination with low temperature/dissolution technique. Another type of imaging strategy is free radical imaging based on Overhauser effect. Thus the free radical imaging in biomedical material through Overhauser effect is denoted as OMRI in this chapter. This chapter discusses the principle of OMR, OMRI scanner, and resonator configuration, and the biomedical application.

Keywords OMRI noninvasive REDOX imaging · Biomedical Overhauser magnetic resonance imaging

Introduction

Overhauser effect is an energy transfer phenomenon between different spin systems [1]. It is known as NOE (nuclear Overhauser effect) in NMR and widely used in molecular structure analysis. It is generally recognized as dynamic nuclear polarization (DNP) process to establish a hyperpolarized spin state to increase sensitivity of magnetic resonance detection of nuclei. This type of DNP approach is also regarded as a promising technique for sensitive detection of NMR/MRI of chemical probes in

K. Ichikawa (✉) · M. Yamato · T. Naganuma
Faculty of Pharmaceutical Sciences, Kyushu University, Fukuoka, Japan
e-mail: ichikawak@niu.ac.jp

L. J. Berliner, N. L. Parinandi (eds.), *Measuring Oxidants and Oxidative Stress in Biological Systems*, Biological Magnetic Resonance 34,
https://doi.org/10.1007/978-3-030-47318-1_11

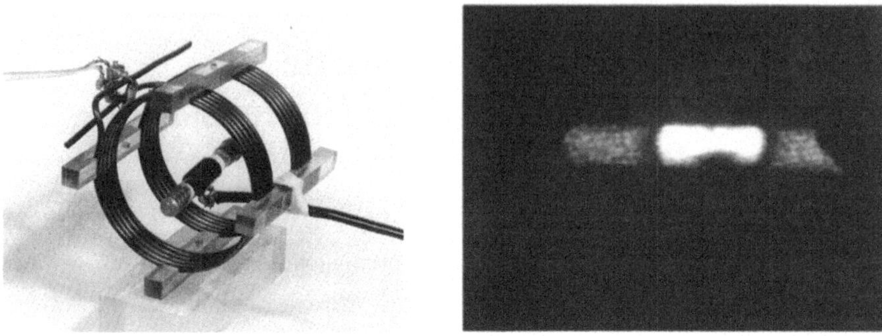

Fig. 11.1 OMRI resonator assembly of first OMRI(PEDRI) resonator (left) and OMRI image of Tempol solution (right). In this case, separate-solenoid NMR receiver coil was used. MRI intensity increased in the middle part of the Tempol tube phantom after ESR excitation. (Modified from Lurie et al. [18])

combination with low temperature/dissolution technique. Combined with low temperature effect on spin population, the hyperpolarization could be more than 10,000 times for ^{13}C nuclei [2], enabling to monitor enzymatic conversion of pyruvate to alanine or lactate in real time. The chemical shift change of NMR-positive nuclei, which is induced upon metabolism of the chemical probe, can be observed in real time depending on the enhanced NMR signals [3].

Other type of imaging strategy is free radical imaging based on Overhauser effect. In 1988, free radical imager based on Overhauser effect was first reported as Proton Electron Double Resonance Imaging (PEDRI; [4]; Fig. 11.1). Overhauser effect polarizes nuclear magnetization by excitation of unpaired electrons, and the nuclear species are not necessarily restricted to proton spin [5]. Thus the free radical imaging in biomedical material through Overhauser effect is denoted as OMRI in this chapter.

Using OMRI, free radical image can be obtained in high spatial resolution since the detection process is exactly the same as MRI, while typical ESR-based imaging provides a few mm of spatial resolution due to its wide spectral linewidth. OMRI has been utilized for measurement of tissue partial oxygen pressure, pH, redox status, etc. [6–9]. Contrast agents, such as nitroxyl or trityl radicals, have been utilized to image in vivo physiology based on change in their ESR spectra according to their interaction with oxygen molecules, redox compounds, etc., and physiological information is calculated from the raw OMRI images.

Principle of OMRI

A brief outline of the principles is presented in this section [10, 11]. OMRI is based on the Overhauser effect that polarizes nuclear spins, resulting in enhanced amplitude of NMR signal, when the ESR transition of the paramagnetic compound is saturated. The degree of enhancement, E of the NMR signal ($I = 1/2$ for proton spin) which is coupled to unpaired electron spin ($S = 1/2$) is given by,

$$E = \frac{I_Z}{I_0} = 1 - \rho f s \frac{\gamma_e}{\gamma_N}$$

where,

$\langle I_z \rangle$: polarized proton intensity
I_0: proton intensity at thermal equilibrium
ρ: coupling factor,
f: leakage factor
s: saturation parameter
γ_e: electron gyromagnetic ratio
γ_N: nuclear gyromagnetic ratio

Coupling factor ρ is close to 0.5, when interaction between electron and proton spins is dominated by dipole–dipole interactions,

$$\rho = \frac{R_1^{IS}}{R_1^{II}} \cong 0.5$$

where

R_1^{IS} : $I - S$ spin relaxation
R_1^{II} : $I - I$ spin relaxation in the presence of S spin

Spin-lattice relaxation of nuclei within the molecules via proton–proton dipolar coupling diminished the polarization. The leakage factor f that accounts for the loss of polarization is sensitive to the motion, and it depends also upon the concentration of electron spins. Leakage factor f is described as follows,

$$f = \frac{R_1^{II}}{R_1^{II} + R_{10}^{I}} = \frac{r_1 c}{r_1 c + R_{10}^{I}} = 1 - \frac{T_1}{T_{10}}$$

where

R_{10}^{I} : inherant relaxivity of I spin
T_1: proton spin-lattice relaxation time in the solution with electron spin
T_{10}: inherent proton spin-lattice relaxation time without electron spin
r_1: relaxivity of electron spin
c: concentration of electron spin

As the concentration of the agents is increased, the leakage factor approaches to unity.

The degree of saturation of the electron spin, s, describes energy absorption of the spin system and has a large impact on the polarization. Saturation factor s

depends on the electron spin relaxation rates of free radical compounds at a given ESR power and formulated from Bloch equation as,

$$s = \frac{\gamma_e^2 B_1^2 T_{1e} T_{2e}}{1 + \left(\Delta\omega T_{2e}\right)^2 + \gamma_e^2 B_1^2 T_{1e} T_{2e}}$$

where

B_1: magnitude of ESR oscillating magnetic field,
T_{1e}, T_{2e}: spin-lattice and spin-spin relaxation times of electron spin, respectively.
$\Delta\omega$: off resonance shift

Enhancement factor E is the equilibrium value of NMR signal enhancement after infinite time of ESR excitation is carried out. In experiments, the degree of E is formed at the rate of T_1, which is described as follows,

$$E - 1 = \left(E_{\text{inf}} - 1\right) \times \left(1 - e^{-\frac{T_{\text{ESR}}}{T_1}}\right)$$

where,

E_{inf}: equilibrium value of NMR signal enhancement after infinite time of ESR excitation
T_{ESR}: duration of ESR excitation

The last part of the equation is called as Quality transfer factor. The polarized spin state returns its thermal equilibrium value at the rate of T_1 after ESR excitation is terminated.

Enhancement factor E can reach maximum values of 110 and 165, respectively, for ^{14}N and ^{15}N nitroxyl free radicals for pure dipolar, and 220 and 330 for scalar interactions in theory. However, many factors in experiments affect polarization, including additional hyperfine interaction between the hydrogen nuclei and the unpaired electron of the free radical compounds. For example, ESR lines of nitroxyl radicals are inhomogeneously broadened due to the presence of the unresolved hydrogen hyperfine, and excitation of one of the nitrogen hyperfine lines will result only in partial saturation. Nevertheless, the enhancement factor can be approximately given by the above equations by irradiating a single ESR line of the nitrogen hyperfine line of the nitroxyl-free radicals.

OMRI Scanner and Resonator Configuration

OMRI instrument configuration is basically the same as that of MRI, with the exception that there is an ESR resonator and corresponding ESR amplifier for electron excitation (Fig. 11.2a).

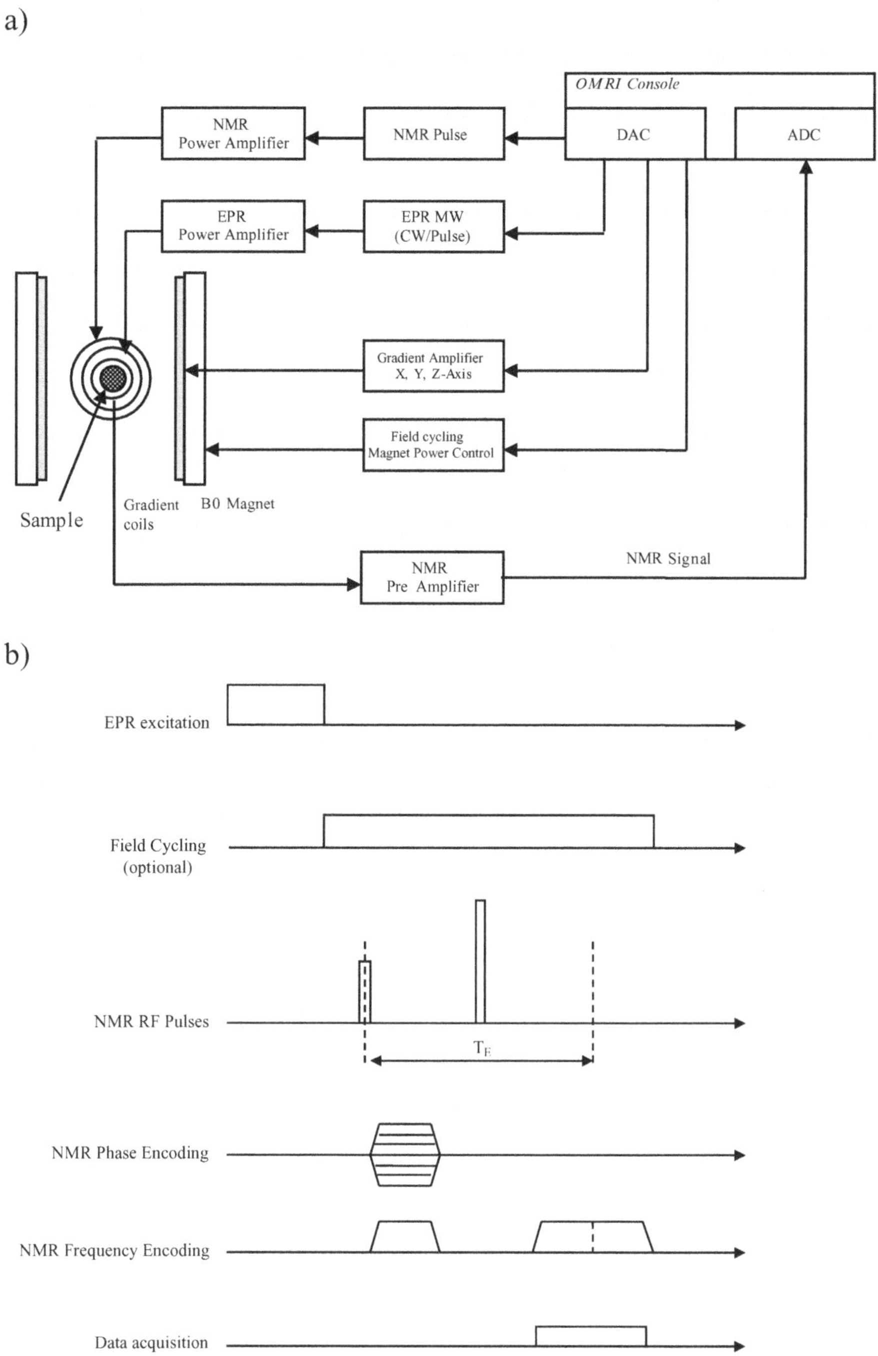

Fig. 11.2 Block diagram of a field-cycling type OMRI scanner (**a**) and OMRI Spin Echo sequence diagram (**b**). Fig. XX Alderman grant resonator for low frequency OMRI

One of major differences between OMRI and MRI scanner is the B_0 configuration. In biomedical OMRI, the ESR frequency requires to be ensured for good ESR microwave penetration, skin depth, to the dielectric biological sample [12]. Thus the magnetic field strength of OMRI is determined by the skin depth of the ESR microwave frequency for the object size of interest. Conventional OMRI scanner for whole-body imaging of small animal utilize 150–450 MHz for ESR excitation, i.e., 5–15 mT, which is less than 1/100 of conventional MRI scanner, unless the area of interest is focused on the surface of live animal, such as skin or shallow region of muscle, and the ESR field can be several hundreds of mT [13]. This is the reason why construction of OMRI scanner is not possible just as an add-on to ESR devices in the clinical MRI system. Small animal MRI scanner is hardly available for this magnetic field ranges and OMRI scanner is typically constructed based on in vivo ESR system, which utilizes B_0 magnet between 5 and 40 mT.

OMRI pulse sequence also requires ESR excitation process in addition to conventional MRI pulse sequences (Fig. 11.2b). Proton T_1 relaxation time ranges between a few to thousands of milliseconds at low field strength, i.e. 5–40 mT. As described in previous section, the time constant of OMRI polarization is proton T_1 and maximum polarization requires five times of T_1, ESR excitation time (T_{ESR}). Thus, one set of OMRI pulse sequence, repetition time (T_R), requires hundreds of milliseconds to a few seconds, while duration of typical MRI pulse sequence is in the range of a few to dozens of milliseconds.

A complementary approach in biomedical OMRI configuration is field-cycling [4, 14, 15]. There are multiple steps in OMRI measurement, including ESR excitation, MRI encodings, and detections, which can be performed either simultaneously or separately. In field-cycling OMRI, proton polarization through ESR saturation is carried out at low field strength, such as 5–10 mT, to ensure good microwave penetration for large animals, which is followed by switching the external magnetic field strength to high field strength, such as 450 mT, for NMR detection. NMR sensitivity increases as B_0 to the power of 1.5 and NMR sensitivity of 5 mT/450 mT field cycling OMRI can be more than 800 times higher than that of fixed-field B_0 configuration at 5 mT, if the relaxation of hyperpolarized to thermal equilibrium during field-cycling time is not taken into account. Potential issues we encounter are high current capacitor banks required to elevate the external field using electromagnet(s) in short transition time, i.e. dozens of ms in instrumentation. Changes in tissue T_1 of the organ of interest in physiological and disease conditions may confuse the analysis of disease models of animals, since hyperpolarized spin state forms and returns to its thermal equilibrium at a rate of T_1 of the organ of interest.

A configuration of biomedical OMRI resonator for field-cycling system is shown in Fig. 11.3. The OMRI resonator is designed for use outside RF shielded room and the resonator consists of local RF shield, ESR transmission coil in Alderman-Grant structure [16, 17], NMR transmission, and receiver coils. Anesthetized animal is mechanically retained in the OMRI resonator and the local RF shield is in cupper mesh structure, which allows to directly monitor animal position and breathing status during OMRI measurement.

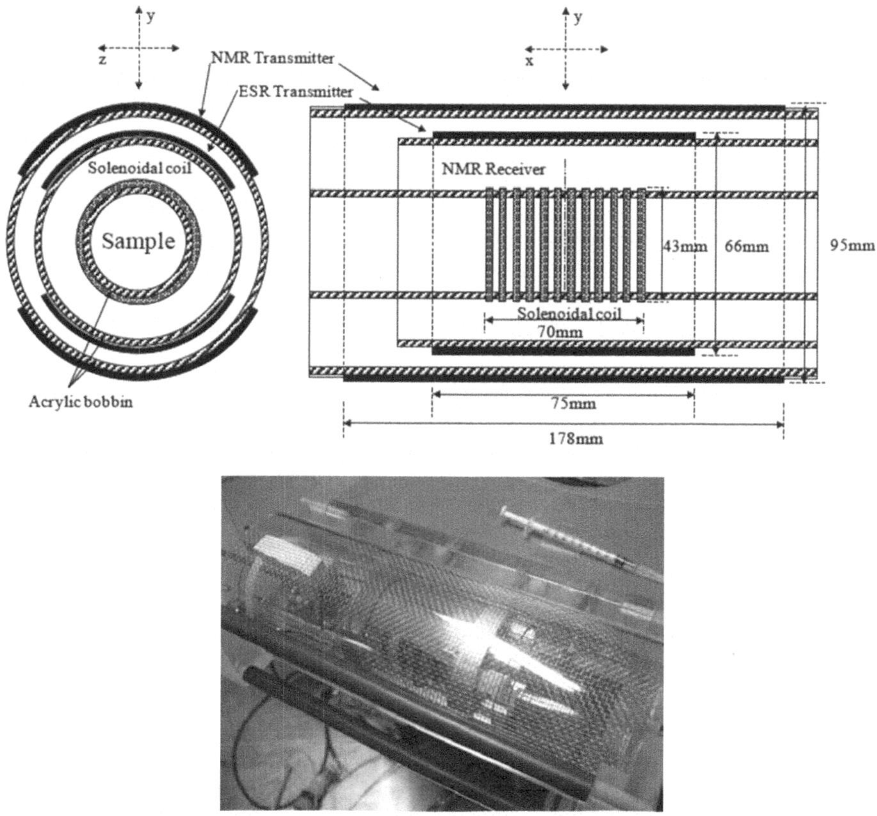

Fig. 11.3 A diagram of field-cycling type OMRI resonator (top) and its photograph (bottom)

Biomedical Application and Conclusions

An example of biomedical application of OMRI is presented, in which the OMRI was utilized for redox status measurement in Parkinson's disease model [6]. Parkinson's disease is a progressive, neurodegenerative disease, and it is suggested that mitochondrial dysfunction induced by oxidative stress may play an important role. The 6-hydroxydopamine (6-OHDA) model in rats was used to study the mechanisms of Parkinson's disease, in which mitochondrial dysfunction is associated with the progress. Nitroxyl-free radical is a group of compounds which is sensitive to redox reactions, and the compounds are reduced enzymatically by complex I and complex II in the electron transport chain of mitochondria. One of the nitroxyl-free radical, 3- methoxycarbonyl-2,2,5,5-tetramethylpyrrolidine-l-oxyl (methoxycarbonyl-PROXYL), was utilized in the study since methoxycarbonyl-PROXYL is redox-sensitive and blood-brain permeable.

Thus the reduction of methoxycarbonyl-PROXYL to corresponding nonparamagnetic compounds was noninvasively visualized in rat brain to monitor changes in redox reaction in Parkinson's disease model (Fig. 11.4). After injection of the methoxycarbonyl-PROXYL solution into rat, the OMRI intensity in rat brain immediately increased as a result of delivery of the molecule over blood-brain barrier. Then the OMRI intensity gradually decreased with time and the decreasing rate was calculated as rate constant of the molecule in reaction with redox system in the brain. As can be seen, the decrease in imaging intensity in the lesioned hemispheres was significantly slower in the 6-OHDA-lesioned rats while the distribution of methoxycarbonyl-PROXYL to the lesioned hemisphere was the same as that to the contralateral hemisphere, which was confirmed with microdialysis measurement. The results suggested that the reductions were due to the conversion from methoxycarbonyl-PROXYL to corresponding nonparamagnetic compounds, rather than to modification in methoxycarbonyl-PROXYL clearance from the brain tissue.

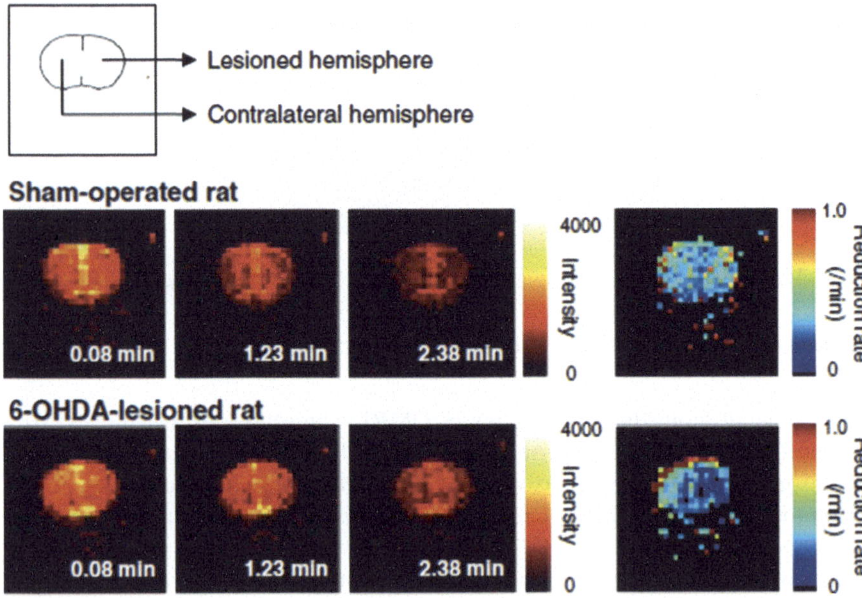

Fig. 11.4 Redox status images with OMRI using a blood-brain-barrier-permeable nitroxyl-free radical as contrast reagent, methoxycarbonyl-PROXYL. Rats were given with 6-OHDA in the right striatum and examined with OMRI 6 weeks after the treatment. OMRI images in the head region were successively obtained to calculate the reduction rates, as redox status (right). (Modified from Yamato et al. [6])

References

1. Overhauser AW. Polarization of nuclei in metals. Phys Rev. 1953;91(2):476.
2. Ardenkjaer-Larsen JH, Fridlund B, Gram A, Hansson G, Hansson L, Lerche MH, Servin R, Thaning M, Golman K. Increase in signal-to-noise ratio of > 10,000 times in liquid-state NMR. Proc Natl Acad Sci U S A. 2003;100(18):10158–63.
3. Nishihara T, Yoshihara HA, Nonaka H, Takakusagi Y, Hyodo F, Ichikawa K, Can E, Bastiaansen JA, Takado Y, Comment A, Sando S. Direct monitoring of γ-glutamyl transpeptidase activity in vivo using a hyperpolarized (13) C-labeled molecular probe. Angew Chem Int Ed. 2016;55(36):10626–9.
4. Lurie DJ, et al. Design, construction and use of a large-sample field-cycled PEDRI imager. Phys Med Biol. 1998;43(7):1877–86.
5. Murugesan R, et al. Fluorine electron double resonance imaging for 19F MRI in low magnetic fields. Magn Reson Med. 2002;48(3):523–9.
6. Yamato M, et al. Overhauser-enhanced magnetic resonance imaging characterization of mitochondria functional changes in the 6-hydroxydopamine rat model. Neurochem Int. 2011;59(6):804–11.
7. Matsumoto S, et al. Simultaneous imaging of tumor oxygenation and microvascular permeability using Overhauser enhanced MRI. Proc Natl Acad Sci U S A. 2009;106(42):17898–903.
8. Samouilov A, et al. In vivo proton-electron double-resonance imaging of extracellular tumor pH using an advanced nitroxide probe. Anal Chem. 2014;86(2):1045–52.
9. Golman K, et al. Dynamic in vivo oxymetry using Overhauser enhanced MR imaging. J Magn Reson Imaging. 2000;12(6):929–38.
10. Benial AM, et al. Dynamic nuclear polarization properties of nitroxyl radicals used in Overhauser-enhanced MRI for simultaneous molecular imaging. J Magn Reson. 2006;182(2):273–82. 133.
11. Ardenkjaer-Larsen JH, Laursen I, Leunbach I, Ehnholm G, Wistrand LG, Petersson JS, Golman K. EPR and DNP properties of certain novel single electron contrast agents intended for oximetric imaging. J Magn Reson. 1998;133(1):1–12.
12. Röschmann P. Radiofrequency penetration and absorption in the human body: limitations to high-field whole-body nuclear magnetic resonance imaging. Med Phys. 1987;14(6):922–31.
13. Tokunaga Y, Nakao M, Naganuma T, Ichikawa K. Construction of 0.15 Tesla Overhauser enhanced MRI. Adv Exp Med Biol. 2017;977:393–8.
14. Lurie DJ, et al. Field-cycled PEDRI imaging of free radicals with detection at 450 mT. Magn Reson Imaging. 2005;23(2):175–81.
15. Naganuma T, Nakao M, Ichikawa K, Utsumi H. Development of a new redox molecular imaging method. Yakugaku Zasshi. 2015;135(5):733–8.
16. Alderman DW, Grant DM. Efficient decoupler coil design which reduces heating in conductive samples in superconducting spectrometers. J Magn Reson. 1979;36(3):447–51.
17. Petryakov S, et al. Modified Alderman-Grant resonator with high-power stability for proton electron double resonance imaging. Magn Reson Med. 2006;56(3):654–9.
18. Lurie DJ, et al. Proton electron double magnetic-resonance imaging of free-radical solutions. J Magn Reson. 1988;76(2):366–70.

Index

L. J. Berliner, N. L. Parinandi (eds.), *Measuring Oxidants and Oxidative Stress in Biological Systems*, Biological Magnetic Resonance 34,
https://doi.org/10.1007/978-3-030-47318-1

O

The manufacturer's authorised representative in the EU is Springer Nature Customer Service Centre GmbH, Europaplatz 3, 69115 Heidelberg, Germany. If you have any concerns regarding our products, please contact ProductSafety@springernature.com

Printed and bound by CPI Group (UK) Ltd, Croydon, CR0 4YY
30/09/2026
02225503-0005